XPONENTIAL 2024

San Diego, California, USA
22 - 25 April 2024

ISBN: 978-1-7138-9933-4

Printed from e-media with permission by:

Curran Associates, Inc.
57 Morehouse Lane
Red Hook, NY 12571

Some format issues inherent in the e-media version may also appear in this print version.

Printed by Curran Associates, Inc. (2024)

Association for Unmanned Vehicle Systems International
2700 South Quincy Street
Suite 400
Arlington, VA 22206
USA

Phone: (703) 845-9671
Fax: (703) 845-9679

www.auvsi.org

Additional copies of this publication are available from:

Curran Associates, Inc.
57 Morehouse Lane
Red Hook, NY 12571 USA
Phone: 845-758-0400
Fax: 845-758-2633
Email: curran@proceedings.com
Web: www.proceedings.com

TABLE OF CONTENTS

Deloitte.

Economic and data perspectives of S&T & R&D portfolio optimization

MAKING AN IMPACT THAT MATTERS since 1845

https://doi.org/10.52202/075106-0001

Problem statement

Achieving the US National Security objectives is underpinned by the ability to successfully transition Science and Technology (S&T) and Research, Development, Testing, and Evaluation (RDT&E) into programs, affordably and at scale.

The 12- to 20-year Department of Defense (DoD) weapon systems development cycle is the subject of numerous efforts including provisions in the National Defense Authorization Act, think-tank studies, and DoD internal reviews. The national security and national defense strategies link the US' ability to accelerate technology into the deployed force.[i] Therefore, the opportunity to accelerate S&T and RDT&E into programs benefits the DoD in terms of force and economic effectiveness.

Factors identified within the paper to help convert R&D into programs include economic feasibility, technological maturity opportunity/cost trade-offs, and program management. Broader factors include developing the government workforce and gaining buy-in from the industrial base.

https://doi.org/10.52202/075106-0001

Overview

The fighting in the Russia-Ukraine war has demonstrated how legacy Soviet-era systems have combined with rapid modifications and are then employed quickly into the front-line operations. Since 2022, we have seen changes in tactics, techniques, and procedures (TTPs) used by threat actors. These TTPs then employ new, modified, and old systems—including using aerial drones with ordnance over troops in contact.

The war in Ukraine is changing how modern wars are conducted. The war now employs drones, military and commercial satellites with imaging and communications, and missile defense systems.[ii] The use of data, even artificial intelligence, plays an increasingly bigger role integrating the entire battlespace. Moreover, the war has produced a relative tactical stalemate, which becomes fertile ground to bring into the conflict technologies that may historically take five to 20 years to mature. It is important to understand the developmental and transition costs that pull RDT&E into operations and then sustain those systems in forward deployed geographies.

The national security threats from nation states highlight the importance of technological pace and risks of parity in warfare. Near peer threats are leveraging a "system of systems" approach integrated across all domains including geography and robotics to degrade US advantages.[iii] Combatant commands and commanders (COCOMS) are engaging in our most forward geographical countries recognizing the employment by adversaries from competition to conflict.

As technological advancement is increasing and the adoption of new/transformative technologies becomes more prevalent amongst near peer competition, the need for converting research into fielded technologies that can be leveraged on the battlefield is paramount. In 2019 DEVCOM reported that 30% of S&T activities transitioned to a project management office in the Army.[iv] If only 30% are transitioned to a project management office, then 70% or more technologies fall into the infamous "valley of death." The "valley of death" is a colloquialism often used withing the DoD to explain a phenomenon where technologies and research languish and are unable to transition into a DoD military program that becomes acquisition/procurement and not R&D.

Deloitte published two previous papers in 2023 that explored the interaction of price and cost effect on traditional systems acquisition across the RD&A portfolio.[v] The 2nd paper focused on how economic and operations/compliance risk can impact Foreign Military (FMS) accounts.[vi]

FMS is potentially a very strong lever for the DoD and the Defense Industrial Base to increase the conversion of S&T/R&D transition due to the >$100B (Army) portfolio for FMS. The downside is that an S&T pipeline, which has 12-20 years to maturity, can have increased and compounding costs associated with each program. In some cases, this could render a program unaffordable. For example, a 5% cost growth per year would equal to 175%-200% increase from inception estimates.[vii]

The aim of the paper is to put forward the ability to increase the probability of success and persistence across the S&T, RDT&E, and program transition; but, more importantly, increase the probability of transitioning the right technologies.

We highlight the impacts of the current governance structure of S&T programs, look at commercial industries and how they approach R&D, evaluate the cost and financing of S&T within the Planning, Programming, Budgeting, and Execution (PPBE) process, and finally look at the role of data within the S&T process. Additionally, the paper will often differentiate between S&T (a subset of RDT&E that focuses on nascent research and development) and RDT&E.

S&T and R&D successful transition relies on three factors:

a) Timeline relevant to operational demand;

b) Technical maturity and feasibility; and

c) Affordability in procurement and sustainment.

https://doi.org/10.52202/075106-0001

RDT&E in context

According to a Congressional Research Service (CRS) report in FY2022, the DoD accounted for 41.1% of all federal R&D appropriations ($65.7 billion of $159.6 billion in FY2022).[viii] The appropriation that DoD uses to fund R&D activities is the RDT&E appropriation. Additionally, RDT&E accounted for ~16% of total DoD obligational authority in FY22.[ix] RDT&E is a significant percentage of the National Security Strategy in terms of pure dollars and total obligations.

RDT&E "as share" of DoD total obligational authority (TOA) has been up and down. The last five years has seen an increase in RDT&E as a share of TOA from FY17 to FY22. The US investment in RDT&E is directly tied to the threat of near peer competition and the rise of foreign powers and the technological pace of change throughout the world. Additionally, the increased investment in RDT&E since 2017 was due in part to synchronize the S&T/RDT&E pipeline to match the national security and military strategy.[x]

It is also instructive to compare the R&D funding for Aerospace and Defense (A&D) companies as depicted in the graphic below. When looking at publicly available 10K documents, we see that the top six A&D companies by revenue over the last ten years have an R&D percentage of top-line revenue that is relatively stable. So as revenue grows for the top 6 A&D Companies, net R&D spend grows proportionally, whereas DoD R&D percentage is outpacing increase in TOA. We can also compare them proportionally to RDT&E funding in the above graphic. At a high level, it seems that the US government is shouldering more of the burden in R&D for defense related technologies from 2017 to 2022.

The RDT&E appropriation is broken out into seven separate budget activities. The DoD considers the first three budget activities (6.1-6.3) as S&T and 6.4, 6.5, 6.6, and 6.7 are considered the development, test, and evaluation components. Going from 6.3 to 6.4 is considered where technology transition happens and where the "valley of death" often occurs.

The graph below highlights the funding of activity in the past FY (FY22) which shows the preponderance of funding is in the 6.4, Advanced Concept Development, where technology becomes a program of record.

FY22 RDT&E FUNDING BY RDT&E BUDGET ACTIVITY[xii]

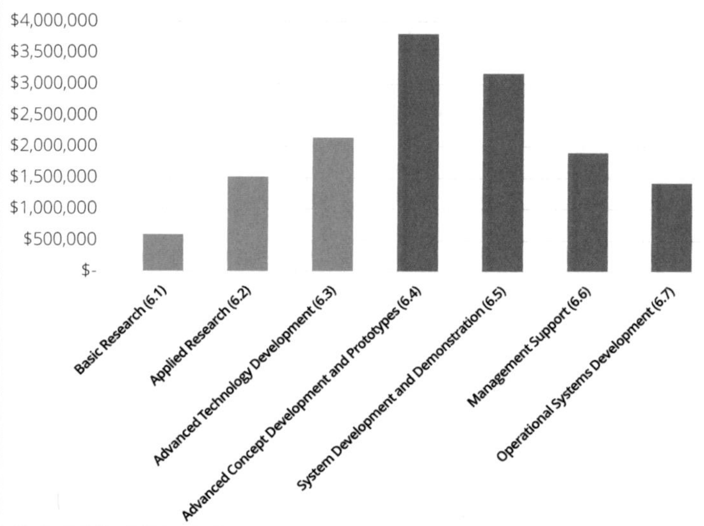

A&D "BIG 6" DEFENSE COMPANIES' REVENUE AS PERCENTAGE OF REVENUE VS. DoD RDT&E SPEND AS PERCENTAGE OF ToA[xi]

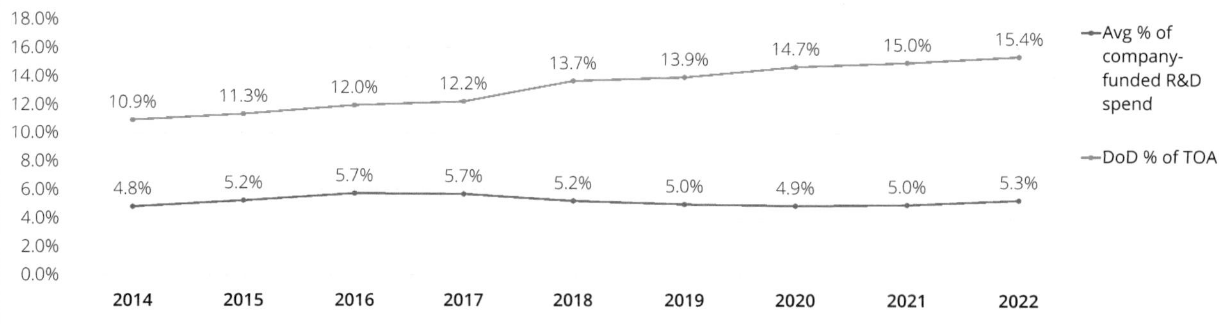

https://doi.org/10.52202/075106-0001

External and internal planning factors create a non-linear path for prototypes and technology to be successfully transitioned to a DoD program. These include market/economic variables such as supply chain and pricing pressures in the economy, risk aversion from the project managers (PMs), and mission requirements. As such, it creates a less efficient pipeline for successful S&T transition where the incentives of various stakeholders, particularly the S&T labs and the PMs, are mismatched.

As the technology transitions between the budget activities the money initially invested (e.g., budget activities 6.1 to 6.2) increases over time due to price and cost factors the longer it stays within S&T and a decision is not made (fielding or killing the research). To move between channels, the cost and overall risk of the technology compounds; inflation is one such factor, but other risk factors include non-standardized data and structure, contracting, design or computational tools, etc.

The process graphic below illustrates the concept from the previous two paragraphs. Transitioning from each budget activity has different risks, and the cost of R&D increases over time. There are also technology readiness levels (TRLs) associated with each budget activity that should be met to be compliant to transition.

Ultimately, to maintain US national security in the interim and long term, the US should invest in research and S&T, but the US should also focus on making sure that the S&T pipeline is synced to the overall R&D pipeline so that technology transition is more successful. The 2023 National Defense S&T Strategy puts it succinctly, "We must take steps to preserve our leadership and counter our competitors who have taken direct aim at this advantage."[xiii]

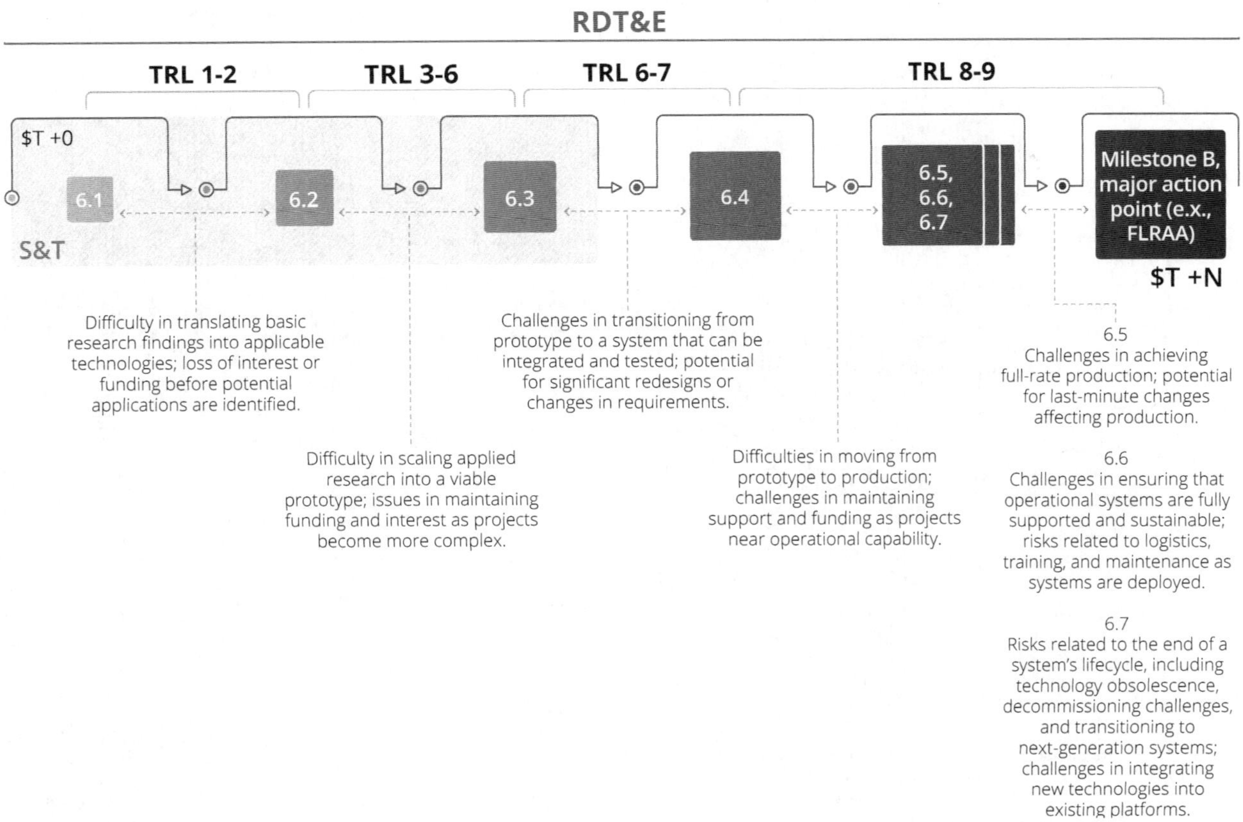

https://doi.org/10.52202/075106-0001

Governance and process

CURRENT S&T GOVERNANCE PROCESS

Currently, more than a dozen technology transition initiatives are overseen by the Office of the Secretary of Defense (OSD) and various military departments.[xiv] These programs offer organized methods and financial support to ease the process of transitioning technologies. They uniformly present opportunities for moving technologies from the S&T portfolio to end-users, including weapon system acquisition programs or military personnel in operational environments.

The 2023 National Defense S&T Strategy lays out plans for coordination between the OUSD (R&E) and Under Secretary of Defense Acquisition and Sustainment to "develop new processes, procedures, and forums to drive close collaboration with the Military Services' and OSD acquisition and sustainment communities to spearhead rapid fielding of capabilities at speed and scale."[xv]

In the context of S&T, transition agreements (TAs) and TRLs are crucial for understanding whether that technology development aligns with strategic and operational needs. TAs formalize partnerships between developers and end-users, defining goals, timelines, and responsibilities to streamline the integration of new technologies into defense systems. Concurrently, TRLs provide a standardized framework to evaluate technology maturity, facilitating informed decision-making and resource allocation. Together, these mechanisms are designed to enhance strategic coordination, risk management, and the timely delivery of capabilities, playing a vital role in maintaining the DoD's technological superiority in a dynamic global threat environment.

While reimagining the governance of S&T presents considerable challenges, the cost of inaction can be far greater. Without proactive measures, organizations risk finding themselves with mature R&D outcomes that, due to escalating costs over time, may become increasingly difficult to afford and implement. This scenario could significantly hinder the DoD's capacity to respond promptly to new threats and sustain its technological superiority in a competitive global environment. Therefore, it's imperative to address these governance challenges head-on to make sure that S&T investments yield timely, cost-effective, and strategically-relevant outcomes.

https://doi.org/10.52202/075106-0001

The graphic below depicts representative S&T and Technology Transition programs across OSD and the military departments to illustrate the need for effective and non-duplicative governance structures.

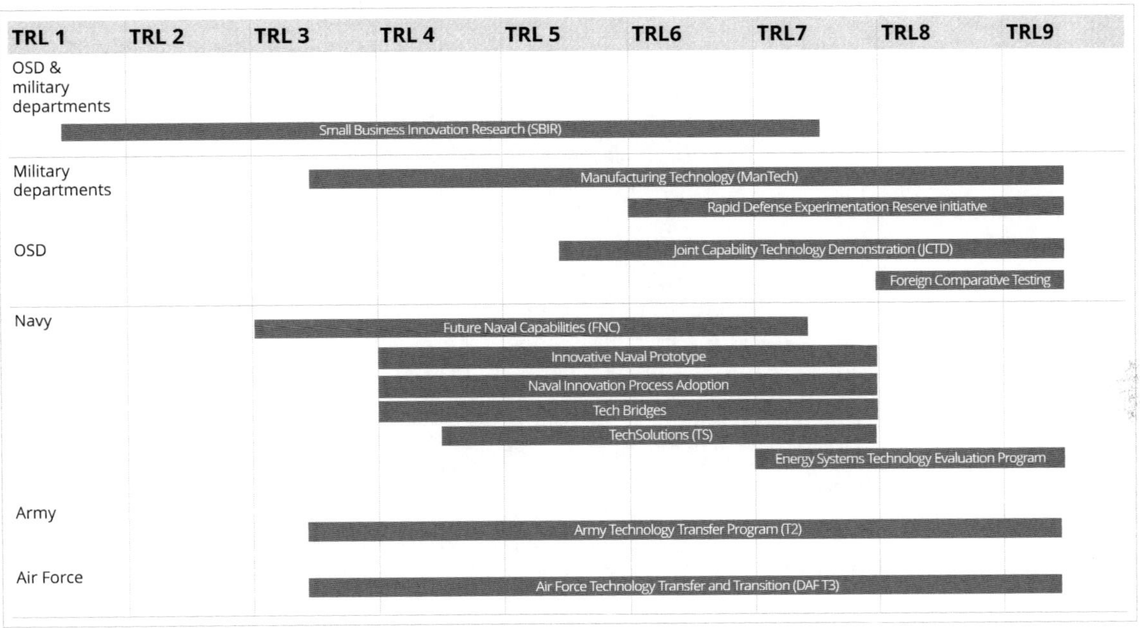

[xvi]

THE NEED FOR EFFECTIVE GOVERNANCE IN S&T

In the context of transitioning S&T efforts to practical applications, the role of governance cannot be overstated. Effective governance acts as a foundational element from the very inception of these initiatives, aiming to guide the transition process with a clear, strategic framework. This involves setting priorities, allocating resources efficiently, and establishing policies and standards that foster innovation while aligning with broader objectives.

By integrating governance early on, stakeholders can navigate complexities and challenges inherent in the transition process more effectively, thereby enhancing the potential for successful integration of new technologies into operational use. This proactive approach contributes to making S&T projects not only visionary but also pragmatic, fostering outcomes that are both innovative and aligned with the needs of end-users, such as weapon systems acquisition programs or military personnel in the field.

Without a well-visioned governance structure, programs can face a lack of cohesive and strategic decision-making processes, ineffective communication channels among stakeholders, and a failure to align S&T initiatives with pressing operational needs and future warfare requirements. This can lead to a suboptimal utilization of resources, resulting in delays, cost overruns, and the inability to rapidly integrate cutting-edge technologies into defense systems.

Moreover, when organizations forgo co-developing governance frameworks with their S&T pipeline, an often-fragmented approach to research and development is observed by leaders, with insufficient collaboration between military, academic, and industry partners.[xvii] This may lead to a disconnect between the development of innovative technologies and their practical application in military contexts, as evidenced by the low conversion rate of S&T projects to operational capabilities.[xviii]

https://doi.org/10.52202/075106-0001

In an era where adversaries rapidly harness new technologies for tactical gains, bureaucratic inertia and inefficient processes within S&T transition pipelines can challenge the United States' responsiveness to emerging threats and its enduring strategic and technological leadership. S&T governance structures are essential to streamline decision-making, foster innovation, and maintain a competitive edge in a fast-evolving global landscape.

GOVERNANCE CHALLENGES IN S&T

Estimating the cost of governance challenges in the DoD S&T transition activities requires a methodical approach to understand the financial impact of misaligned projects and inefficiencies. The framework for this analysis considers the total federal spending in DoD S&T and the percentage of projects that are not aligned with strategic goals or existing programs.

First, consider the total federal spending in DoD S&T. This figure represents the comprehensive budget allocated for science and technology activities within the DoD. For the sake of this analysis, we can assume the total federal spending in DoD S&T for a given fiscal year is set at the FY23 budget request of $16.5 billion.[xix]

Next, determine the percentage of projects that are not aligned with strategic goals or existing programs. This misalignment indicates governance challenges, as it reflects a lack of strategic focus and integration with the DoD's broader objectives and operational needs. Misaligned projects are less likely to transition successfully into operational capabilities, thereby representing inefficiencies and wasted resources. Suppose based on previous GAO estimates of independent R&D projects, that 38% of DoD S&T-related projects are not aligned with strategic goals or existing programs.[xx]

In this example, if the total federal spending in DoD S&T is $16.5 billion and 38% of the projects are not aligned with strategic goals or existing programs, the cost of misaligned spending due to poor governance would be $6.27 billion each fiscal year.

This $6.27 billion represents the financial impact of governance challenges in terms of misaligned S&T projects within the DoD. This figure is a conservative estimate, as it does not account for other potential inefficiencies such as duplicative research efforts, administrative overheads associated with misaligned projects, or the opportunity costs of not investing these resources in more strategically aligned initiatives.

The cost of governance challenges in DoD S&T transition activities can be significant, underscoring the importance of strategic alignment, effective project management, and efficient use of resources to optimize the impact of federal spending in defense-related science and technology. With an increased legislative trend of capping the discretional defense budget authority, understanding these gaps and their associated costs to the US taxpayer is of increasing importance.

R&D lessons from commercial industries

To improve the S&T conversion rate, the DoD can draw insights from commercial R&D governance, commercialization strategy and partnerships that foster innovation, growth, and market competitiveness. Given the fast-paced nature of change and technological development in advanced industries, streamlining the R&D process is just as important as increasing net R&D expenditure to yield results for the stakeholders.

Perhaps no industry is as invested in successful R&D commercialization than the pharmaceutical industry. Between 2000 and 2020, the pharmaceutical industry had the highest average R&D budget as a percentage of net revenues (19%), with the semiconductor, technology hardware, and software industries lagging at an average at 15%.[xxi]

The importance of effective R&D governance is highlighted when looking at the pharmaceutical industry's R&D spend and drug yield. Despite a continuing upward trend in R&D allocation, the industry still faces challenges with negative R&D productivity, suggesting that greater R&D spend may not necessarily result in increased commercialization success rate.[xxii] The R&D success rate from Phase I clinical trials to final approval was consistently approximately 10% over various study periods from 1963 to 2020, and the rate did not increase over the years.[xxiii]

THE STAGE GATE METHOD AND REAL OPTIONS ANALYSIS

The Stage Gate method, also referred to as phase gate process, is an industrial commercial project management technique that breaks up a project into shorter phases according to milestones, which must be met sequentially to move on to the next milestone.[xxiv] These frequent checkpoints in early stages of development that will determine if a project should be pursued further or not can mitigate long term R&D project risks in two ways.

First, these "gates" allow for early identification of errors and resolution of discrepancies that can save time and resources—both financial and human resources—from being wasted. Implementation of this methodology can help to reduce increasing R&D project costs over time, including inflation-related costs, and make sure that new weapons systems are in the hands of warfighters in a timely manner amid escalating geopolitical tensions.

Second, frequent communication and direct inputs from the stakeholders in this customer-focused innovation methodology allow R&D project leaders to accurately identify and understand the needs of the client.[xxv] Furthermore, it holds the project leaders accountable and prevents workstreams from becoming siloed from evolving operational needs or maturing to a point that timely commercialization is unattainable. Early assessment of a product's theoretical viability, constant evaluation of actual viability during development and reorientation of the trajectory to meet the desired objective can help companies ensure that their investments are directed towards R&D workstreams that are most likely to reach commercialization.

By adopting the stage gate method, DoD may have more opportunities to address R&D problems as early as possible and prevent unnecessary investment in projects that are not scalable or feasible.

The graphic below demonstrates different stages of the Stage-Gate model where activities take place and five gates where decisions to either continue or end the project are made.

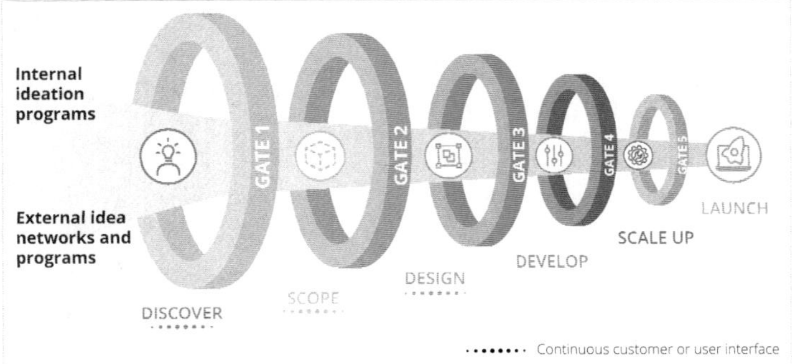

Source: Stage Gate International[xxvi]

https://doi.org/10.52202/075106-0001

A similar alternative to the Stage Gate method is the Real Options Analysis (ROA), which is a financial approach that assigns value to investment situations that are "non-financial," such as R&D that is yet to be commercialized.[xxvii] Industry players use ROA to provide opportunities, such as increasing manufacturing capacity at a factory or developing new technology, with risk-adjusted strategic values that can inform investment decision-making process. This financial modeling approach could help inform DoD's R&D selection decisions by front loading risk identification and estimating the rate of commercialization and return before making any investment.

USE OF AI AND DIGITIZATION OF DATA FOR R&D

In addition to the above project management and decision-making techniques that can guide DoD's R&D framework, there are concrete examples of how commercial industries are accelerating their R&D timetable.

Companies in the industry are also pursuing digitization of data both from and beyond clinical trials to enhance the R&D process. Specifically, investments in automation and predictive analytics help streamline drug development process with improved accuracy, and incorporation of real-world evidence (RWE) such as digital health and insurance records can help pharmaceutical companies gather more data points to increase the efficacy of new drugs under development.[xxviii]

TALENT RECRUITMENT AND DEVELOPMENT

Talent recruitment and development are also critical components of R&D that can accelerate growth in critical and emerging technologies—semiconductors, biotechnologies, artificial intelligence and machine learning—as part of the DoD's and the broader US government's long-term national security objective.[xxix] Following the nationwide prioritization to develop advanced semiconductor ecosystem in the country, many industry leading associations are investing in STEM education for students and standardizing technical curriculum for the future workforce.

For example, the Semiconductor Industry Association (SIA) is partnering with a non-profit organization focused on robotics to educate and mentor young students with a goal to develop a strong pipeline for the semiconductor industry.[xxx] By identifying and nurturing students into the niche area of semiconductors, the industry builds itself a workforce with prerequisite skills necessary for its field.

Fostering talent development can be more efficient and effective when there is a collaboration with government stakeholders. Just as how the National Semiconductor Technology Center (NSTC) is aimed to coordinate public-private partnership for semiconductor R&D, a more active role of R&D focused Defense Industrial Base Consortium (DIBC) for the DoD may be able to foster more resilient innovation in S&T and RDT&E.

https://doi.org/10.52202/075106-0001

Linking S&T R&D with PPBE and cost analysis

A successful S&T R&D pipeline should maintain at least one constant: supporting a *common good within resource constraints.*[xxxi] Successful R&D inherently benefits the organization and citizenry whom fund it, providing meaningful and enduring value. Portfolio Managers may rely on several specific analysis methods when developing effective and efficient R&D pipelines.

a. Feasibility and ROI analysis: Portfolio managers should develop models to verify predicted total cost through the lens of workforce planning tools, resourcing, and equipment cost. This permits determination of affordability and increases confidence in fielding estimates.
b. Impact assessments and prioritization: Portfolio managers should employ robust and validated methods for scoring and prioritization of alternatives and the impact of selective funding decisions on out-year mission capabilities.

Evaluating an effective R&D pipeline through qualitative constraints such as product fit, capability, and regulatory guidelines may be the tip of the iceberg in evaluating pipeline growth, but cost/impact modeling and financial management can provide portfolio managers multiple options to obtain and objective, evidence-based viewpoint of S&T program feasibility.

The DoD PPBE process provides a framework for evaluating Courses of Action (COAs) through assignment of quantified values and predicted impact of requirements realization.[xxxii] Leading practices such as firmly-articulated requirements and benefits within the Program Objective Memorandum (POM) promote buy-in to S&T and R&D investment prioritization, and assist both stakeholders and decision-makers to view investments through the lens of this framework.

Proper emphasis on supporting resources including manpower and equipment is necessary in evaluating project effectiveness. Using forward-looking workforce planning tools and historical supporting equipment data are crucial in supplementing project information. The Office of Personnel Management's (OPM) five-step Workforce Planning model[xxxiii] evaluates specific project requirements as well as training requirements, gap reduction strategies, and strategies for identifying potential barriers. The financial cost of resourcing should also be evaluated to avoid underutilization/overutilization of resources. Historical equipment costs for similar projects can be used with an inflation factor to estimate future costs.

Portfolio managers striving to deliver on the public and organizational good objective may encounter difficulty evaluating the pipeline for R&D projects based on traditional corporate finance metrics such as future cash flow potential, return on assets, and discounted cash flow. Current gaps between project potential and ultimate outputs most often relate to project buy-in, commitment to delivery, and funding gaps. Scoring techniques used in the PPBE process mixed with key topic areas for evaluating project effectiveness may bridge this gap.

The PPBE Reform Commission has outlined the need for consistent and equitable treatment of DoD RDT&E appropriations to "protect...early-stage investments but also highlight technologies in need of continued coordination efforts between labs and users to ensure early-stage research responds to operational needs."[xxxiv] Portfolio managers can achieve consistency and equity

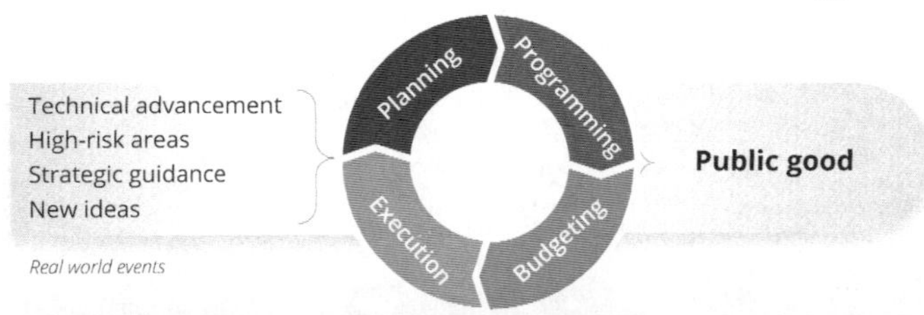

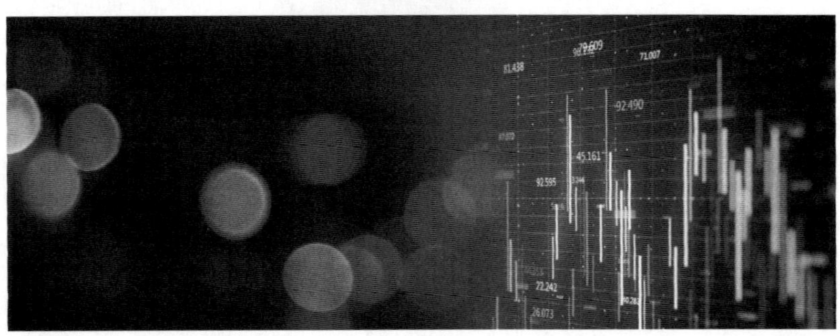

https://doi.org/10.52202/075106-0001

of evaluation as prescribed by the Commission by applying validated data and evidence-based inputs to an objective scoring regime that includes a wide selection of stakeholders and subject matter experts to inform the process. A framework of this type may include some or all of the following parameters:

- **Assessment for innovation and uniqueness.** Assessing the level of innovation and uniqueness to identify truly cutting-edge technologies with a clear potential to deliver a competitive advantage in national security.
- **Market potential.** Consideration should be given to the size and growth potential of the market for a technology. Thorough analysis to determine scalability and commercial viability of the product or solution beyond its immediate application to national security is essential, as commercial adoption enhances viability for economic feasibility and long-term sustainment.
- **Financial due diligence and revenue model.** If adopting or modifying a nascent commercial technology that is on the critical path of an S&T/R&D pipeline, an examination of the technology-owning vendor's revenue model is necessary to understand how that vendor generates income and whether the product, and possibly the vendor, are sustainable. Evaluating financial health of that company would ideally include reviewing financial statements, cash flow, profitability, and other metrics.
- **Leadership and expertise on the management team.** Evaluation of the capabilities and experience of the management team at a scientific research laboratory, national lab, University Affiliated Research Center (UARC), or Federally-funded R&D Center (FFRDC) is an often-overlooked element when making S&T investments. The team's ability to execute the business plan and navigate challenges is a critical factor in the investment decision.
- **Intellectual property (IP) and patents.** Typically in S&T projects, all IP is owned by the government; even though IP rights are increasingly being shared to encourage commercialization of novel technologies. According to Stevenson-Wydler Act, Federal Laboratories are permitted to issue exclusive license for patents.[xxxv] The number of patents or citations, however, is not directly (if at all) indicative of the value of S&T output for end-user warfighting capabilities such as survivability, lethality, and sustainability.

Example framework:

CRITERIA	PARTICIPANT #1	PARTICIPANT #2	...	PARTICIPANT #X
MANPOWER	→ 5	↑ 8		→ 5
EQUIPMENT	↓ 1	↑ 8		↓ 3
INNOVATION AND UNIQUENESS	↑ 9	→ 4		→ 5
MARKET POTENTIAL	↓ 1	EXAMPLE SCORING DATA		8
FINANCIAL DD/REVENUE MODEL	→ 4	↑ 8		→ 4
LEADERSHIP AND EXPERTISE	↓ 1	↓ 0		→ 3
IP/PATENTS	→ 6	↑ 10		↓ 1

Note: Criteria may be assigned weightage by performance indicators by priority.

Greater insight may be gained by performing an assessment survey across S&T investments since the start of the Global War on Terror in 2001, and the systems, Programs of Record, and fielding success that followed. This historical review would capture investments with a returned value beyond the typical POM cycle, for which standard DoD KPIs and metrics are not readily applicable.

Additional outputs of a successful S&T may be considered, including the impact on policies, on-the-job training of junior researchers and scientists, and recruitment for the future workforce, all of which are difficult to capture through traditional financial metrics.

Perhaps, a binary metric of scientific success would be an effective method: Was a conclusion reached? Did the research add to the body of knowledge, and did it expand the possibilities for future work? Did it confirm that a certain line of research should not be pursued further, at least until certain other problems or obstacles upon which the research is dependent are solved? A research effort that does not push technology forward is not necessarily a failure, particularly when seeking truly pivotal, breakthrough technology advancements.

Success, therefore, may also be more appropriately measured at the portfolio level, rather project/program level, which leads to the necessary step of setting a threshold for how transformational a selected line of research must be to deserve investment, and how much.

This would also permit planners to evaluate *integration across that portfolio* as a key performance indicator (KPI). Specifically, looking across a portfolio of S&T investments, would their internal scientific activities and outputs (if some or all were successfully transitioned into R&D) support each other technologically; or are they disparate, indicating that the research organization has potentially become too diversified or strayed from their core competencies.

https://doi.org/10.52202/075106-0001

The critical role of data management in DoD's RDT&E success

The final dimension we explore within the RDT&E pipeline is data and data management. In R&D, the imperative for robust data management and analytics cannot be overstated. These elements are not merely auxiliary; they are the bedrock upon which informed decision-making, rapid innovation, and the seamless transition of technologies from conceptual frameworks to operational dominance are built. The complexity of modern military operations and the relentless pace of technological advancements necessitate a data-centric approach, wherein the efficacy of S&T initiatives is significantly amplified through strategic data management and analytical processes.

The DoD and associated organizations face significant challenges in the realm of data management and analytics for R&D. These challenges notably include siloed datasets, inconsistent data standards, and the underutilization of data assets. These issues not only impede the pace of technological innovation but also hinder the effective transition and scaling of new technologies that are crucial for maintaining the US technological edge in national security.

Despite the wealth of data at the DoD's disposal, there's a pervasive issue of underutilization. This underutilization stems from a combination of factors, including the lack of comprehensive data management strategies, inadequate data literacy among staff, and technological limitations in data analytics tools. Consequently, valuable insights that could potentially drive innovation and improve operational efficiencies remain untapped, significantly impacting the DoD's ability to respond swiftly and effectively to emerging threats and opportunities.

The challenges associated with data management practices directly impact the efficiency and effectiveness of S&T/R&D projects. Siloed datasets and inconsistent standards not only slow down the R&D process but also contribute to the so-called "valley of death" for many promising technologies. This phenomenon, where technologies fail to transition from research to operational use, is exacerbated by the underutilization of data, as potential applications and improvements for these technologies often go unrecognized.

 https://doi.org/10.52202/075106-0001

In response to these challenges, it's imperative to adopt broad data management and analytics strategies. Such strategies should aim to break down data silos, standardize data management practices across the DoD, and leverage advanced data analytics tools to fully utilize available data assets. By addressing these challenges head-on, the DoD can enhance the transition rate S&T research and development into scalable programs, thereby bolstering US national security in an increasingly competitive global landscape.

To facilitate this integration, adopting frameworks and methodologies that promote data decentralization, federated governance, and treating data as a strategic asset is crucial. These principles, inspired by leading practices in data management architectures, such as the concepts like the Unified Data Reference Architecture (UDRA), provide a comprehensive blueprint for embedding data-driven insights throughout the technology development lifecycle.[xxxvi]

Data decentralization within S&T entails distributing data management responsibilities, so data is managed by those with the closest operational insights. This can enhance the relevance, quality, and accessibility of data, empowering project teams with the autonomy to manage their data efficiently. This autonomy fosters an environment conducive to rapid innovation, where data-driven insights can be integrated into the development process, from conceptualization to fielding.

Federated governance supports this decentralized approach by instituting a unified set of data management policies, standards, and protocols across S&T and its partners. This aims to create an environment where data management practices may vary across teams and projects but all adhere to a common framework that facilitates data interoperability, security, and quality. Such governance structures enable effective data sharing and collaboration, not just within the DoD but also with academia and industry partners, creating a synergistic ecosystem where collaborative innovation thrives.

https://doi.org/10.52202/075106-0001

Action plan

Based on our experience, to increase the S&T and overall RDT&E conversion rate, S&T practitioners need to mitigate risk, apply governance best practices, conduct comprehensive technology assessments/ analysis, align to requirements, and incorporate standard data management structures.

What undergirds these approaches is developing training to standardize the skills for S&T project lead and technology sponsors. Providing the RDT&E workforce with the requisite training to get the right technologies adopted as programs of record can ultimately help the US achieve its national security objectives and compete with near peers.

The following are common business issues across the RDT&E process that are candidates for optimization to increase the US ability to field novel—and specifically the right technologies—required on the modern battlefield.

Step 1: Apply governance leading practices.

In the realm of Technology Transition governance, an effective framework is predicated upon three essential steps that collectively facilitate a structured, accountable approach to transitioning technologies.

1. Drafting a foundational Technology Transition Agreement (TTA) that establishes the initial governance framework and sets the stage for all subsequent activities. This would include the customization of the TTA, tailored to the unique characteristics of the technology, its developmental trajectory, and the attributes of the providing entity, so governance considerations are woven into the fabric of the transition process.
2. Planning to facilitate the technology's seamless integration from S&T or R&D into its subsequent operational phase, a process governed by strategic oversight to achieve a successful "design-in." This includes the introduction of a broad governance model, embedding structured processes and oversight mechanisms to guide the entire transition period to confirm that each phase adheres to established governance protocols.
3. Rigorous evaluation process, employing specific criteria endorsed by the DoD to assess the technology's readiness and suitability for integration into the lifecycle, a testament to the critical role of governance in achieving technological excellence and operational readiness.

Central to this process is the strategic coordination of planning and estimation within the PPBE/POM framework, an exercise steeped in governance principles to achieve alignment with overarching strategic goals and fiscal discipline.

Step 2: Integrate trade-off analysis, cost modeling, impact modeling, and defining key metrics for success early in the S&T process.

Key objectives for any S&T project are to gain knowledge, make decisions on the technology based on the knowledge gained from research, and transition the right technologies into programs of record. To achieve better S&T outcomes, the S7T community should consider integrating technologies, or portfolios of technology, earlier in the POM cycle to help drive requirements and alignment to eventual program sponsors. This requires developing measure of success and instituting criteria that support sponsor outcomes early in the S&T process.

By leveraging management science techniques and aligning to the POM process, S&T professionals can be better positioned to assess, prioritize, and transition the right technologies effectively. S&T professionals can then determine, in a standardized methodology, which technologies and outcomes are well-suited to enhance their portfolio.

https://doi.org/10.52202/075106-0001

Step 3: Develop data management leading practices starting with data governance and common infrastructure.

Deloitte's data management and analytics services cater to the needs of the DoD, federal, and state governments with a focus on scalability and precision. Our approach leverages a scalable platform for data integration, grounded in UDRA, to help organizations address complex data challenges efficiently. We provide a unified governance solution that enables you to manage across diverse data landscapes, alongside analytics and AI services built on scalable infrastructure for in-depth data processing and analysis. Deloitte's secure data sharing framework also features advanced access controls, facilitating strategic decision-making and innovation in national security with an emphasis on operational efficiency and data-driven insights.

Operationalizing these principles requires a detailed understanding of the roles and services that support data management, so data-driven insights inform every stage of the technology development process. By prioritizing a data-centric approach, S&T professionals can overcome the identified challenges, enhancing their ability to rapidly develop and deploy innovative technologies. This strategic emphasis on data not only accelerates technological advancement, but also strengthens the DoD's ability to maintain its technological edge in an effort to remain well-positioned to address the evolving challenges of national security in an increasingly digital and data-driven era.

FINAL THOUGHTS

A recent S&T portfolio initiative has been making news across the defense industry and government. The program known as Replicator has been established to field and deliver thousands of multi-domain autonomous systems by August 2025 to hedge against influence and military buildup in INDOPACOM.[xxxvii] This initiative's core focus is to get innovative and critical technologies in the hands of the warfighter faster, at scale, and responsibly. They intend to do so through a repeatable and standard process.

Now more than ever, the ability to align requirements early in the POM cycle, establish sound governance and accountability, leverage management science techniques to evaluate technologies and get the right technologies to programs of record, and establish a unified data management structure is critical to the success of innovative initiatives.

https://doi.org/10.52202/075106-0001

Deloitte services

a. Economic modeling and valuation

Deloitte's specialists build economic models that can help the US government assess the financial implications of existing and potential drivers and to formulate tailored optimization strategies. Deloitte's economic specialists conduct modeling, scenario-based analysis, and simulation exercises. These services help clients understand the implications of industry trends for the US government and assess potential risks as well as develop possible options to mitigate and potentially eliminate those risks.

b. Planning, Programming, Budgeting, and Execution (PPBE)

Deloitte's clients encompass DoD's PPBE process across several verticals, and our service offerings span many components of the budgeting cycle. Our professionals have subject matter experience throughout the PPBE process and are very familiar with cost tracking and workforce planning methodologies. Additionally, Deloitte's Government and Public Services clients include those with dated budget model tools, and our professionals can provide software implementation and process improvement COAs across industry and government.

c. Quantitative risk and decision analysis

Deloitte's clients include DoD services, federal, and state governments. Our professionals range from DoD specialists with 40+ years of specialized DoD major defense acquisition experience to financial economists and data science analysts. Deloitte's specialists may also apply the methodology to planning, PPBE; S&T; and R&D transition. The analysis would include a forecast of the inflation factors, cost estimates, and would provide mitigation recommendations for large investment in DoD R&D.

d. Supply chain and industrial base

Deloitte's specialists are equipped with state-of-the-art tools like CentralSight™ which can illuminate and separate components of the supply chain from tier 1 prime contractors through tier 2-x. When the supply chain illumination data analysis and visualization tools are combined with Deloitte's business analysis of individual companies, valuable insights are generated that can drive decision making within the government.

f. Commercial aerospace and defense leader

Deloitte is a leading presence in the A&D industry, providing services to greater than 95 percent of the Fortune 500 A&D companies through more than 1,500 practitioners. Many of our practitioners have worked in the A&D industry and/or the military prior to joining our organization and have deep industry experience across many sectors. The level of experience and connectivity of these practitioners allows us to provide insights and observations regarding the latest in commercial concepts, technology, and operating procedures to help our clients address their most pressing issues.

https://doi.org/10.52202/075106-0001

Principal authors

TERRY BOYD
Managing Director in Risk & Financial Advisory, Government and Public Services
tboyd@deloitte.com
+1 571 225 6119

JOHN BURKE
Specialist Executive in Risk & Financial Advisory, Government and Public Services
joburke@deloitte.com
+1 256 426 2099

MATT CRIMI
Manager in Risk & Financial Advisory, Government and Public Services
mcrimi@deloitte.com
+1 256 476 6188

PETE SILVER
Manager in Deloitte Consulting LLP, Government and Public Services
pesilver@deloitte.com

JAMESON MASCARA
Consultant in Deloitte & Touche LLP, Government and Public Services
jmascara@deloitte.com

ANDREW MOTON
Consultant in Deloitte Consulting LLP, Government and Public Services
amoton@deloitte.com

SUZIE KIM
Analyst in Deloitte & Touche LLP, Government and Public Services
suzkim@deloitte.com

SOHAN DATE
Consultant in Deloitte Consulting LLP, Government and Public Services
sodate@deloitte.com

ALEX REGUEV
Consultant in Deloitte & Touche LLP, Government and Public Services
areguev@deloitte.com

https://doi.org/10.52202/075106-0001

ENDNOTES

i https://www.cto.mil/wp-content/uploads/2023/05/2023-NDSTS.pdf

ii https://www.washingtonpost.com/opinions/2023/09/18/ukraine-war-drone-military-technology/

iii https://media.defense.gov/2023/Oct/19/2003323409/-1/-1/1/2023-MILITARY-AND-SECURITY-DEVELOPMENTS-INVOLVING-THE-PEOPLES-REPUBLIC-OF-CHINA.PDF

iv https://www.dau.edu/sites/default/files/2023-11/AcqDays_Topic%208_Tech%20Transition%20Brief%20-%20FINAL.pdf

v 2023_June_DoD Inflation Eminence Paper_Deloitte POV.pdf

vi 2023_Security Assistance and Foreign Military Sales Pressing FY24-30 Considerations.pdf

vii https://sgp.fas.org/crs/natsec/R44711.pdf

viii https://sgp.fas.org/crs/natsec/R44711.pdf

ix https://sgp.fas.org/crs/natsec/R44711.pdf

x https://www.gao.gov/assets/gao-13-286.pdf

xi https://www.asafm.army.mil/Budget-Materials/#fy-2018

xii https://www.asafm.army.mil/Budget-Materials/#fy-2018

xiii https://www.cto.mil/wp-content/uploads/2023/05/2023-NDSTS.pdf

xiv https://www.gao.gov/assets/gao-13-286.pdf

xv 2023 National Defense Science & Technology strategy, n.d. https://media.defense.gov/2023/May/09/2003218877/-1/-1/0/NDSTS-FINAL-WEB-VERSION.PDF

xvi https://www.gao.gov/assets/gao-13-286.pdf

xvii "2013 Annual Report: Actions Needed to Reduce Fragmentation, Overlap, and Duplication and Achieve Other Financial Benefits Report to Congressional Addressees." 2013. https://www.gao.gov/assets/gao-13-279sp.pdf.

xviii Jonathan Specht, "Topic 8 – Tech Transition" Defense Acquisition University, https://www.dau.edu/sites/default/files/2023-11/AcqDays_Topic%208_Tech%20Transition%20Brief%20-%20FINAL.pdf

xix FY2023_Budget_Request_Overview_Book.pdf (defense.gov)

xx GAO-20-578, DEFENSE SCIENCE AND TECHNOLOGY: Opportunities to Better Integrate Industry Independent Research and Development into DOD Planning

xxi David Austin and Tamara Hayford, Research and Development in the Pharmaceutical Industry, Washington, DC: Congressional Budget Office, April, 2019, https://www.cbo.gov/publication/57126

xxii "$6.16 Billion Per Drug Approval: Almost Half of Big Pharma Companies Hit Negative R&D Productivity," Forbes, August 8, 2023, https://www.forbes.com/sites/alexzhavoronkov/2023/08/09/616-billion-per-drug-approval-almost-half-of-big-pharma-companies-hit-negative-rd-productivity/?sh=205267475447

xxiii Ryan Kimmitt and Marcela Vieira, "Time and Success Rate of Pharmaceutical R&D," Geneva Graduate Institute, July, 2020, https://www.knowledgeportalia.org/r-d-time-and-success-rate; Tohru Takebe, Ryoka Imai, and Shunsuke Ono, "The Current Status of Drug Discovery and Development as Originated in United States Academia: The Influence of Industrial and Academic Collaboration on Drug Discovery and Development," Clinical and Translational Science vol. 11, no. 6 (July 2018): 597-606, 10.1111/cts.12577

xxiv "Stage Gate process: How to prevent project risk," Asana, May 31, 2023, https://asana.com/resources/stage-gate-process

xxv Scott J. Edgett, "The Stage-Gate Model: An Overview," Stage-Gate International, https://www.stage-gate.com/blog/the-stage-gate-model-an-overview/

xxvi Edgett, "The Stage-Gate Model: An Overview."

xxvii Martin Haugh, Introduction to Real Options, Columbia University, https://www.columbia.edu/~mh2078/FoundationsFE/RealOptions.pdf

xxviii https://www.ncbi.nlm.nih.gov/pmc/articles/PMC9299990/

xxix "NIST Seeks Input on Implementation of National Standards Strategy for Critical and Emerging Technology," National Institute of Standards and Technology, September 7, 2023, https://www.nist.gov/news-events/news/2023/09/nist-seeks-input-implementation-national-standards-strategy-critical-and

xxx "SIA, FIRST® Announce Partnership to Strengthen U.S. STEM Workforce," Semiconductor Industry Association, May 4, 2022, https://www.semiconductors.org/sia-first-announce-partnership-to-strengthen-u-s-stem-workforce-2/

xxxi National Science Foundation

xxxii Congressional Research Service

xxxiii Office of Personnel Management

xxxiv PPBE Reform Commission Interim Report - https://ppbereform.senate.gov/wp-content/uploads/2024/03/Commission-on-PPBE-Reform_Full-Report_6-March-2024_FINAL.pdf

xxxv Congress.gov

xxxvi DoD Office of the Assistant Secretary of the Army for Acquisition, Logistics, and Technology. Unified Data Reference Architecture, Washington D.C.; Department of Defense 2023, https://sam.gov/opp/f39ecc3623c14642be9531618e6953b3/view

xxxvii https://defensescoop.com/2024/02/14/diu-consults-pentagon-policy-shop-replicator-technology/

https://doi.org/10.52202/075106-0001

AUTONOMOUS AERIAL DRONES CONNECTING PUBLIC SAFETY: OPPORTUNITIES AND CHALLENGES FOR THE FUTURE

Raymond Sheh,[1] Donald Harriss,[2] and Karen Geappen[3]

Reliable voice and data communications are critical in supporting public safety responders during incidents and disasters and will become more important as ground robots and Internet-of-Things (IoT) sensors for public safety applications proliferate. Uncrewed Aircraft Systems (UAS), consisting of an Uncrewed Aerial Vehicle (UAV) and associated systems, are valuable tools for public safety personnel to reduce risk to the public and themselves. They improve the service provided to the community by delivering rapidly deployable "Eyes in the Sky."

It is a natural progression for UAVs to be more closely integrated into the communications of public safety operations. For example, they can provide critical communication links in public safety operations and connect and provide information directly to nearby personnel, devices, and perhaps even the public. UAVs have the potential to fill a vital technology gap where existing infrastructure is damaged or non-existent and where structures or topography make traditional radios and satellite communications ineffective.

In this paper, we discuss the unique capabilities that the closer integration of UAS into public safety communications brings to current and future public safety missions, the challenges users may face in a UAS deployment, and the regulatory and ethical considerations that come with the use of UAS. This discussion dovetails with work that the U.S. National Institute of Standards and Technology (NIST) Public Safety Communications Research (PSCR) Division's Uncrewed Aircraft Systems (UAS) portfolio is undertaking to drive the development of UAS for public safety, with an emphasis on reliable communications, responsible cybersecurity, and artificial intelligence (AI) risk management. This paper will reference related data points from the recently released NIST Cybersecurity Framework (CSF) 2.0 and Artificial Intelligence (AI) Risk Management Framework (AI RMF).

[1] Uncrewed Aircraft Systems Research Lead, Public Safety Communications Research Division, National Institute of Standards and Technology, 100 Bureau Drive, Gaithersburg MD 20877, USA
[2] Uncrewed Aircraft Systems Technical Lead, Public Safety Communications Research Division, National Institute of Standards and Technology, 325 Broadway, Boulder, CO 80305, USA
[3] Independent Researcher, PO Box 174, Bull Creek, Western Australia, Australia, 6149

https://doi.org/10.52202/075106-0002

Introduction

Public safety personnel rely on the integrity and availability of communications for their safety and that of the broader public. This reliance is especially true in dynamic response situations and unstructured environments, such as wilderness rescue, wildland and urban fires, and active shooter situations. Personnel in the field need accurate and relevant real-time information from various sources and locations to inform actions, such as a fire actively jumping a fire break. Incident commanders need to know how a situation evolves in real-time and need to be able to look to their personnel in the field and "Eyes in the Sky" to provide situational awareness of personnel, property, and other assets. There are many documented cases where communication delays and breakdowns in data delivery have cost lives, property, or unrecoverable damage to the local ecosystem.

Infrastructure disruptions can further complicate disaster management by preventing the use of cellular networks and other fixed radio infrastructure. Point-to-point systems, such as Land Mobile Radio (LMR), are limited by the chosen protocols, the laws of physics, and the laws of information theory pertaining to the quantity and quality of information LMR can transmit. These factors define an upper limit as to what is possible for a given number of people spread over a given area, in a given environment, and the use of a given amount of radio frequency spectrum.

Uncrewed Aircraft Systems (UAS), consisting of an Uncrewed Aerial Vehicle (UAV, often also called a drone) and associated systems and software[4], are increasingly employed in public safety missions. UAS are becoming potential solutions to shortfalls in public safety communications. Their rapidly decreasing cost, increasing functionality, and ease of use make UAS valuable assets for many public safety agencies, whether as an "Eye in the Sky" or for more advanced operations[5].

Today, most public safety UAS operations have two-way data communications only between the UAV and operator control unit (OCU, often a handheld controller). The OCU receives commands from the operator(s) and transmits these commands to the UAV. In turn, the OCU receives data from the UAV, such as imagery, videos, and telemetry (e.g., location and heading). The OCU may also receive other information from the UAV, depending on the UAV's sensors.

While not usually considered part of communications, the UAV, and sometimes OCU, will receive information from systems such as the Global Positioning System (GPS) and broadcast information such as Remote Identification (Remote ID) to meet FAA and other regulatory requirements.

This standalone UAS model was suitable for situations where the UAS was only an "Eye in the Sky," whereby the imagery had to be mediated through personnel at the OCU, such as the pilot or payload operator, and then perhaps the incident commander, before being available to other personnel in the field.

However, as the quality of data from UAS has improved, along with their ability to operate with increasing levels of autonomy, requiring information from the UAS to pass through the singular OCU can present a bottleneck to getting timely information to personnel in the field. Additionally, most UAS designs are closed or proprietary, making it difficult to communicate

---

[4] In this document, we make a clear distinction between the UAV, the flying vehicle itself, and the UAS, which includes the UAV and its associated support equipment and systems, such as antennas, operator control unit (OCU), communications infrastructure, and so-on.

[5] While this document primarily focuses on Small Uncrewed Aircraft Systems (sUAS) that are in common deployment across public safety agencies, the principles are largely applicable to UAS in similar roles, regardless of size.

https://doi.org/10.52202/075106-0002

with other endpoints in the field. For example, personnel in the field usually cannot receive information directly from the UAV. The UAV also usually cannot receive data from IoT sensors in the field.

Furthermore, a UAV, particularly one capable of loitering for more extended periods, is often in a much better position to communicate with personnel in the field than fixed or mobile ground-based stations. It is quite possible that the only equipment that is in the right place to transmit critical information to a particular person in the field might be the UAV.

Imagine if the UAV could integrate and participate more fully in the response team's communications ecosystem. Instead of only communicating operationally with one OCU, it behaves more like another team member in the communications network. Perhaps it could decide to directly transmit a subset of its data to personnel within range without needing to go through the OCU. Perhaps the UAV could act as a relay, using its extra height to link personnel to each other and to the incident commander, who might be on the other side of a ridge and be limited as to with whom they can directly communicate. Multiple UAVs could even form a mobile ad-hoc network (MANET)[6], a type of wireless network that can reconfigure itself as the UAVs and other nodes move in and out of range of each other. The network dynamically routes information between nodes that cannot directly communicate with each other by finding intermediate nodes through which data can be relayed. Such a network can maintain communications across a changing operational environment as network nodes on vehicles, UAVs, and personnel move in and out of range of each other.

The principle is simple enough. The underlying technology exists and has been demonstrated to work. Indeed, with the increasing levels of autonomy of UAS, particularly those operating in clear sky and away from obstructions, a UAS can truly be a team player. It can take commands from different people at various times while performing autonomously as an "Eye in the Sky" and as a communications relay. However, as with all new technology, the details matter. It is not enough to prove that a system works. The system needs to be appropriately, reliably, safely, and ethically integrated into public safety operations.

This paper will first provide background in communications, UAS, cybersecurity, autonomy levels, and risk management. We will then describe how new, upcoming technologies relating to autonomous, highly connected UAS can benefit public safety and delve into the associated risks, potential regulatory interactions, ethics, and social acceptance.

Background

Communications in Public Safety

First responder autonomous robots, like UAS, require secure, priority communication capabilities. This requirement presents technical challenges such as allocation of services, attention to cybersecurity, and use of AI data. Prior knowledge of public safety communications is required to understand the challenges presented by these new technologies. Public safety uses dedicated wireless interoperability frequencies for first responder events and communications. These frequencies include various channels in the 700MHz and 800MHz bands and interoperability channels in the Very High Frequency (VHF, 30 MHz - 300 MHz) and Ultra High Frequency (UHF, 300 MHz - 3 GHz) bands. These channels are divided into dedicated talk groups for specific purposes. Over time, many of these channels have moved from analog to digital communication methods for better bi-directional communication. However, switching

---

[6] For the purpose of this discussion, we make the distinction between MANET and the broader category of Mesh Networks that include pre-defined, fixed networks.

https://doi.org/10.52202/075106-0002

from analog to digital has yet to solve scaling problems, especially where data-rich technologies require additional bandwidth. To better use these limited resources, a frequency spectrum segment from the 700MHz range has been allocated to deploying dedicated broadband cellular communications for first responders. This allocation includes enhancements for greater coverage, interagency collaboration, security, and data applications for situational awareness. New communication tools like UAS and data-rich methodologies, such as AI, require digital communication networks that go beyond traditional systems. Industrial, Scientific, and Medical (ISM) frequencies are used for most UAS communications. The ISM band has opened up new opportunities for innovation and accessibility in wireless communication, leading to the growth of various technologies that rely on wireless communication. However, there are not any dedicated spectrum offerings for UAS operations; first responder activities must compete with non-responder civilian traffic on the ISM spectrums. New deployment systems, such as Drone as First Responder (DFR)[7] systems, which we will discuss in more detail, can take advantage of cellular broadband communications for prioritized, responder-focused networks.

First responders' operational demands are growing quickly, and traditional networks cannot provide reliable communications. Modern facilities are moving towards connected dispatch systems using AI-driven applications to orchestrate remote command centers and autonomous assets while monitoring boots-on-the-ground operations. To support these diversified, interoperable systems, networks will require more bandwidth and be able to handle multiple traffic types.

Flexible, scalable, and seamlessly integrated deployable broadband networks can meet the increasing demand for reliable, high-performance connectivity. These networks leverage AI-driven, advanced routing, software-defined networks and meshing protocols to ensure uninterrupted connectivity despite network failures or congestion. This level of service allows for robust applications that use AI at the user network edge in near real-time. It is good practice for organizations to achieve reliable communications by establishing end-to-end agreements among stakeholders to ensure the safe and secure use of networked systems and AI-driven applications. Building a reliable, sustainable ecosystem requires collaboration across industries and adopting common standards and protocols to enable interoperability and seamless integration of different first responder applications.

Current Use of UAS in Public Safety

Public safety personnel predominantly use UAS to obtain visual information, usually through conventional (visible light) cameras but also increasingly through thermal imagery and 3D mapping. UAVs are not only used in lieu of crewed aircraft but also in new applications, such as flying low under trees or into buildings, where it is impossible to use crewed aircraft. While the situations vary between different public safety applications, in general, the goal is to put eyes onto a situation from a perspective where it is difficult, dangerous, or impossible to otherwise gather that information.

In the general case, the UAS is deployed by public safety personnel at, or close to, the response location. The deployment may be by a dedicated UAS response team or by personnel who have other roles but also maintain proficiency and equipment for deploying UAS as necessary. In such use cases, the UAV communicates directly with an OCU, generally in an ISM band, although some communicate in specific public safety bands. The OCU sometimes has a

[7] King, S., Major, s., Maccollum, M. "Drone as First Responder Programs: A New Paradigm in Policing," *MITRE*, 2023-08-11, https://www.mitre.org/news-insights/publication/drone-first-responder-programs-new-paradigm-policing

https://doi.org/10.52202/075106-0002

communications link, usually via a cable, wifi, or cellular service, to allow the UAV to communicate with other systems. These systems are often collaboration tools[8] like Team Awareness Kit (TAK)[9], DroneSense[10], or DroneDeploy[11]. These tools allow this information to be shared with other team members. Some of these services also provide the ability to control the UAV through the OCU.

We are increasingly seeing the use of DFR, which refers to a class of UAS that aims to have the UAV deploy to the response location independent of the responding personnel. The UAV is usually pre-stationed in weatherproof pods on the roofs of buildings in the service area, from which they launch and where they land without physical human intervention. They are generally controlled through a control center that is neither in proximity to the UAV nor the response location. Due to current regulatory requirements, there may be a person in the proximity of the UAV pod whose role is to maintain situational awareness of the airspace around which the UAV flies. With the advent of automated airspace monitoring technologies, this restriction is loosening.

Decoupling the flight of the UAV from the responding personnel requires a new paradigm in communication. The UAV is no longer near a dedicated OCU with which it communicates. Instead, systems use a deployment pod, which may have colocated antennas for communication with the UAV. The UAV is also in a dedicated service area, which allows the UAV to communicate with known communications infrastructure, such as 4G and 5G cell phone towers. We are starting to see UAVs that communicate with this infrastructure. For example, several DFR-focused UAVs have one or more cellular SIM cards and the appropriate cellular modems to connect to 4G and 5G infrastructure.

We are also starting to see UAVs for public safety use that can participate in Mobile Ad-hoc Networks (MANET). This allows the UAV to dynamically discover and communicate with other public safety entities, such as vehicles and people equipped with compatible MANET equipment, as they move in and out of range.

These new communications paradigms bring immense opportunities for greater reliability and safety, as well as new use cases. However, they also bring considerable technical, regulatory, safety, and ethical challenges to consider. The purpose of this document is to present some of these opportunities and challenges, to give stakeholders greater visibility, and to allow them to make informed choices regarding the development and adoption of such technologies.

Cybersecurity

For the purpose of this paper, we use the definition of cybersecurity used in NIST Special Publication 800-53 Rev. 5[12]:

"Prevention of damage to, protection of, and restoration of computers, electronic communications systems, electronic communications services, wire communication, and electronic communication, including information contained therein, to ensure its availability, integrity, authentication, confidentiality, and nonrepudiation."

---

[8] Any mention of commercial products is for information only; it does not imply recommendation or endorsement by NIST.

[9] Team Awareness Kit Product Center, "TAK Product Lines," accessed 2024-04-03, https://tak.gov/products

[10] DroneSense, "DroneSense Home Page," accessed 2024-04-03, https://www.dronesense.com/

[11] DroneDeploy, "DroneDeploy Home Page," accessed 2024-04-03, https://www.dronedeploy.com/

[12] Joint Task Force, "NIST Special Publication 800-53 Revision 5 Security and Privacy Controls for Information Systems and Organizations," *National Institute of Standards and Technology (NIST)*, 2020-12-10, https://doi.org/10.6028/NIST.SP.800-53r5

This is a broad definition that deliberately encompasses both a wide variety of assets and a wide variety of impacts on an organization.

Particularly important to note, when it comes to increasingly connected UAS, is that cybersecurity is not just limited to people breaking into the UAS to steal information. As technology has become more accessible, a wider variety of adversaries have proliferated, with a correspondingly wide variety of motivations to cause damage.

Cybersecurity risks have also increased in both variety and severity as critical systems become more interconnected. Attackers can enter through a seemingly unrelated system and find ways to jump between networks, an action known as "moving laterally." For example, several well known hacks of big corporate networks started with the attackers compromising systems of related organizations, such as contractors or suppliers, and then using the access that those organizations had to break into their actual target. In this context, all organizations in the network have a stake in protecting each other.

While the wording of this definition focuses on equipment, the human factor is also critical to consider in cybersecurity, both in terms of attacks coming from inside an organization, as well as attackers stealing an individual's credentials. This may be as simple as guessing things like passwords or password reset details or as complex as crafting specific, directed "phishing" emails that trick the individual into handing over important information. Two-factor or multi-factor authentication is often touted as a solution to this problem by requiring the user to prove in multiple ways that they are who they say they are. For example, the system might ask for both a password and a one-time code that is emailed to them. Of course, in the context of connected UAS, multi-factor authentication has to be chosen carefully. If the UAS is being used by public safety as part of an effort to restore communications, requiring cell phone access to receive a code to launch the UAV might not be a good solution.

There is much guidance available in the cybersecurity space. This includes:

- The NIST Privacy and Risk Management Frameworks.
- NIST SP 800-53 Rev. 5, Security and Privacy Controls for Information Systems and Organizations.
- NIST SP 800-171 Rev. 2, Protecting Controlled Unclassified Information in Nonfederal Systems and Organizations[13].
- The Criminal Justice Information Services (CJIS) Security Policy[14].
- International Organization for Standardization (ISO) and International Electrotechnical Commission (IEC) ISO/IEC 27001:2022 Information security, cybersecurity, and privacy protection[15].

However, there is very little in the way of comprehensive guidance available in the UAS space. As we illustrated above and will discuss further later in this document, there are

[13] Ross, R., Pillitteri, V., Dempsey, K., Riddle, M., Guissanie, G. " NIST SP 800-171 Rev. 2 Protecting Controlled Unclassified Information in Nonfederal Systems and Organizations," *National Institute of Standards and Technology (NIST)*, updated 2021-01-28, https://doi.org/10.6028/NIST.SP.800-171r2

[14] CJIS Information Security Officer, "CJISD-ITS-DOC-08140-5.9 Criminal Justice Information Services (CJIS) Security Policy Version 5.9," U.S. Department of Justice, Federal Bureau of Investigation, Criminal Justice Information Services Division, 2020-06-01, https://www.fbi.gov/file-repository/cjis_security_policy_v5-9_20200601.pdf

[15] International Organization for Standardization (ISO) and International Electrotechnical Commission (IEC), "ISO/IEC 27001:2022 Information security, cybersecurity and privacy protection, Information security management systems, Requirements," https://www.iso.org/standard/27001

https://doi.org/10.52202/075106-0002

considerable dangers to blindly applying cybersecurity guidance to UAS without considering the unique aspects of the application.

Specific guidance regarding the management of cybersecurity risk for Uncrewed Systems is still in its relative infancy. Organizations that have issued guidance that relates to Uncrewed Systems include:

- The Cybersecurity & Infrastructure Security Agency (CISA), part of the U.S. Department of Homeland Security (DHS), has issued guidance that focuses on the procurement and use of Uncrewed Aircraft Systems[16] rather than the development, manufacture, or broader integration into wider organizational systems and processes.
- Both the Defense Innovation Unit (DIU) of the U.S. Department of Defense (DoD) and AUVSI have issued guidance on the procurement of UAS based on the National Defense Authorization Act (NDAA)[17]. DIU maintains a "Blue UAS" list[18] while AUVSI maintains a "Green UAS" list[19], which list the UAS that they have evaluated according to the NDAA criteria. This guidance is primarily focused on geopolitical, cybersecurity, organizational, and supply chain considerations of the UAS manufacturer rather than its integration into the organization of the end user's broader systems and risk management.
- The NIST Information Technology Laboratory (ITL)[20] produces guidance on a wide variety of cybersecurity topics, including the aforementioned resources. While it does not have an Uncrewed Systems-specific focus, it does focus on applications such as Automated Vehicles[21]. Much of the guidance within the Cyberphysical Systems and Internet of Things (IoT) topics is also highly relevant to Uncrewed Systems.

As we will discuss later, we have established a working[22] group that aims to adapt these resources to the UAS space to better fill the gaps in available guidance to stakeholders, including end users, manufacturers, developers, researchers, and regulators.

Artificial Intelligence

For this discussion, we will use the ISO/IEC 22989:2022[23] definition of AI:

"... a technical and scientific field devoted to the engineered system that generates outputs such as content, forecasts, recommendations or decisions for a given set of human-defined objectives."

Within this broad definition, we will also discuss the following:

---

[16] Cybersecurity & Infrastructure Security Agency (CISA), "Be Air Aware," accessed 2024-04-03, https://www.cisa.gov/topics/physical-security/unmanned-aircraft-systems

[17] House Armed Services Committee, "NDAA - National Defense Authorization Act," accessed 2024-04-03, https://armedservices.house.gov/ndaa

[18] Defense Innovation Unit, "About Blue UAS", accessed 2024-04-03, https://www.diu.mil/blue-uas

[19] Association for Uncrewed Vehicle Systems International (AUVSI), "Green UAS," accessed 2024-04-03, https://www.auvsi.org/green-uas

[20] National Institute of Standards and Technology (NIST), "Information Technology," accessed 2024-04-03, https://www.nist.gov/information-technology

[21] National Institute of Standards and Technology (NIST), "NIST Automated Vehicles Program," accessed 2024-04-03, https://www.nist.gov/programs-projects/nist-automated-vehicles-program

[22] National Institute of Standards and Technology (NIST), "PSCR UAS Working Group, Cybersecurity and AI Risk Management for Uncrewed Aircraft Systems (UAS) in Public Safety," updated 2024-02-27, https://www.nist.gov/ctl/pscr/pscr-uas-working-group

[23] International Organization for Standardization (ISO) and International Electrotechnical Commission (IEC), "ISO/IEC 22989:2022 Information technology, Artificial intelligence, Artificial intelligence concepts and terminology," accessed 2024-04-03, https://www.iso.org/standard/74296.html

- Machine Learning (ML) describes AI systems whose behavior depends also on observed input data.
- Statistical ML describes ML systems whose behavior depends specifically on the statistical properties of the observed input data.
- Deep Learning (DL) describes Statistical ML systems whose behavior depends on fitting highly complex, randomly initialized sets of equations to the observed input data.

It is essential to understand that various types of AI have different properties. DL is very topical right now and has some very well-known properties, such as requiring immense amounts of data. However, other forms of ML, such as regression techniques, use task-related data differently and may require less of it. Some DL and ML systems use foundational models and transfer learning, whereby the system first learns to make decisions relating to a different task and then uses that experience to help it learn to make the desired decisions more quickly.

There are other forms of AI that are not ML and have yet other properties. Classical planning techniques do not rely on the statistical properties of data and instead solve problems expressed in much more mathematical terms. Reasoning systems, in contrast, make use of large amounts of background knowledge, such as knowledge of physics, combined perhaps with a limited amount of task-related data, to make their decisions.

This distinction is important because different forms of AI expose stakeholders, including users, the organization, and the community, to different risks. However, these may only be obvious when detailed information about the underlying forms of AI is known. These distinctions can significantly affect how risks associated with the deployment and use of the system should be managed. As UAVs become more autonomous, with less human oversight, and as these new communications paradigms become more common, the management of these risks becomes even more critical.

For example, if a user disagrees with a decision that an AI system makes, their ability to find out who was correct, and why the decision was made, varies considerably depending on the underlying form of AI used in the system. In some cases, it is obvious that the AI system made a mistake, such as the autonomous UAV clearly not reacting to an obstacle that appears in its sensors. In other cases, it is less obvious, such as when the UAV takes an unexpected route that still reaches its destination. Disagreements like this are not necessarily a sign of failure in the system. After all, one of the primary reasons for using AI is for the system to perform better than humans, so it should not be surprising that the AI system might sometimes make decisions that humans might disagree with in the moment.

Some AI systems are relatively transparent and able to explain their decisions in a way that suitably trained humans can understand in an actionable manner[24]. If the system was correct in its decision, it might help users understand how it came to its decision, and the factors it considered. Systems based on classical planning and, to some extent, regression often have such properties.

However, many AI systems, particularly ones based on statistical ML, are not so transparent. This lack of transparency may be because the system makes its decisions in a manner that is too complex, or otherwise too difficult, to be meaningfully interpreted, as is the case in DL systems. In other cases, it may be because the system has not arrived at its decision in a way that has a meaningful interpretation, such as systems that develop their decision-making models through random trial and error.

---

[24] Sheh, R., Monteath, I., "Introspectively Assessing Failures through Explainable Artificial Intelligence," *IROS-17 Workshop on Introspective Methods for Reliable Autonomy (IMRA)*, 2017

https://doi.org/10.52202/075106-0002

Along similar lines is the ability to correct errors that are encountered. The new applications enabled by highly connected UAVs often run with reduced human oversight. This makes human judgment less effective as a control on systematic errors and increases the importance of correcting detected errors in a verifiable manner. Once again, AI systems vary considerably in their ability to be repaired and to check that the repair fixes related errors without causing new errors.

Levels of Autonomy

When discussing opportunities and risks associated with autonomous UAS, it is useful to distinguish the various levels at which decisions are made within the UAS, if those decisions represent autonomy or automation, and if those decisions are made on the UAV or with the assistance of other parts of the UAS such as the OCU. In this section, we will describe some illustrative examples from currently deployed UAS, loosely guided by the NIST Autonomy Levels for Unmanned Systems (ALFUS)[25]. As a new generation of connected UAS applications run with reduced human oversight, it is imperative to understand which decision-making processes, at the various levels of autonomy, were previously supervised and where systems are now expected to be trusted to "do the right thing."

Control Autonomy. The vast majority of UAVs in use by public safety are autonomous to some degree at the control level. The operator will command a direction in which to fly, rotate, ascend, or descend. An autonomous controller in the UAV takes this command, combines it with sensor data, and turns it into motor commands that fly the UAV in the desired direction, accounting for the wind, variations in the UAV's behavior due to age, wear and tear, damage, and any added accessories. This type of autonomous controller deals with very simple tasks in a moderately complex environment of wind and varying system characteristics. The human provides the commanded direction but does not provide the UAV with assistance in determining how to spin the motors to achieve that commanded direction.

This contrasts with other UAVs, such as traditional model aircraft or acrobatic drones, which rely almost entirely on the operator to deal with such variations. These UAVs use very simple analytical controllers and simple sensor data and processing to determine how to spin the motors in response to the operator's commands.

More recent public safety UAVs have an additional autonomy layer that makes use of AI, often including at least some DL, to detect obstacles around the UAV and determine if the UAV should stop or fly around the obstacle, overriding the commanded direction of flight from the operator. These UAVs deal with the same simple task of "fly in a given direction" but now account for a more complex environment where there are obstacles with which to collide.

On top of these lower-level autonomy layers, most deployed UAVs receive their commanded flight direction from a human operator. Examples of these flight-level commands might be "fly forward at a moderate speed" or "turn left." Some are controlled by an OCU that is much like that of a model aircraft, with two joysticks that the operator uses to command the UAV to move in a particular direction and at a particular speed. It is assumed that the operator is handling these higher levels of task and mission complexity while the lower levels handle, and abstract away, some of the environmental and system complexities.

[25] Ad Hoc ALFUS Working Group Participants, "NIST Special Publication 1011-II-1.0, Autonomy Levels for Unmanned Systems (ALFUS) Framework Volume II: Framework Models Version 1.0," *National Institute of Standards and Technology (NIST)*, 2007-12-28, https://doi.org/10.6028/NIST.sp.1011-II-1.0

Task Autonomy. Most UAVs deployed in public safety applications also have some basic automation features to which the operator can delegate simple flying tasks, such as flying orbits or a grid pattern. Automation of these task-level behaviors is distinguished from autonomy in that automation is generally a set of pre-defined or pre-computed commands and does not respond to unexpected situations with the intent of continuing its task. They can only handle static and usually relatively simple environments. Most orbit or grid pattern features on UAS, be they running directly on the UAV or elsewhere in the UAS, such as the OCU, will generally stop if they encounter high winds or an unexpected obstacle (that the lower-level autonomy cannot handle transparently).

Fly-to-Point and Return-to-Home task-level behaviors can vary in their level of complexity. Early systems were purely automated, whereby the UAV would ascend to a given height and then fly directly towards a defined location, such as the denoted fly-to point, takeoff point or OCU location. Other Fly-to-Point and Return-to-Home features can be elaborate autonomous behaviors, which can fly around obstacles based on a 3D map, either provided or generated by the UAV or other parts of the UAS. DFR systems typically exhibit autonomous behaviors at this level, with the operator, usually in a command center, providing the fly-to points.

Some UAVs have a degree of landing autonomy that allows the UAV to vary its landing location depending on the suitability of the terrain. Such systems make extensive use of AI to characterize the terrain and assess the risk of landing on a bad surface versus the risk of running out of power. Some of these assessments can be quite involved, particularly if they also account for other agents, such as animals that may appear at the intended landing site. Such systems can potentially override the human operator. This capability is crucial for systems where there is a significant latency between the operator and the UAV, such as in the case of novel communications technologies that might improve reliability and range but at the expense of increased latency. In such cases, this additional latency may make it impossible for the operator to quickly intervene to address unexpected events.

There are autonomous goal-directed behaviors that are increasingly seen on UAS deployed by public safety. In such systems, the location that the UAV should fly to is determined on the fly by (usually) onboard sensing, with the human operator providing some specific mission information. An example of such an advanced task-level behavior is a UAV that autonomously tracks and follows an operator-designated object of interest that is moving relative to it through the use of its onboard sensors, such as its cameras. Such a system may need to deal with substantial mission and environmental complexity with minimal human intervention. This is especially the case if it must handle the object of interest being temporarily hidden behind other objects and, thus, needs to decide how to re-acquire the object.

Mission Autonomy. Finally, there are systems that can autonomously generate whole mission-level, complex, multiple-stage, dynamically adaptive flight plans based on data collected in real time. For example, some systems can autonomously fly an initial simple grid pattern over a complex environment, such as a vehicle accident, to build a rough 3D map. While the UAV is still in the air, it can analyze the rough map to understand the location of various objects in the environment and then autonomously develop a flight plan and fly that plan to gather more detailed data to generate a comprehensive 3D map. As it does this, it might continue to analyze its map and adjust its mission plan accordingly, perhaps even scheduling time to return to its base to replace its battery. In such a case, the human operator might only tell the system the boundaries of the location to map and the desired level of fidelity; the system makes all subsequent decisions by itself.

https://doi.org/10.52202/075106-0002

As we will discuss later in this document, the integration of additional communications roles into autonomous UAS will require additional levels of autonomy that will build on those mentioned above.

SAE Levels of Autonomy. It is important to distinguish between the levels of autonomy mentioned above, which divide autonomy into levels of control abstraction and complexity, and levels of autonomy that describe the relative need for human intervention in the system as a whole. While there is often some correlation between the two, they represent different concepts. Perhaps the most commonly cited is outlined by the Society of Automobile Engineers (SAE) International in Taxonomy and Definitions of Terms Relating to Driving Automation Systems for On-Road Motor Vehicles J3016_202104[26] (J3016). This reference is often called the "SAE Levels of Autonomy." Although J3016 was originally developed in the context of self-driving cars, these levels are often adapted for other forms of autonomous systems, including UAVs.

The J3016 levels range from Level 0, meaning no automation, to Level 5, meaning full automation. Of course, a Level 0 vehicle can still have autonomy at the lower levels, as we refer to them in this document. For instance, according to J3016, a Level 0 vehicle might still have autonomous systems that sense obstacles and apply the brakes or adjust the steering to avoid an accident.

It is also essential to consider the complexity of the application and environment when adapting J3016 to other applications. For example, a Level 5 autonomous vehicle, according to J3016, is specific to vehicles driving in a "conventional" manner on publicly accessible roadways. The Level 5 vehicle may not know how to drive forward along a four-wheel-drive trail, nor is it expected to know how to back up a large trailer on a roadway (even if it may be capable of doing so).

When applying SAE Levels of Autonomy to describe, for example, UAVs used as communications relays or as systems that otherwise provide information to multiple personnel on the ground, it is vital to do so in the context of a well-defined mission and environment. A UAV that only needs to fly to and hover at a fixed location during the day, in good weather, with a clear path between its launch/land point and that fixed location in the sky, could claim Level 5 autonomy with fairly modest capabilities. In contrast, a UAV that must maintain relay communications to multiple mobile ground stations in a mountainous wildfire scene with challenging terrain, smoke, and wind would need very advanced capabilities to be considered Level 5 Autonomous. In this scenario, it would need sensing and AI capabilities to determine the best location to be in as the ground stations moved, avoid terrain and other obstacles, and respond to changes in local weather due to the fire itself.

Management of Risk

NIST SP800-39 defines risk as follows:

"A measure of the extent to which an entity is threatened by a potential circumstance or event, and typically a function of: (i) the adverse impacts that would arise if the circumstance or event occurs; and (ii) the likelihood of occurrence."

Risk management is further defined as:

"The process of identifying, assessing, and responding to risk."

---

[26] Society of Automobile Engineers (SAE) International, "Taxonomy and Definitions for Terms Related to Driving Automation Systems for On-Road Motor Vehicles J3016_202104," 2021-04-30, https://www.sae.org/standards/content/j3016_202104/

Risk management is not necessarily (or even usually) about purely minimizing a set of risks associated with a given activity. Rather, it is more nuanced and includes such tasks as enumerating and characterizing the different risks associated with the activity relative to similar alternatives, comparing them to the risk tolerance of the given application in consideration of the benefits, including possible reduction in other risks, determining controls that are worth applying, and deciding if the residual risk of that activity is acceptable.

The public safety community is more aware than most of the need for risk management. Public safety personnel routinely make on-the-spot decisions as to the risks to take in the moment, accounting for a wide variety of unusual situations, be it the decision to enter a burning building or to drive at high speed the wrong way down a busy street. Coupled with this is an innate sense of factors that manage this risk, many of which are pre-prepared and some of which are societal. For example, when public safety personnel decide to get to the burning building quickly by driving the wrong way down a street, society participates in this risk management by getting out of the way when we see the flashing lights and hear the sirens.

As already defined, risk management requires identifying, assessing, and responding to risk. These three tasks are usually well-defined for well-understood domains, such as driving at high speed the wrong way down a busy street. Even with the advent of vehicles with increasingly autonomous features, such as adaptive cruise control on public safety vehicles, regulations regarding minimum levels of performance usually allow traditional risk assessments to continue to be relevant[27], in large part because the human is still responsible and maintaining situational awareness. There are second-order risks associated with humans beginning to rely (perhaps too much) on some of these assistive autonomous features; however, thus far, these have not risen to a significant enough level to induce a major change in risk management culture.

The different levels of autonomy and the associated implementations of AI significantly influence the options available for managing risk. This is particularly the case in new domains where there are not commonly accepted standards of performance, requirements, or even a common language with which to express requirements and capabilities. This results in an inability to appropriately identify the risks, assess their impacts, and respond to them accordingly. Where risks are underestimated, personnel are likely to unknowingly take excessive risks. In contrast, over-estimating risks can result in not deploying capabilities or deploying other capabilities that may be less appropriate, shifting risk to other parts of the operation.

In the process of increasingly integrating UAS into public safety communications infrastructure, it is critical that public safety is able to appropriately identify the associated risks, assess their impacts, and respond to them. While these new technologies provide new capabilities and reduce other risks, they can impart new risks to existing communications infrastructure.

Governance for Uncrewed Systems

Governance is the collection of systems of processes, mechanisms, and rules by which an organization is controlled and operates. It ensures that the people within the organization are held accountable when moving towards organizational objectives. Appropriate governance bridges across the organization as well as up and down its hierarchy. It is a traceable collection of artifacts linking organizational objectives with processes, procedures, and technology. It also needs to flow in both directions, with approvals informing teams and metrics reported to determine effectiveness. The focus on organizational objectives amalgamates what is important to the

[27] The behavior of new technologies in other vehicles in the environment is a different question, as evidenced by recent problems with public safety interactions with autonomous vehicles and electric vehicle fires.

https://doi.org/10.52202/075106-0002

organization, such as social considerations, regulatory requirements, technology, and organizational skills and capabilities.

Governance is a key pillar when it comes to risk management including the use of UAS for public safety response. It is necessary to have in place governance that defines and helps personnel responding to incidents to make the decision to deploy UAS, the actions they should take when using UAS, and when to recall UAS. When an organization is tasked with operating in a real-time environment where there are human safety aspects, governance is critical in ensuring that all personnel operate knowing what others are doing, even with degraded capabilities. For example, in a corporate setting, not having a continuity plan for the loss of network connectivity can be inconvenient, whereas in an emergency response situation, not having network connectivity can be the difference between life and death.

By creating and operationalizing appropriate governance, an organization achieves risk management in two aspects: organizational and operational. First, the organizational aspect is with knowing that the UAS is used in accordance with organizational objectives. It will support and enhance and not inadvertently hinder a mission. Defining what it can and can't be used for buffers the organization from aspects related to inappropriate deployment and use. It allows for monitoring of the UAS as an appropriate tool in the organization's arsenal and fine-tuning of its capabilities, configuration and deployment within operation sets.

The second is to manage the risks associated with the deployment and use of the UAS in the field by establishing pre-determined and understood operating bounds, expectations and processes for all personnel involved. It removes ambiguity and assumptions between operators who may not be able to directly communicate with each other. It allows incident coordinators and commanders to have clear expectations when directing and coordinating between teams. By having such aspects already documented, precious time is saved in briefings and response set-ups. It also saves precious time and avoids confusion during an incident response when likely disruptive events (such as loss of communications) can occur as there are pre-defined and understood continuity and work-around processes.

To achieve good eventual governance of UAS, the governance process must start when debating whether and what systems to acquire. A clear goal must be the alignment of usage with the organization's key objectives in public safety. Capabilities and applications described in this paper and others that are not able to be covered that are relevant to an organization should be considered when assessing the bounds an organization sets for UAS, which then feed into acquisition requirements. Once selected and procured, governance artifacts such as policies, plans and procedures are needed to support and implement the governance. By assessing capabilities, application and alignment with objectives prior to acquisition and commissioning, it ensures the UAS is fit for purpose and the personnel can consistently make decisions regarding the use of UAS, all personnel have pre-defined knowledge and expectations of the UAS, and everyone is clear on when to use and when to disengage UAS from the field.

Opportunities and Challenges

As introduced earlier, we will consider two mutually compatible forms of integration of UAS into public safety communications. The first is enabling the UAV to communicate information it has gathered from its sensors, perhaps with some processing, directly with personnel in its vicinity, rather than needing to first transmit this through its OCU. The second is using the UAV as a communications relay, receiving information and transmitting it onward.

https://doi.org/10.52202/075106-0002

UAV-to-Personnel Communications

Some existing UAVs are able to communicate with personnel in the field (and, in the future, other devices such as robots and IoT sensors). Direct UAV-to-personnel communications enable a variety of novel public safety use cases. As previously discussed, at its simplest, this allows personnel in the vicinity of the UAV to observe what the UAV is observing without those personnel needing to communicate with the OCU. This is often the case for analog video transmissions, where any appropriately configured device can also observe the video stream.

Direct UAV-to-personnel communications also enable more complex use cases. For example, in a fast-moving response situation, it may be advantageous for personnel on the ground to call for an autonomous UAV to make an observation, such as looking in a window or providing an overhead view. This use case involves the ad-hoc, dynamic assignment of a UAV to personnel requesting particular data. While the person making the request might be considered the UAV operator, they are providing relatively abstract, mission-level instructions (such as "look in that window" or "give me an overview of this area") rather than low-level flight commands (such as "fly forward" or "turn left"). The person is also only in command of the UAV for a short part of the flight.

One could imagine a fleet of UAVs stationed or in flight in the response area, waiting to service such requests, somewhat like a temporary, on-scene DFR system. While waiting, they might even perform background tasks, such as acting as communications relays or building more detailed 3D maps of the environment.

However, existing widespread methods pose some challenges to more general adoption. In this section, we will describe how such communications usually happen with most currently available UAS. We will then outline how this is being achieved with some of the latest systems and systems of the future.

Existing UAV to Personnel Communications. Having the UAV communicate with other personnel is perhaps the simplest use case, and, in fact, many early public safety UAVs and even some current smaller and lower cost UAVs have such a capability. These UAVs broadcast their video feeds, similar to how a radio or television station might broadcast to anyone with a receiver. The UAV's imagery can be received by any device that is using the same frequency and, if the broadcast is scrambled, has the descrambling key. The video feeds through such systems tended to be lower in resolution and were subject to interference that degraded image quality. However, they did have the advantage of gradually degrading image quality as signal quality decreased rather than suddenly cutting out, as was the case with legacy digital systems. Nowadays, such analog systems are limited to smaller and lower-cost UAVs, particularly ones for indoor use, as these systems tend to be smaller and lighter. The sudden variations in radio link quality in indoor environments due to obstructions can also make the gradual degradation in image quality an advantage. For example, a skilled operator may still be able to make enough sense of a severely degraded image to fly the UAV back to where there is a good signal.

Most modern, larger outdoor UAVs are fully digital systems that communicate to their paired OCUs through a data connection. Using such a data connection is more like receiving video through an on-demand Internet streaming service, where the communication is directed at one receiving entity. In some cases, this is based on conventional 802.11 (WiFi) protocols. Others, such as DJI Ocusync, use their own proprietary methods. Unlike analog systems, which broadcast a signal without any feedback confirming that the signal was received, these digital systems generally establish a two-way data connection between the UAV and its dedicated OCU and often use the same connection for control and transmission of other information, such as telemetry.

https://doi.org/10.52202/075106-0002

Digital systems are more secure and allow for the reliable transmission of greater amounts of data. They can take advantage of advanced underlying radio systems that can account for errors due to the environment, such as interference from other devices in the same radio band and radio reflections from buildings and other structures. They also provide ways to authenticate the identity of the receiver and log who has access to, and control of, the UAV. These data links can be properly encrypted, using techniques that, unlike analog scrambling, require significantly greater resources to compromise.

Some newer and more advanced digital systems also have the advantage of being able to dynamically adapt the amount of data used according to both the needs of the mission and the quality of the communications link. For example, when the UAV is stationary and has a good signal, the picture changes slowly so the UAV can transmit a higher-resolution picture to the OCU. As the signal quality degrades, the UAV can adapt and, depending on the situation, either transmit lower resolution imagery that is still sufficient for control or further reduce the frame rate if the priority is resolution. This brings back some of the advantages that analog systems have, by degrading the image as the signal becomes too poor to support high quality imagery, rather than cutting out completely.

However, by switching from a broadcast to a connection-based protocol, most modern UAVs lose the ability to send the data to multiple receivers with one transmission. Instead, it is typical for the UAV to communicate directly with the OCU, which then transmits the data onwards. The OCU is usually bandwidth-constrained, as it communicates via a cellular modem. It usually transmits the data onwards to a streaming server on the Internet, which then sends the stream to multiple receivers. When an incident commander and other personnel at the site of the response receives a video feed from a UAV on their mobile device, that video is likely to have taken the following path, even if the UAV and operator are right next to them.

- The UAV transmits the video to the OCU via its dedicated radio link.
- The OCU transmits that video via the cellular network to a streaming server hosted by a cloud services company.
- The incident commander's personal device, along with the personal devices of other personnel, make separate connections to the streaming server via a cellular network to download and display the video.

Multiple data transmissions are not a major issue in places with good signal coverage. However, where cellular infrastructure is already strained, or where there are only a few temporary cellular towers present, multiple transmissions present a real challenge, both for the UAS and for other users who need to communicate over the same airwaves. Furthermore, this also means that the UAV can only communicate with users who are either within range of the OCU or the cellular network. In a response situation, it is possible that public safety personnel on the ground have no signal from the cellular network or the OCU and yet are in proximity to the UAV.

UAV to Cellular Network Communications. UAVs are starting to become available with cellular modems that can connect to one or more existing cellular network providers. Data from the UAV is sent directly to the cellular network rather than through a single OCU. In many urban applications, and particularly for applications like DFR, the service area has reliable 4G or 5G data coverage. This makes cellular networks an attractive option for use cases where it is safe to assume that a reliable network is present, enabling Beyond Visual Line of Sight (BVLOS) operation anywhere in the covered area. This is particularly advantageous in areas with taller buildings that would block the signal between the UAV and a single OCU.

Even in situations where the network may be stressed, direct-to-cell network communications can still have an advantage over routing communications via the OCU. The primary advantage is avoiding the need to maintain a radio connection with the OCU, which may be impossible, or require much higher levels of power due to distance or buildings. For DFR applications in urban environments, where there may be many small base stations serving a large population, it is quite possible that the closest cellular tower may be closer than the OCU. Thus, the UAV may require less power to connect to the cellular network than to connect directly to the OCU.

A second-order advantage of having the UAV communicate directly with the cellular network is that it can relieve some of the congestion of the airwaves in the vicinity of the OCU. This may be co-located with other emergency response personnel who also require significant quantities of bandwidth, both on the cellular network and in the ISM or dedicated public safety frequencies within which the UAVs operate. Having the UAV communicate directly to the cell towers, and potentially different towers than the ones near the OCU, can reduce contention for this limited resource, especially if the connection through the cellular network back to the OCU is shared among personnel in the area.

One of the primary disadvantages of having the cellular network as the primary communications channel for the UAV is that of latency. As the imagery must now travel from the UAV to the cellular network, to a server somewhere on the Internet, and then to the OCU, most likely via another cellular connection, significant delays can occur between an event occurring in the view of the UAV and the operator seeing it on their screen. Similarly, the operator's commands will also be delayed in being received by the UAV. Worse yet, this delay can vary over time, sometimes rapidly, due to network congestion, particularly in networks where public safety traffic does not have priority.

This delay, particularly if it varies randomly, greatly increases the difficulty and risk involved in having the operator perform tasks that require a timely response, such as avoiding obstacles or reacting to wind and other disturbances. Therefore, such UAVs will need to rely more heavily on autonomy to operate safely. In particular, they will need to avoid or adapt to fast-changing situations that require actions that are infeasible for an operator to address through a high-latency connection. Understanding the implications of such communications systems on AI risk is an important part of assessing the trade-offs of adopting AI technology.

The use of cellular networks also changes the failure modes associated with the communications between the UAV and OCU. On the one hand, the range of the UAV is likely to be increased for a similar amount of power required, as the UAV can connect to towers that are closer to it than the OCU or to towers that are not obstructed by buildings or other structures. On the other hand, the involvement of additional systems in the communication between the UAV and OCU increases the probability of communication dropouts that personnel on the ground have very little control or visibility over. Addressing this risk likewise requires a UAV that is capable of reliably and autonomously handling fast-changing situations while the operator is unable to communicate with it.

Finally, there are potential problems associated with the fact that current cellular networks are simply not expecting to connect to user equipment (UE) that is flying in the sky. Cellular networks are highly optimized to cover specific areas of service, generally on the ground and in buildings. This optimization ranges from the antennas, which can be highly directional to targeted at these service areas, to the algorithms used to hand connections off between towers based on when a particular piece of UE is likely able to see another tower, and the calculations that the network makes to determine how much power to use to communicate with a given piece of UE in a given location.

https://doi.org/10.52202/075106-0002

While UAVs flying in the service area can communicate with the towers, the fact that they are flying in the sky, in places that the cellular network does not expect them and was not optimized for, can cause problems. For example:

- Cellular networks closely manage power levels used to communicate with UE based on specific assumptions to avoid unnecessary interference and optimize bandwidth. These calculations can be incorrect for UAVs that appear in locations that the network does not expect.
- The antennas on cell towers can be highly directional and are typically pointed downward slightly as it expects UE to be at ground level. This can result in poorer-than-expected reception for a UAV that is flying higher than the tower.

UAV Direct Communications. There are also increasing options for UAVs with digital communications to directly communicate with personnel and devices in the vicinity. In its simplest case, multiple devices, be it multiple OCUs or a combination of OCUs and other devices, can communicate directly with the UAV. In applications that require only a low amount of bandwidth, such as the UAV gathering data from IoT sensors in the field or receiving commands from users, this is generally straightforward.

Where there is a requirement for higher bandwidth, such as streaming video, there are several options available. The simplest is to have the UAV make a direct connection to each recipient of the video. This allows each recipient to have a unique stream but requires significant bandwidth and computation power on the part of the UAV, as it must send the data multiple times. Alternatively, it may require that the image quality be degraded due to limited bandwidth and computation.

More advanced systems may be able to broadcast or multicast the same video stream to several recipients in the field. This means that each recipient receives the same video, but it also means that the UAV needs to send the video stream only once.

UAV as Communications Relays

In addition to the UAV communicating directly to other entities for the purpose of its own mission, UAVs can be used to relay communications from other entities. At its simplest, the UAV could be a fixed communications repeater, with connections to the two people or devices that need to communicate. This setup takes advantage of the UAV's elevated position to "bounce" the signal around an obstacle or over a distance that would otherwise prevent the two people from directly communicating. In such an application, the UAV would simply need to station itself at a fixed location and perhaps return to its base occasionally to replace batteries or refuel. Two UAVs can also "take turns" so that one is always in the correct location while the other is returning for a battery change or refueling. Alternatively, a tethered UAV could stay in one location for an extended period of time. Such a paradigm can involve relatively little autonomy, especially if a human operator is closely monitoring the mission.

A significantly more complex example would be the use of multiple UAVs, each equipped with MANET radios, that communicate with each other and with multiple radios on the ground, be it personnel or devices. Such a system can be an advantage, both in increasing the possible communications and coverage range and also in increasing available bandwidth. This can be an advantage in a congested radio environment, even if the people and devices could otherwise communicate with each other. This is because if a radio needs to run at high power to reach the desired receiver, it takes up that frequency across a large area of space, much like a person shouting across a room interferes with all of the conversations in that room. In contrast, a radio running at lower power, talking to a nearby UAV, that then relays that data onwards to the next closest UAV and so on to the receiver, allows other nearby communications to occur on the same frequency, much like how many people can talk quietly in a room without interfering with each other, even if some of them are passing messages along.

More advanced use cases include using a combination of technologies. For example, in a location that has no cellular network coverage due to remoteness or damage to infrastructure, it is possible to place a cellular network antenna on a large UAV to which conventional cell phones can connect. The UAV can then connect to a MANET or other type of network that provides connectivity to the outside world[28]. The UAV becomes a "flying cell tower", providing coverage to public safety in the critical initial period of a response before additional temporary structure is deployed.

While such a system can be effective with fixed locations for the UAVs, in the future, tight integration between advanced autonomy on the UAVs and the radio systems can also allow the UAVs to dynamically station themselves in the locations where they can be most effective given the changing communications needs of the network, the movement of the radios on the ground, and the need for the UAVs to take turns returning home for battery changes.

Applications and Benefits. To analyze the benefits and risks of using UAS as part of a MANET or as a data ferry, let's give an example using a high-altitude mountainous recreation area. Many of these locations are often remote and lack the infrastructure to provide data or voice communications. Typically, these are purposefully designated and protected areas, such as national parks and wilderness areas, oriented to preserve the natural environment. As such, the buildout of any infrastructure is limited or prohibited. Not only are broadband cellular communications unavailable, but first responder LMR is often limited or non-existent in these areas. As such, responders typically depend on VHF wireless communications to relay messages and information.

First responders operating in wilderness areas contend with a variety of incidents, such as search and rescue missions, drownings, exposure to elements, medical emergencies, criminal activities, and natural or human-made events. Because of this, the first responders require specialized skills, training, and communications tools, such as UAS. UAS aid in search and rescue missions and can also relay communications in difficult terrain. In-flight UAV with data ferry and/or MANET capabilities can enhance information exchange for responders and incident commanders in extreme environments. With enough resources, MANET provides network redundancy, load balancing, reliability, and quality communications. It is able to do so dynamically, as nodes, such as UAVs, vehicles, and personnel with radios, move in and out of range of each other.

The interconnectedness provided by MANET allows responders to share resources that one node can access across all nodes in the network. For example, a UAV with a MANET node and a cellular modem can use its extra height to reach a cellular network tower and obtain internet access. In that case, cellular access can then be provided to personnel, UAVs, and other equipment in the vicinity who may be unable to access that tower. This allows personnel to communicate more widely and access crucial resources such as databases, weather reports, and Common Operating Picture (COP) tools. It also enables more advanced AI features on UAVs and other equipment that benefit from access to the internet, be it to access additional data and computation or share the data with other personnel not in the immediate vicinity.

---

[28] Knight, R. "Restoring Communication After a Disaster," *Inside Unmanned Systems*, 2023-12-13, https://insideunmannedsystems.com/restoring-communication-after-a-disaster/

https://doi.org/10.52202/075106-0002

MANET also brings with it new risks beyond those of traditional wireless networking systems. For example, a variety of Denial of Service attacks are possible on MANET that can inhibit all network communications. This is particularly problematic for attacks that target the dynamic reconfiguration capabilities of the network. Latency can also be an issue, as data must be passed from one node to the next, each adding some delay. This can be exacerbated with the use of UAVs, where a UAV leaving the network to recharge might mean that the network needs to re-route data through longer paths. Worse yet, if the UAV happens to have been the single path between two sections of the network, it can result in the network splitting in two, at least until other nodes can span the resulting gap. To an external user, it may not be obvious that a single node might be the only path between two parts of the network and that it should be protected. Due to its additional workload, this single node may also deplete its battery more quickly.

In situations where MANET is not feasible or practical, such as where there are not enough UAVs or other nodes to maintain continuous coverage, a UAV can be used as a data ferry to collect information by periodically flying over an area and relaying data back to mission commanders. Although this approach can still accomplish most of the aforementioned applications, it may not provide real-time services.

A data ferry is a good solution for applications, such as periodically gathering data from IoT sensors, that may not be time sensitive but require regular updates. The periodic nature also means that low powered sensors need only connect and broadcast data to the UAV when it is nearby, reducing its overall power use. In a public safety context, this can be particularly valuable for a future generation of deployable sensors in places such as unstable buildings or wildland fires. The lack of a continuous connection may also be an advantage in an adversarial situation, such as in law enforcement, as it becomes much more difficult for an attacker to use the periodic connectivity to enter and attack other systems.

Data ferry solutions also carry risks that include single-point-of-failure of the UAV, the lack of real-time connection, and the lack of bandwidth. For applications where the data is sensitive, the potential for the UAV to fall into malicious hands is also a concern. Due to the significantly larger quantities of data that might be carried by a data ferry UAV, particular care needs to be taken to encrypt data at rest (when stored on-board the UAV) as well as in flight (during transmission), and to properly authenticate legitimate devices and users.

Managing AI Risks

To fully and appropriately take advantage of these novel communications paradigms, it is necessary for the UAVs to exhibit at least task, if not mission, levels of autonomy. For example, even if the mission is to fly to and stay in one location, increased communications delays between the UAV and OCU due to the use of MANET or cellular networks means that the UAV must be able to autonomously handle fast-changing situations and respond to situations like other flying objects in the area (including birds), sudden gusts of wind, or unexpected obstacles in the path. More advanced use cases require one or more UAVs to decide, within their mission area, the best locations where they can be most useful, perhaps making decisions that even human operators would not have the information to make.

The NIST AI Risk Management[29] Framework (AI RMF)[30] and Playbook[31] are tools that can be used to help manage the risks associated with transferring more of these responsibilities onto AI systems. While profiles that directly relate the AI RMF to the UAS for Public Safety application are the topic of future work, the core playbook is still a useful resource to guide the risk management process. In particular, it highlights where potential risks lie and facilitates more informative discussions with vendors by focusing on the key aspects of risk associated with these systems.

Adopting the AI RMF can also provide organizations with renewed visibility into other risks that may exist elsewhere in their operations with which they may not be aware. Much of the AI RMF is not specific to AI, particularly when it comes to governance.

The AI RMF divides the management of AI risk into four functions: Govern, Map, Measure, and Manage.

The Govern function focuses attention on cultivating an appropriate culture of risk management across the development, deployment, evaluation, and acquisition of systems that include AI. It also outlines processes, policies, procedures, and documents for the management of these risks and provides structure that helps align AI risk management functions with those of the organization. This includes human factors such as training, legal issues, and consideration for the unexpected, such as incident detection, continuity, response and recovery when an AI or similar system is embedded into a process. Public safety organizations who wish to deploy increasingly autonomous UAS may not have had to integrate AI concerns into their existing organizational risk management policies. The Govern function helps to provide structure to assist with this integration.

The Map function contextualizes the framing of risk relating to the AI system. It helps to enumerate the many different processes involved in the system, including both the UAS and systems that connect to it, and to determine where AI components may exist, their effect, and their scope. It also helps users to identify possible interactions between these systems, AI and otherwise, that may pose additional sources of risk.

The Measure function highlights the tools that are available to assess the various risks surfaced by the Map function. It also provides insight into what these measurements mean and where the state-of-the-art might be deficient.

Finally, the Manage function takes the identified and measured risks and helps to determine appropriate actions to take based on the outcomes of the Govern function. Public safety organizations are particularly well placed to adopt the Manage function, for many of their activities are generally considered too risky for the general public. It is generally a bad idea to run into a burning building or drive a large vehicle the wrong way down Main Street. Public safety personnel sometimes have to manage the risks associated with doing such activities to avoid a potentially bigger risk of not undertaking those activities. The Manage function, informed by the Map and Manage functions, helps to develop policies that enable this nuanced approach, whereby riskier activities are permitted in appropriate situations, with suitably controlled risk.

---

[29] National Institute of Standards and Technology (NIST), "NIST Trustworthy & Responsible AI Resource Center," accessed 2024-04-03, https://airc.nist.gov/

[30] National Institute of Standards and Technology (NIST), "NIST AI 100-1 Artificial Intelligence Risk Management Framework (AI RMF 1.0)," 2023-01-26, https://doi.org/10.6028/NIST.AI.100-1

[31] National Institute of Standards and Technology (NIST), "NIST AI RMF Playbook", accessed 2024-04-03, https://airc.nist.gov/AI_RMF_Knowledge_Base/Playbook

https://doi.org/10.52202/075106-0002

Managing Cybersecurity Risks

Like the AI RMF, the NIST Cybersecurity Framework (CSF) 2.0[32] is designed to help organizations understand and manage their cybersecurity risks. As UAS become increasingly connected, and in novel ways, they expose organizations to new risks that they may not have been aware of. Furthermore, the increasing connectedness and advances in technology more generally also means that organizations are exposed to a much wider range of cybersecurity threats. Adversaries are no longer individuals or entities who are trying to break into a specific organization. Ransomware targets anyone who values their data or system more than their money. Hactivists target anyone who can provide visibility to their cause. Any number of anarchy-focused individuals and groups simply want to cause disruption regardless of target. Frameworks like CSF 2.0 highlight the space that organizations need to consider and provide structure to guide the management of these risks.

Unlike the preceding AI discussion, it is likely that organizations adopting highly connected, increasingly autonomous UAS will already have cybersecurity policies and regulations, such as the Criminal Justice Information Services (CJIS) Security Policy, NIST SP 800-53 Rev. 5 Security and Privacy Controls for Information Systems and Organizations, or NIST SP 800-171 Rev. 2 Protecting Controlled Unclassified Information in Nonfederal Systems and Organizations. However, it is very likely that the existing policies do not account for the novel requirements, capabilities, and risks associated with highly connected, increasingly autonomous UAS (or, indeed, UAS or cyberphysical systems in general). Most cybersecurity risk management policies are geared towards IT systems in an office environment, perhaps with some limited exceptions for field use.

The CSF 2.0 is not another cybersecurity checklist. Rather, in the context of integrating these new technologies, it provides a systematic, structured way of analyzing an organization's existing and desired security postures, measures that might be considered to control the risks associated with these new technologies, and how security policies might be adapted to accommodate these while continuing to appropriately manage risk.

For example, highly connected UAS are increasingly being connected to critical systems, such as Computer Aided Dispatch (CAD), Geographical Information Systems (GIS), and Common Operating Picture (COP) systems, either directly or through intermediaries. Compromising critical systems can result in very significant disruption and risk to public safety organizations, even if the attackers do nothing beyond forcing the system offline[33].

They also connect to local systems that enable real-time streaming of videos and other information to the incident commander and other personnel. Such links are increasingly expected by personnel and management as the benefits of these technologies become clear. However, these systems often do not satisfy the cybersecurity policies of the networks to which they connect. Some of the unique aspects of highly connected UAS that often do not fit well with conventional cybersecurity policies include the following.

---

[32] National Institute of Standards and Technology (NIST), "NIST Computer Security Resource Center," accessed 2024-04-03, https://csrc.nist.gov/

[33] Rector, K. "Hack of Baltimore's 911 dispatch system was ransomware attack, city officials say," *The Baltimore Sun*, updated 2019-08-07, https://www.baltimoresun.com/2018/03/28/hack-of-baltimores-911-dispatch-system-was-ransomware-attack-city-officials-say/

- Lack of traditional cyber confidentiality controls. An emphasis on the availability and real-time processing on potentially limited hardware resources, resulting in less focus or ability to implement traditional cyber controls such as encryption and authentication.
- Non-standard operating system setup. The computer systems in the UAV, OCU, and other components of these advanced UAS are often embedded computers, running operating systems based on Android or Linux. However, these are often heavily customized and proprietary.
- Proprietary network protocols or messages. Similar to customized and proprietary operating systems, many such devices utilize proprietary network communications to maximize performance across potentially unreliable connections. This inhibits the ability for traditional network controls, monitoring, and securing.
- Security updates. Due to the high level of customization and the need for soft or hard real-time operation, updates to the software and operating system require extensive testing. This means that security updates for the underlying operating system are likely to be delayed or unavailable from the manufacturer and difficult or impossible for the end user to apply themselves.
- Inability to run management software. These computer systems are often heavily locked down by the manufacturer and cannot run other software. Safe physical operation of the system relies on the manufacturer having certainty about the performance and reliability of the system, which precludes the running of software that the manufacturer has not tested.
- No access for organizational network security personnel and retention of manufacturer's default access tokens. For the same reason, administrative access to these computer systems is often not available to the organization's network security personnel to prevent changes that may affect the safe operation of the UAS. While the network security personnel may be IT and security experts, they may not necessarily be experts in real-time system control or the UAS's internal safety measures. The manufacturer may also retain a level of administrative access tokens on the UAS for maintenance in warranty or repair situations which can't be changed by the organization.
- No physical access. The purpose of a UAV is to go where personnel cannot; therefore, by definition, it will spend much of its operating life outside the direct physical possession of organization personnel.
- Security theater. The novelty of the field makes it easier for better known security risks, which themselves may or may not have some element of legitimacy, to become over-emphasized. This draws attention and resources away from other security concerns that may be less well understood. For example, security controls relating to country-of-origin can help address the issue of state-sponsored adversaries. However, there are still many cybersecurity concerns that affect all systems, regardless of country-of-origin. An over-emphasis on country-of-origin controls can result in these other risks being de-emphasized. At the very least, it is necessary to acknowledge that even if the well known security issues are controlled, there are still risks across the entire system that need to be otherwise mitigated or accepted.

Organizations with security policies that do not include the unique needs of UAS and other cyberphysical systems end up having these systems function under exemptions, formal or otherwise. This results in ad-hoc, or no, oversight or visibility into these systems by the organization's cybersecurity team.

Relaxing these security policies to such systems without increasing actual cybersecurity risk exposure can be a difficult process. For every control that is loosened, additional controls need to

https://doi.org/10.52202/075106-0002

be put in place to manage that risk. For example, adopting UAS may require additional segregation of networks to make it more difficult for a malicious actor to move laterally from a compromised UAS into other critical systems.

However, it is vital that this be done in such a way as to maintain the benefits of these advanced UAS and their novel communications capabilities; otherwise, the organization risks falling into a "worst of both worlds" situation. The CSF allows organizations to analyze how these novel systems can fit within their existing risk management. It provides assistance for the development of procedures and policies for their appropriately risk-managed use and deployment, helping to identify where these additional controls might be placed to manage risk while allowing the benefits.

Governance Issues for Connected UAS

Governance relates to the human processes and expectations around using UAS within the operating objectives for public safety response. The challenge is to be appropriate to the intended application of UAS and not miss any aspects brought on by the novel application and use of UAS. Governance concerning UAS, when done correctly, is something that is used as one of the many inputs into the system - so that when it is operational, the personnel using it know the system performs in accordance with their processes and procedures.

In terms of governance, a starting point is to think of an intelligent UAS as a "new team member," who is unfamiliar with existing teams, processes, practices, and expectations. This is reasonable particularly if the new AI capabilities of the UAS allow it to take over decision making tasks that were previously the responsibility of a human. However, there are some significant differences to consider. For example, there are basic assumptions around common sense, ethical standards, and cultural values that may need to be made explicit. On the other hand, there are concerns that do not apply to AI systems, such as the need for organizational processes to be "fair" to the AI system.

It is important to create governance artifacts that document these governance decisions. This ensures that the UAS is appropriately specified and configured (in verification and validation testing) and that the UAS is provided with appropriate inputs and used in the appropriate manner, consistent with these governance decisions. These artifacts are also important for incident and activity review, where opaque AI may preclude easy root cause analysis without the ability to analyze the initial inputs and explicit instructions.

For example, a UAS may be directly communicating with personnel and making determinations on what information to disseminate from its potentially vast set of information. Governance would drive the policy that dictates how much, when and to whom information is disseminated. An inappropriate policy creates the risk that information is disseminated to the wrong people, or is irrelevant or incomplete for their situation.

Governance may be at a more abstract level and provide Incident Commanders the flexibility to determine how the UAS should disseminate its information dynamically. However, too much flexibility runs the risk that the UAS is asked to perform beyond its capabilities. From the personnel's point of view, governance sets their expectations on the information available to them during an activity. If the UAS behaves in a manner inconsistent with governance, such as due to situations that were unforeseen in the process of developing the governance, user expectations may not be met.

Another example of a factor to consider when developing governance for autonomous systems that disseminate information is the need to consider the human effect of information. Responders may receive information of a distressing nature, such as news that their properties or loved ones have been affected by the incident to which they are responding. Previously, such information might have been disseminated by a human, who could overlay judgment and take additional measures to address the distress that this may cause. For example, they may choose an appropriate time when support is available. Automated dissemination of such information, by default, has no such filter. An influx of information can also distract personnel or cause information overload. While it is possible to autonomously curate information, inappropriately curated information can also lead to incorrect decision making. Appropriate governance is necessary to develop consistent organizational (and, by extension, system) policies to address these issues.

The use of UAS as part of the communications infrastructure also presents new governance challenges. While UAS bring considerable advantages to a wide variety of public safety applications, they also bring significant risks for which the management is still immature. Governance decisions need to be made to determine when it is appropriate to use such systems and if the resources necessary to control the risks are appropriate. If the UAS is performing other activities in addition to being a communications relay, how is it going to decide which activity to continue performing if there is contention? For example, a UAS may be tasked with keeping watch on a moving firefront. To continue performing this task, it may need to move from its position and disrupt communications. It is sometimes hard enough for a human to make a decision where the consequences either way are either unknown or can be catastrophic.

At an organizational governance level, it is important that there are checks and balances to ensure underlying plans and procedures align with the overarching organizational policies and rules. This is not just for new technologies like highly connected UAS, but also for all forms of technology and is generally good governance practice anyway. Governance artifacts should be harmonized or, at least, compatible with each other and fit with other existing processes. Even without the introduction of new technology, governance artifacts can become a patchwork, written at different points in time and by different departments. The introduction of new and novel technology can exacerbate this patchwork. Two risks in particular arise from incompatible procedures across an organization.

The first results from confusion, where it is unclear which procedures might apply in a given situation. For example, it can be unclear what procedure might apply to a cellphone that is also used to control a UAS. It might be a personal or corporate managed device that is temporarily connected to the OCU for controlling the UAS. Alternatively, it may be a cellphone that is built into the OCU from the factory. The user may not even be aware that the OCU contains a cellphone.

The governance stack has to be, at the same time, consistent and yet also aware of the risks associated with the cellphone as part of the UAS and connected to the network. Problems can arise if the policies and procedures for cellphones were developed without visibility into the organization's UAS program and that cellphones used in the UAS program may have very different requirements to those used elsewhere in the organization. Worse yet, such policies will not account for additional risks due to these cellphones being connected to more systems than is normal. Consider the scenario where a cellphone is connected via the OCU to a UAV. The UAV might then be serving as a node in a MANET carrying other traffic. Even if the intent is for the cellphone to be logically separated from the traffic on the MANET, it is necessary for governance to be aware of the additional risks. This may include an adversary being able to exploit a vulnerability to move laterally between the two systems, an attack that has already been

https://doi.org/10.52202/075106-0002

demonstrated in motor vehicles[34]. Alternatively, a configuration problem in one could affect the availability and performance of the other.

Secondly, the novel use of new technology could require deviations from well-established existing security processes. It may be impossible to perform important tasks with new technology while following existing policies and procedures. This incompatibility may be due to aspects of the underlying technology, design choices made by the developer, existing policies and procedures from a time when these technologies did not exist, and/or because of the way in which the policies and procedures were written.

Exceptions to existing policies and procedures may increase risk, particularly where they are made informally and not subject to proper risk management and controls. Exceptions can also affect organizational risk culture. The precedent might also cause users to make assumptions and generalizations that if the policy could be "broken" for UAS, they could also break it for all systems.

In the case where the incompatibility is due to the manufacturer's design decisions, particularly ones that reduce security, it is preferable to write into specifications the need for the system to be compliant with appropriately secure policies and procedures. These specifications eliminate the need for the exception. For example, suppose a UAS might not satisfy an organization's data encryption requirements for a given application. It would be reasonable to have the manufacturer add the necessary encryption, for the organization to move to a system that does have the requisite encryption, deliberately decide that the additional risk is acceptable (perhaps subject to additional controls), or to forgo the capability altogether.

There may be instances where exceptions are required, such as for a UAV that is particularly small or lower powered and where adding the necessary encryption affects its performance in an unacceptable manner. In this case, UAS specific governance must include a risk assessment process to justify its need, measures that could be taken to otherwise control the risk, approval at the appropriate level of authority to accept any residual risk, clear scope of when it is acceptable, and appropriate logging and monitoring. Periodic review of the exception and repeating the risk assessment and approval should also occur as part of overall risk management, including evaluating if technology has progressed to the point that it may be worth transitioning to a different system that does not require the exception.

Conclusion

As UAS become more connected with people and systems beyond just the operators, so too do the challenges of appropriately understanding and managing both the benefits and the risks of employing highly connected UAS. Cybersecurity and AI risks, in particular, tend to be poorly understood and managed. In this paper, we have discussed some of the background of communications and UAS in public safety. We also provided an overview of AI, autonomy, and governance as it applies to UAS and the public safety use case.

We then presented some of the opportunities and challenges in these new, highly connected, increasingly autonomous UAS. We focused on two categories of use cases. The first were use cases where the UAV communicates mission data, such as videos and commands, directly with personnel and other systems in its vicinity rather than going through the OCU. The second were use cases where the UAV serves to relay communications between other people and systems. While there are many other types of use cases, these two encompass a wide range of applications.

---

[34] Greenberg, A., "Hackers Remotely Kill a Jeep on the Highway—With Me in It", *Wired*, 2015-07-21, https://www.wired.com/2015/07/hackers-remotely-kill-jeep-highway/

We describe some of the specific applications within these categories, followed by specific challenges in cybersecurity and AI risk management, and how frameworks such as the NIST Cybersecurity Framework 2.0 and the NIST AI Risk Management Framework can be helpful tools. We also discuss governance issues that should be considered when conducting a risk-benefit analysis of the use of these systems.

Our efforts in this space are just beginning; much work will be done in the future. We have established a working group to develop guidance and other resources for managing cybersecurity and AI risks of UAS in public safety applications. This effort aims to help all stakeholders, including manufacturers, vendors, developers, researchers, trainers, regulators, insurers, lawyers, and politicians, to understand this space better.

The initial focus of this effort is on the development of resources to help fire chiefs, police chiefs, and other public safety managers ask questions of vendors and other technical people, including guides that help them understand the responses and the ensuing effects on risk management. Much of the information that end users need to make the decisions discussed in this document is not provided by default to end users before or after deployment. Indeed, much of this information is only currently known to those who are directly developing the software systems and may be several steps removed from the end users. Having end users better informed when asking questions of their vendors is an initial step in improving cybersecurity and AI risk management.

We are also developing corresponding resources for other stakeholders in the UAS for Public Safety ecosystem. We will particularly focus on manufacturers and developers to help them better understand the reasons why end users need this information and allow them to proactively surface this information from their organizations and software supply chains. This will also help inform the design of new systems and improve their ability to be appropriately and transparently risk-managed.

Finally, we are developing free, open-source resources for educators, with an initial focus on graduate and advanced undergraduate students in engineering and law. The goal is to make it easier for educators to incorporate cybersecurity and AI risk management into their teaching, with UAS in public safety as an example use case. The hope is to improve the future management of cybersecurity and AI risk further, not just in terms of highly connected UAS for public safety but also in society in general.

https://doi.org/10.52202/075106-0002

AUTONOMOUS CYBERSECURITY AND AI RISK MANAGEMENT FOR UNCREWED SYSTEMS: CHALLENGES AND OPPORTUNITIES USING THE NIST FRAMEWORKS

Raymond Sheh,[1] Karen Geappen,[2] Donald Harriss[3]

Uncrewed systems are increasingly connected and automated. With these increases come significant changes in the cybersecurity and Artificial Intelligence (AI) risks associated with such systems. This is particularly important in safety-critical applications, such as public safety, security, infrastructure inspection, hazardous materials handling, transportation and delivery, and entertainment.

This paper provides an overview of uncrewed systems and the ways in which increasing connectivity and automation pose new and increased cybersecurity and AI risks. We provide some examples of these risks and suggest examples of controls that may be employed to manage these risks, with reference to the NIST Cybersecurity Risk Management Framework (CSF) 2.0 and the NIST AI Risk Management Framework (AI RMF). We also outline our working group's efforts to produce detailed, actionable guidance and other resources to help stakeholders better manage these risks, particularly for the uncrewed aircraft systems (UAS) for Public Safety application.

INTRODUCTION

Uncrewed systems, such as uncrewed aircraft systems (UAS), are increasingly widely used in critical applications such as public safety, security, infrastructure inspection, hazardous materials handling, transportation and delivery, and entertainment. Previous generations of commonly available uncrewed systems have been mostly human-controlled. Artificial Intelligence (AI) provides basic control of the motors and sensors in response to simple commands, such as to move in a given direction. Such systems also have limited integration into other critical software systems. While cybersecurity risks have always been a reality, they were usually limited in scope with realized risks only impacting the individual system.

Systems like Drone as First Responder (DFR) and autonomous patrol and inspection systems are now frequently deployed. These have increasing levels of autonomy and connectivity with critical software systems, such as emergency dispatch and digital twins as well as interconnection with other uncrewed systems. Certainly, highly connected, intelligent systems are not new. Specialized industries and the military have had a wide variety of such systems for several decades, often pushing the forefront of the technology. However, technology leaps come with a significant

[1] Uncrewed Aircraft Systems Research Lead, Public Safety Communications Research Division, National Institute of Standards and Technology, 100 Bureau Drive, Gaithersburg MD 20877, USA

[2] Independent Researcher, PO Box 174, Bull Creek, Western Australia, Australia, 6149

[3] Uncrewed Aircraft Systems Technical Lead, Public Safety Communications Research Division, National Institute of Standards and Technology, 325 Broadway, Boulder, CO 80305, USA

https://doi.org/10.52202/075106-0003

cost in custom research and development, in addition to the policies, procedures, and other controls to manage risk. The results have historically been difficult, or impossible, to apply widely to mainstream users and other stakeholders in the uncrewed systems ecosystem due to specialization, classification, and other restrictions on dissemination.

There are few resources that help mainstream stakeholders apply existing language, standards, metrics, governance, and risk management frameworks to manage the novel cybersecurity and AI risks posed by these increasingly connected and intelligent uncrewed systems. Contextualizing cybersecurity and AI risks related to the adoption of such technologies can support users in making informed choices for effective risk management. Proper risk management can remove barriers to adoption, improve efficiencies, reduce organizational risk exposure, and increase return on investment.

This paper aims to provide an overview of cybersecurity and AI risk management and outline some of the challenges to managing risks posed by highly connected and intelligent uncrewed systems. We provide some salient examples of how these risks can manifest themselves, and we end with some actionable suggestions that organizations can take to improve their risk management. While we will be mainly using UAS[4] and public safety as the example modalities and applications, the general principles can be applied across other applications, including ground and water vehicles.

BACKGROUND

Definitions

There are various competing definitions for the key terms of cybersecurity and artificial intelligence. While there is no universal definition for these terms, we will use the following definitions for this discussion.

Cybersecurity. For this discussion, we use the definition of cybersecurity used in NIST Special Publication 800-53 Rev. 5[5]:

"Prevention of damage to, protection of, and restoration of computers, electronic communications systems, electronic communications services, wire communication, and electronic communication, including information contained therein, to ensure its availability, integrity, authentication, confidentiality, and nonrepudiation."

We will delve into three critical factors that users must consider to understand the definition and scope of cybersecurity in general and for uncrewed systems in particular. The first is identifying the assets and interests to be protected. This may be physical, but it may also be something more abstract, such as privacy, integrity, trust (in the system, data, or operation), reputation, continuity of operation, or financial, regulatory, or legal exposure. The second factor is identifying the entities from which the asset or interest must be protected. The third involves assessing what the potential

---

[4] In this paper, we will make the distinction between the Uncrewed Aerial Vehicle (UAV), which is the equipment that flies, and the overall Uncrewed Aircraft System (UAS), which includes the UAV and all of the systems that directly support and integrate with its operation, such as the Operator Control Unit (OCU), flight planning software, maintenance and logistics software, and interfaces to collaboration and Common Operating Picture (COP) systems.

[5] Joint Task Force, "NIST Special Publication 800-53 Revision 5 Security and Privacy Controls for Information Systems and Organizations," National Institute of Standards and Technology (NIST), 2020-12-10, https://doi.org/10.6028/NIST.SP.800-53r5

 https://doi.org/10.52202/075106-0003

damage might be. Identifying these three critical factors informs priorities for applying cybersecurity.

In the past, industry attention was mostly focused on attackers looking to take over control of systems and/or gain access to protected information. Such attackers still exist; however, the modern threat landscape, coupled with a higher degree of interconnection between systems, requires considering a wider variety of potential attackers and their targets.

For example, as uncrewed systems become more tightly interconnected with other critical systems, such as dispatch, collaboration, and geographical information systems (GIS), the uncrewed system can become an entry point for pivoting into the overall network of the organization or partner organizations. An attacker might have no interest in the uncrewed system except that it might be the least-protected device with access to the network.

As technology becomes cheaper and more widely available, it is easier for a more diverse range of adversaries to attack critical systems. Also, as critical systems become more interconnected, it becomes easier for adversaries to find paths through the network to those systems, paths that might not even be known to the owners of those systems. For example, the threat of ransomware exists against anyone who attackers think will pay to regain access to their data or systems. At the same time, "hacktivists" might only care that their actions will result in disruption and publicity regardless of the actual value of their target. There are also individuals and groups who simply want to cause disruption regardless of the target.

As we will discuss later, this changing threat landscape, in terms of both advancements in uncrewed systems and changes in threat actors, dramatically changes the risk calculus for many users of uncrewed systems who might have previously considered themselves an unlikely target.

Artificial Intelligence. Artificial Intelligence, or AI, is a term that has very different meanings in different contexts. ISO/IEC 22989:2022 defines AI as "a technical and scientific field devoted to the engineered system that generates outputs such as content, forecasts, recommendations or decisions for a given set of human-defined objectives.". This definition reflects the broad scope of AI, although it also encompasses a lot of traditional software systems that one might not immediately consider to be AI, particularly in the popular media. For our discussion, we will focus on dividing up the space of AI in two ways.

First, AI systems can be divided by how much visibility humans have into their inner workings. They can tend towards being more transparent or more opaque in terms of the ability of humans to understand, at an actionable level, within a reasonable timeframe, and with reasonable resources, how it works and why it made a given decision. This distinction is critical for discussing risk management as much of traditional risk management relies on predicting what a system might do and performing meaningful root cause analysis.

Second, AI systems can be divided into the ways in which they make their decisions and the sources of knowledge used to make those decisions. Much of the recent discussion in popular media has focused on deep learning (DL). It is important to understand that there are many other forms of AI with different capabilities, benefits, and, more importantly, risks and risk management techniques.

For example, DL is a type of statistical machine learning (statistical ML). DL depends on fitting highly complex, randomly initialized sets of linear equations, with some non-linear elements, to the observed input data. Other forms of Statistical ML fit equations to the observed input data that are less complex and less random. Such systems might not be able to represent the same complexity as DL, but they might also be easier for humans to understand and the system may behave more consistently.

Statistical ML is a type of machine learning (ML) that bases its behavior on the statistics of the data it sees. While statistical ML is the dominant form of ML, there are forms of ML that also take in other forms of information, such as mathematical models of the system, and use the observed data to augment these models. Such systems might require more knowledge of how the system works, such as a knowledge of physics, but can also behave more predictably and require less training data.

ML is a type of AI that takes in, processes, and stores some form of information from the task or environment and uses this stored information to influence the system's future behavior. Its behavior is not only dependent on its programming but also on historical information with which it was presented. Other forms of AI, such as classical planning, reasoning, logic programming, and expert systems, generally take input from outside the task to inform its behavior.

For example, we might choose two different ways to teach a system how to catch a ball, based on observing its trajectory at the moment that it is thrown. An AI system based on DL would observe the ball being thrown at it many times, learn to predict where the ball is going to land based on what it sees of the initial trajectory, and then learn how to move to that location. This is similar to a human learning a skill by practicing it repeatedly.

Instead, an AI system based on reasoning, rather than ML, might have background knowledge that includes the equations that represent physical properties such as ballistics, gravity, and inertia. It might infer from this knowledge how the ball will move given its observed initial trajectory and calculate where it should move to catch the ball. It represents the task in an analytical manner rather than modeling it based on observation.

Both approaches have benefits and risks. For example, the DL system will naturally account for more complex aspects of the environment, such as air resistance and variations in the wind, that are incorporated into its observations of the flight of the ball. In contrast, the reasoning system might not have background knowledge that includes these factors. Instead, the reasoning system might only account for these as random errors.

In contrast, the DL system is opaque. For example, it does not represent the concept of "gravity" in a way that a human can adjust. If the system were taken to the Moon or Mars with a different gravitational field, it would have to relearn how to catch the ball. Worse yet, as there is no way for a human to tell the system that gravity has changed, it will be impossible to tell if there are any remnants of its behavior that still rely on Earth's gravity. In contrast, the gravity term in the reasoning system could be adjusted explicitly, and the system is likely to perform in an appropriate manner.

Taking risk management designed for a system with one type of AI and applying it to a system using a very different type of AI, even if it performs a similar task, can easily lead to a situation where benefits are foregone and risks are ignored. Blindly adopting risk management plans becomes even more complicated in systems that are hybrids of these techniques. For example, some systems might be analytical in nature but also include a learning component to account for complexities in the world about which it does not otherwise know. Hybrid AI systems can lead to the best of both worlds, but they also risk becoming the worst of both worlds if the risks are not appropriately managed.

Autonomy in Uncrewed Systems

Uncrewed systems vary considerably in how they make use of autonomy and automation. For this discussion, we refer to autonomy as systems that make decisions independently based on their own sensing and computation. Automation, on the other hand, refers to systems that execute

https://doi.org/10.52202/075106-0003

predefined instructions and are only capable of simple deviations from the instructions in response to unexpected events. An example would be a robot that stops when it encounters an obstacle.

Many uncrewed systems are remotely operated with a single human operator directing the vehicle's movements through an Operator Control Unit (OCU). For example, many UAVs are still controlled by humans using basic direction commands like forward, backward, left, right, climb, descend, and yaw.

This commonality can be somewhat deceiving; while a wide variety of UAVs might take the same basic commands, they can vary in the amount of autonomy they apply to follow these commands. Increasing levels of autonomy usually make it easier for the operator to control the UAV. They move the UAV's behavior closer to what the operator expects from an ideal system. This autonomy handles low-level control of the UAV and leaves the operator to handle the movements required to achieve the task at hand. For example, even modest UAVs now incorporate Global Positioning System (GPS) and cameras to determine their position and help them to automatically counteract wind. This allows the operator to fly the UAV as if it were operating in still air, transparently delegating the task of dealing with wind to the autonomy of the UAV.

From a risk management perspective, this example of autonomously dealing with changing wind conditions presents both benefits and risks as compared to systems without it. Clearly, such a system reduces the risk associated with an operator, particularly one who is less experienced, losing control of the UAV if they are insufficiently skilled to respond to changing wind conditions. This is particularly important near obstacles, both because of the increased risk of collision and the more complex air currents around obstacles that even experienced pilots can have trouble managing.

However, it also presents a significant risk, especially if the pilot is unaware of what the autonomy is doing, the limit of its capabilities, and from where the autonomy gets its information. For example, UAVs are mechanically limited in how fast they can fly relative to the air around them based on factors such as their aerodynamics and the power of their motors. Complex maneuvers close to these mechanical limits can pose risks to the stability of the UAV. An operator might command the UAV to fly forward slowly relative to the ground, but if there is a strong headwind, the UAV might actually be flying fast and close to its mechanical limits relative to the air. As the UAV is autonomously compensating for the headwind, the pilot may not even be aware that the UAV is close to its mechanical limits. If the operator then commands a complex maneuver, thinking that the UAV is still moving slowly and far from its mechanical limits, the UAV may be at greater risk of loss of control than the operator would expect.

UAVs, and uncrewed systems in general, have a wide variety of forms of autonomy beyond simply accounting for variations in the environment. Uncrewed systems can vary in complexity in their abilities to perform tasks such as driving or flying to a given point and have different capabilities for autonomously addressing unexpected events on their way to their commanded destination. For example, they may:

- Take a straight-line path.
- Drive or fly around unexpected obstacles.
- Plan a variety of paths and choose the one that they deem to be the best based on a wide variety of factors such as power consumption, the riskiness of the different paths (by some measure), and so-on.

Even more complex levels of autonomy include mission-level decision making and behavior, such as finding and tracking an object of interest or coordinating with other entities.

Autonomy at task and mission levels presents particular challenges to risk management because of the wide variety of ways in which they can be implemented and the unique risks posed. Proper risk management requires information from vendors and developers that often does not flow to the end users outside of highly integrated sectors such as defense. Short of acquiring the requisite information, end users can at least be informed of the potential risks for which they should be watching and managing.

The aforementioned levels of autonomy are split across levels of abstraction loosely based on the NIST Autonomy Levels for Unmanned Systems (ALFUS)[6]. In the uncrewed systems community, levels of autonomy are also divided based on how much human supervision is required. There are loose correlations between the two different ways in which levels of autonomy are defined, but they do represent very different concepts.

One popular definition is from the Society of Automotive Engineers (SAE) Taxonomy and Definitions of Terms Relating to Driving Automation Systems for On-Road Motor Vehicles J3016_202104 (J3016)[7], often simply referred to as the "SAE Levels of Autonomy." This six level system ranges from level 0, where the operator must devote focused attention to the operation of the system, through to level 5, where the operator need not be present or on call.

It is important to note that the SAE levels of autonomy are designed specifically for roadgoing vehicles on publicly accessible roads and associated locations like parking lots. Specifying a level of autonomy for a system is only meaningful when associated with a definition of its nature, task, and environment. For example, a level 5 autonomous car is not expected to perform autonomously on an off-road trail or to back up a large trailer. Similarly, it is generally meaningless to state that a UAV has or requires "Level 5 Autonomy" if the environment and task are not specified.

Good Risk Governance for Uncrewed Systems

Governance is the collection of systems of processes, mechanisms, and rules by which an organization is controlled and operates. It ensures that the people within the organization are held accountable when moving towards organizational objectives. Appropriate governance bridges both horizontally and vertically across the organization. It is a traceable collection of artifacts linking organizational objectives with processes, procedures, and technology. It also needs to flow in both directions, with approvals informing teams and metrics reported to determine effectiveness. The focus on organizational objectives amalgamates what is important to the organization, such as social considerations, regulatory requirements, technology, and organizational skills and capabilities.

Governance is a key pillar when it comes to risk management. The aim of governance in risk management is to identify and manage relevant risks to the organization achieving its objectives. When done well, it sets the boundaries for acceptable and unacceptable behavior and risk-taking, provides definitions to remove ambiguity, and enables an organization to have clarity and coherence across all its subject matter expert teams.

---

[6] Ad Hoc ALFUS Working Group Participants, "NIST Special Publication 1011-II-1.0, Autonomy Levels for Unmanned Systems (ALFUS) Framework Volume II: Framework Models Version 1.0," National Institute of Standards and Technology (NIST), 2007-12-28, https://doi.org/10.6028/NIST.sp.1011-II-1.0

[7] Society of Automobile Engineers (SAE) International, "Taxonomy and Definitions for Terms Related to Driving Automation Systems for On-Road Motor Vehicles J3016_202104," 2021-04-30, https://www.sae.org/standards/content/j3016_202104/

https://doi.org/10.52202/075106-0003

Good governance becomes even more critical when an organization uses AI or uncrewed systems. By definition, AI implies that elements of the decision are taken out of humans' hands (and minds). Similarly, by definition, uncrewed systems implies that human oversight of the system is reduced. Furthermore, this reduction of human oversight is often also accompanied by greater connectivity and access to other systems, as the AI and uncrewed systems directly access mission data that humans would have normally handled. Good governance is required to manage the risks of activities that humans do not directly oversee.

Effectively, the technology becomes a risk multiplier by opening up different attack vectors in an organization and increasing the number of exposed assets and interests, including the aforementioned abstract assets, such as reputation and legal exposure. This makes risk management control that can cross-cut organizational teams and hierarchy even more important. However, like any risk management control, incorrect application of a control can cause more damage than no control.

Risks that should be covered by governance are those that would hinder the organization from achieving its goals. This generally revolves around protecting an organization's assets and interests, both tangible and intangible, including those of connected systems. Adherence to regulations and legal requirements can be seen as either an intangible reputational risk or a tangible risk related to finances, exposure to legal action, or if applicable, a license to operate.

Many frameworks can be used for the various levels and domains that need risk management governance. The NIST AI Risk Management Framework and the Cybersecurity Framework 2.0, discussed later in this document, may support an organization in establishing operating behaviors and processes. In the context of new technologies and novel applications, the process layer of governance is where we find gaps with published guidance. These gaps include relevant and materially useful metrics that need to be relayed up the governance hierarchy to establish overall organizational risk.

Existing Relevant Frameworks, Standards, and Guidance

As awareness of cybersecurity and AI risks becomes more widespread, various frameworks and guidance are now available to help organizations better manage these risks. Most current guidance is general and not written with uncrewed systems in mind. However, it can provide a good starting point in developing appropriate risk management. In this section, we discuss some salient examples of recent frameworks, standards, and guidance relevant to uncrewed systems.

Cybersecurity Frameworks and Guidance. In early 2024, NIST released the updated Cybersecurity Framework (CSF) version 2.0, providing a set of outcomes across six cybersecurity functions for reducing cybersecurity risks. The functions are Identification, Protection, Detection, Response and Recovery with a cross-cutting function of Governance. The outcomes are industry and organization-agnostic, and therefore, the CSF does not define the implementation of the outcomes. Instead, the framework suggests relevant references, such as NIST Privacy and other frameworks, for guidance on implementation within functions. Examples of resources that the CSF can assist in contextualizing include but are not limited to:

- The NIST Privacy and Risk Management Frameworks.
- NIST SP 800-53 Rev. 5, Security and Privacy Controls for Information Systems and Organizations.
- NIST SP 800-171 Rev. 2, Protecting Controlled Unclassified Information in Nonfederal Systems and Organizations.
- The Criminal Justice Information Services (CJIS) Security Policy.

- ISO/IEC 27001:2022 Information security, cybersecurity and privacy protection.

This contextualization allows organizations to more easily and comprehensively assess their current cybersecurity posture relative to the posture that they wish to adopt and identifies steps to move forward.

The CSF is outcomes-based and technology-agnostic, making it relevant to all technologies. However, without domain-specific guidance for uncrewed systems, best practice implementation to achieve these outcomes is left to the organization, particularly for more unusual aspects of these highly connected, intelligent systems. For many organizations, adapting this general guidance to the specific challenges of uncrewed systems can be difficult.

Specific guidance regarding the management of cybersecurity risk for uncrewed systems is still in its relative infancy. Organizations that have issued guidance that relates to uncrewed systems include:

- The Cybersecurity & Infrastructure Security Agency (CISA), part of the U.S. Department of Homeland Security (DHS), has issued guidance that focuses on the procurement and use of Uncrewed Aircraft Systems[8] rather than the development, manufacture, or broader integration into wider organizational systems and processes.
- Both the Defense Innovation Unit (DIU) of the U.S. Department of Defense (DoD) and the Association for Uncrewed Vehicle Systems International (AUVSI) have issued guidance on the procurement of UAS based on the National Defense Authorization Act (NDAA)[9]. DIU maintains a "Blue UAS" list[10] while AUVSI maintains a "Green UAS" list[11], which list the UAS that they have evaluated according to the NDAA criteria. This guidance is primarily focused on geopolitical, cybersecurity, organizational, and supply chain considerations of the UAS manufacturer rather than its integration into the organization of the end user's broader systems and risk management.
- The NIST Information Technology Laboratory (ITL)[12] produces guidance on a wide variety of cybersecurity topics, including the resources mentioned earlier in this section. While it does not have an uncrewed systems-specific focus, it does focus on applications such as Automated Vehicles[13]. Much of the guidance within the cyber-physical systems and internet of things (IoT) topics is also highly relevant to uncrewed systems.

Often, it is difficult to predict how an organization will use novel technologies and how they will create and implement the necessary cybersecurity policies and other controls to protect organizational assets and interests. The CSF 2.0, focusing on risk management outcomes, provides

---

[8] Cybersecurity & Infrastructure Security Agency (CISA), "Be Air Aware," accessed 2024-04-03, https://www.cisa.gov/topics/physical-security/unmanned-aircraft-systems

[9] House Armed Services Committee, "NDAA - National Defense Authorization Act," accessed 2024-04-03, https://armedservices.house.gov/ndaa

[10] Defense Innovation Unit, "About Blue UAS", accessed 2024-04-03, https://www.diu.mil/blue-uas

[11] Association for Uncrewed Vehicle Systems International (AUVSI), "Green UAS," accessed 2024-04-03, https://www.auvsi.org/green-uas

[12] National Institute of Standards and Technology (NIST), "Information Technology," accessed 2024-04-03, https://www.nist.gov/information-technology

[13] National Institute of Standards and Technology (NIST), "NIST Automated Vehicles Program," accessed 2024-04-03, https://www.nist.gov/programs-projects/nist-automated-vehicles-program

 https://doi.org/10.52202/075106-0003

a useful reference even when domain-specific guidance has not yet caught up. However, leveraging these resources requires a high level of understanding of these novel technologies.

Applying existing frameworks to new applications requires an ongoing, iterative effort by all stakeholders with a keen understanding of established acceptable risk tolerances and how to meaningfully apply them in situations outside of the original use case. Organizations need to be flexible when dealing with cutting-edge, connected, and autonomous systems. Standard security practices and past experiences might not always apply perfectly. This is especially true in such a fast-moving field.

AI Frameworks and Guidance. Society has been aware for a long time that AI poses risks. Indeed, American writer and professor of biochemistry, Isaac Asimov, coined the "Three Laws of Robotics." Many of his stories, such as those in his collection "I, Robot"[14], revolve around how simplistic characterizations of robots with AI result in unintended consequences. When developing guidance on the use of AI, it is easy to fall into the trap of generalizing based on an understanding of how humans, or fictional systems, make decisions. Managing the risks of real AI systems requires considerable nuance and understanding. Unfortunately, this understanding can be very hard to obtain, especially with a supply chain of software and hardware that does not expose the necessary information about included AI systems to the end user.

Several recent efforts have generated frameworks, guidance, policies, and regulations to facilitate the management of AI risk. These are vital in helping organizations to generate more thorough, nuanced, technically actionable, and ethically meaningful policies and procedures. Perhaps most prominent of these are the NIST AI Risk Management Framework (AI RMF)[15], the U.S. Department of Energy (DoE) AI Risk Management Playbook (AI RMP)[16], and the European Union Artificial Intelligence Act. In this discussion, we will focus on the AI RMF as a tool to help organizations better characterize their current AI risk, determine their acceptable level of AI risk, and provide guidance to manage these risks better.

Like the various cybersecurity frameworks, the AI RMF is general and not focused on uncrewed systems. It divides the problem of analyzing an organization's exposure to AI risk through four functions: Govern, Map, Measure, and Manage. Much like the CSF 2.0, it is not a checklist. Rather, it provides a high-level overview that helps organizations better understand their risks and apply other guidance, resources, and policies to manage risks.

The guidance landscape for AI risk management is much less mature than it is for cybersecurity. The existing guidance focuses more on risks associated with the use of AI for decision-making and content generation. This includes such issues as privacy, fairness, ethics, and intellectual property. Government and standards organizations need to provide more guidance that specifically assists organizations with managing the AI risks unique to uncrewed systems.

Uncrewed System specific guidance and resources. The NIST Public Safety Communications Research (PSCR) Division has established a working group to produce guidance and other resources to assist organizations with managing cybersecurity and AI risk associated with UAS,

---

[14] Asimov, I., "I, Robot," *Gnome Press*, 1950-12-02

[15] National Institute of Standards and Technology (NIST), "NIST Trustworthy & Responsible AI Resource Center," accessed 2024-04-03, https://airc.nist.gov/

[16] U.S. Department of Energy (DOE), "DOE AI Risk Management Playbook (AIRMP)," accessed 2024-04-03, https://www.energy.gov/ai/doe-ai-risk-management-playbook-airmp

focusing on public safety applications[17]. The goal of this effort is to work with a wide variety of stakeholders to bridge the gap between general cybersecurity and AI risk management guidance, as well as the specific risks and challenges of uncrewed systems in general and UAS in particular. Although still in its early stages, PSCR researchers developed some preliminary resources, which we will discuss further in this document.

Factors in Managing Cybersecurity and AI Risks for Uncrewed Systems

A comprehensive treatment of the management of all risks is beyond the scope of this paper. This section will highlight a few lesser-known factors to consider when managing cybersecurity and AI risks for uncrewed systems.

Hidden Risks of a Common User Experience

One of the most significant sources of risk for uncrewed systems and intelligent, highly connected systems is their interaction with humans. The risks of an overly complex, unique user interface are well known in terms of cognitive load, errors, and user acceptance.

Having a new system behave in a way that resembles the system it replaces is often seen as reducing risk, as users are already familiar with the previous system and are less likely to make mistakes. However, if the new system has very different underlying risks, the users and the broader organization may not be aware of or understand the need for a change in their risk management. They may, therefore, continue to follow risk management procedures and policies that are inappropriate to the new system. This wastes resources on unnecessary risk management and leaves new risks unmanaged.

A common user experience can hide a change that increases an existing risk, even if the change ostensibly reduces risk. For example, a new UAV that autonomously compensates for wind is often considered to reduce risk. Even in windy conditions, to the operator the new UAV behaves as the previous one did in calm air. However, the new UAV only does so until the point that the wind speed becomes too great. If the user experience of the new UAV is the same as the previous UAV, this poses a risk. The operator was not compensating for the wind and may have therefore been unaware of it. When the autonomy begins to fail, the operator has much less time to react. A common user experience can also hide new risks. For example, failure in the autonomy system of the new UAV due to sensor or GPS errors may require the operator to intervene suddenly and in situations that would not be expected given their experience with the previous UAV.

Nevertheless, there are solutions. For the example of high wind, some UAVs have displays that show an estimate of the surrounding wind speed and warnings if mechanical limits are being approached. However, these are only useful if the operator is aware of these warnings and pays attention to them. This can be challenging if the operator's cognitive load is already high. In such a situation, policies, procedures, and training need to accompany the rollout of the new system, even if the user interface stays the same.

Attack Surface Challenges for Uncrewed Systems

Highly connected, intelligent uncrewed systems present additional attack surfaces compared to more traditional IT systems. A system's attack surface refers to the sum of all points, known as attack vectors, in the system where an unauthorized agent can perform malicious actions. Using the

[17] National Institute of Standards and Technology (NIST), "PSCR UAS Working Group, Cybersecurity and AI Risk Management for Uncrewed Aircraft Systems (UAS) in Public Safety," updated 2024-02-27, https://www.nist.gov/ctl/pscr/pscr-uas-working-group

https://doi.org/10.52202/075106-0003

analogy of a secured room in a building, a physical attack surface might include the doors, windows, walls, floor, and ceiling. Analyzing a system's attack surface ensures we consider how an unauthorized agent may enter. To continue the analogy of the secured room, while placing good locks on the doors and windows might be an obvious measure, good attack surface analysis might also highlight the need to check that an unauthorized person cannot enter the room through the ceiling or ductwork.

Like many IoT systems, uncrewed systems are mobile computing devices operated in public venues, often beyond an organization's logical or physical boundaries. Operating outside of the organization's usual boundaries exposes these devices to external and environmental threats that require greater situational awareness and cyber threat analysis. Furthermore, embedded computer equipment within uncrewed systems can be non-standard due to the need for specialized hardware and software to control the vehicle. The embedded equipment is often impossible to replace or upgrade and might run software that is impossible for central IT services to manage, making them blind spots in traditional cybersecurity monitoring. This void increases cybersecurity risk as uncrewed systems become more connected to critical organizational systems. Uncrewed systems can also threaten other organizational IT systems to which they are connected. This is particularly the case if their integration is ad-hoc and requires formal or informal deviations from IT security policy. As we discuss later, these new, less visible attack surfaces may require careful enumeration and active re-evaluation of security and privacy controls across the rest of the organization to determine how connected they may be to the Uncrewed System.

As discussed previously, a variety of general cybersecurity controls and guidance exist. They are still relevant for uncrewed systems but need to be adopted with a more complex attack surface in mind. For this discussion, we will focus on NIST SP 800-53. This document is used across the US Government for security and privacy controls for information systems (with the equivalent for non-government systems being NIST SP 800-171). Various other countries have equivalent frameworks and/or standards, but the core principle is a set of implementable control targets for information systems.

We will also use, as an illustrative example, the Drone as First Responder, or DFR, application that is becoming more common in public safety. While DFR can be implemented in a wide variety of ways, a common model is for one or more UAS to be pre-deployed on rooftops around the service area within automated unattended weatherproof pods. When required, the UAS launches autonomously from its pod, performs its mission with minimal human intervention, returns to its pod, and then charges, ready for its next mission. The DFR is often connected to other systems within the public safety organization, including the computer-aided dispatch (CAD) system, GIS, and collaboration tools. These connections to other critical public safety systems introduce risks that need to be addressed.

The controls in the NIST SP 800-53 can be customized to meet the diverse requirements of public safety organizations and stakeholders and are implemented as an organization-wide process. The control catalog consolidates requirements from various sources, including mission and business needs, laws, executive orders, directives, regulations, policies, standards, and guidelines. It also includes security and privacy controls from a functional perspective by breaking down topics into focused control families; see the full list in the table below.

ID	FAMILY	ID	FAMILY

AC	Access Control	**PE**	Physical and Environmental Protection
AT	Awareness and Training	**PL**	Planning
AU	Audit and Accountability	**PM**	Program Management
CA	Assessment, Authorization, and Monitoring	**PS**	Personnel Security
CM	Configuration Management	**PT**	Personally Identifiable Information (PII) Processing and Transparency
CP	Contingency Planning	**RA**	Risk Assessment
IA	Identification and Authentication	**SA**	System and Services Acquisition
IR	Incident Response	**SC**	System and Communications Protection
MA	Maintenance	**SI**	System and Information Integrity
MP	Media Protection	**SR**	Supply Chain Risk Management

Table 1: The families of controls in NIST SP 800-53.

We can first look outwards from the UAS towards the systems with which it communicates. To gain perspective on how these control families fit into a first responder IT system, one must break technology into individual tangible assets, policies, and processes to identify cybersecurity requirements. For example, UAS are no longer simple two-device systems consisting of an OCU and flight vehicle but are now a complex, connected ecosystem. This new model consists of a connected interface from the UAS and/or OCU to an internet-connected device, such as a smartphone, cellular modem, or router on a 4G or 5G LTE network. The cellular endpoint may then connect to multiple remote systems. It may also connect to one or multiple local and/or remote first responder dispatch systems; each dispatch system may then connect to multiple networks that provide other response services.

Many DFR systems are hybrid and include services requiring connections to external vendors and cloud-hosted systems. The number of entry points, hardware, software, network, and human interfaces exponentially increases for each connected system. Software sources and supply chains are also potential attack surfaces that must be considered in first responders UAS. Accordingly, the number of responsible parties involved in the operation and maintenance also increases with each connected system. The complex threat landscape heightens as each provider exports functions to another provider, and so on.

Next, as we continue to develop organizational risk policy, we look inwards at how the UAS operates to see where other parts of the attack surface might lie. For example, the UAV will have receivers for other radio signals, such as GPS and Automatic Dependent Surveillance – Broadcast (ADS-B), which allows the UAV to locate and identify other aircraft in the vicinity. While an

https://doi.org/10.52202/075106-0003

attacker is perhaps unlikely to be able to use these parts of the attack surface to penetrate the rest of the organizational network, this attack surface is still important to consider in terms of the security of the mission and the data gathered. For example, an attacker might spoof the GPS on the UAV[18], causing it to think that the collected data is from somewhere else, resulting in incorrect data ingested into the organization's system.

The UAS also possesses a wide variety of computers, most of which do not run conventional operating systems. For example, the UAV itself would have at the very least a main embedded CPU, perhaps a GPU for offloading machine vision and other AI tasks, and one or more embedded controllers that take care of lower-level systems like directly commanding the motors, coordinating the behavior of the battery, and configuring the software-defined radios. These systems run their own software, which can be changed through legitimate software updates or potentially malicious activity, and yet, as previously mentioned, they cannot be managed by traditional IT cybersecurity infrastructure.

This presents a problem for some traditional organizational IT policies, which mandate that, for example, all IT systems be centrally managed. When an IT policy that forbids unmanaged devices collides with the reality that the organization needs to deploy DFR, which often does not integrate into conventional central IT management systems, there is the danger that the new attack surfaces introduced by the DFR will simply be ignored. Instead, a more prudent organization would re-evaluate its IT policy, characterize these and other new attack surfaces, and deliberately allow these new devices onto the network while employing other forms of risk management.

Where is your AI?

A key step in managing AI risk, as highlighted in the Map function of the AI RMF, is to determine where AI presents a risk. For modern uncrewed systems, this is a non-trivial task, made all the more challenging by the often opaque, proprietary, and unusual nature of computing equipment in many of these systems. The complex nature of software supply chains, where even the vendor communicating with the end user may not be fully aware of the nature of the AI within their product, makes identifying risks more challenging.

Properly controlling this risk requires transparency on the part of the entire supply chain. In the absence of transparency, the first step is for all stakeholders to begin asking the right questions. Analyzing the Uncrewed System at each level of autonomy, from the lowest control levels to mission levels of autonomy, can assist in enumerating the places where AI might exist in a system. New features, particularly those that come with software updates and promise the ability to adapt behavior or optimize performance, often also have AI components. As we discuss in the last section, Summary and Future Work, we are also developing guidance and resources to help stakeholders to better understand where AI may be present in their systems and the ways to manage their risks.

Failures, Correctability, and Root Cause Analysis

Along with enumerating the different forms of autonomy versus automation at various levels, as we did in Section 2, it is also useful to classify the forms of autonomy at each level according to how transparent or opaque the underlying AI is, using our aforementioned definitions. This discussion becomes a sliding scale, where automation could be considered the most transparent

[18] Veillette, P. "The Serious Threat Of GPS Spoofing: An Analysis," Aviation Week, 2023-10-09, https://aviationweek.com/business-aviation/safety-ops-regulation/serious-threat-gps-spoofing-analysis

form of autonomy, autonomy based on transparent AI being the next most transparent, and autonomy based on opaque AI being the least transparent and least conducive to root cause analysis.

Knowing why an autonomous system performs a specific action and being able to predict what it might do can be vital in safety-critical applications. Sometimes a trade-off must be made. In some but certainly not all situations, autonomy based on opaque AI can outperform that based on transparent AI over some set of environments and tasks. This is because the environment and/or task contains important complexities that could not be modeled analytically or with more transparent techniques.

The obvious trade-off is that we do not know why it does or does not work. Less obviously, we also have a much more difficult time finding other situations where similar failures may happen. This difficulty poses a unique risk management challenge. A central requirement of risk management is the ability to perform root cause analysis on failures, determine situations where similar failures with the same root causes might occur, and put in place controls to manage that risk in the future.

For example, an autonomous vehicle might exhibit a behavior whereby it fails to see certain objects on the road. For more transparent AI systems, it may be possible to trace the reason for this decision and determine with some degree of certainty the extent of situations where similar errors might occur. Ideally, the system's behavior could be corrected in a verifiable manner. Even if this is not feasible, knowing the extent of the possible similar errors that the system could make enables other risk management controls to be applied.

While more opaque systems can be difficult or impossible to trace, many of them are able to learn from such mistakes. For example, they could feed the sensor data generated from the incorrectly missed objects back into the learning system so that the next time they are encountered, the system is more likely to perform the correct action. On the surface, this seems like a reasonable risk management step. However, it is important to know the extent to which such a "patch" corrects the underlying problem and that the "patch" has not caused new errors elsewhere in the system. For AI systems that trend towards the more opaque end of the spectrum, the answers to these questions can be impossible to know.

From a risk management perspective, it would be unwise to simply patch the behavior of the system for those specific objects. Without understanding exactly why the system could not see those objects, it will be impossible to know for sure what other objects it might also miss. This risk is one that is particularly unique to opaque AI systems.

Appropriately managing the risk of more opaque forms of AI typically requires additional controls beyond those used for more transparent systems. Examples include monitoring for such changes in the system's behavior (especially systems that learn and adjust their behavior regularly), with extensive testing and roll-back of policies, as warranted. Such controls are unlikely to be possible without extensive resources and the cooperation of significant parts of the software and data supply chain. It is an open question as to whether it is appropriate to deploy more opaque AI systems if resources and cooperation are unavailable.

As technology progresses, policies around the use of AI with different levels of transparency cannot be universal or static. The goal should be to manage risks relative to each other. On the one hand, it may be appropriate to take on a new, possibly less well-understood risk in the short term if the trade-off is a dramatic improvement in performance that reduces risk elsewhere. However, this decision needs to be flexible. If future systems can yield the same improved performance but with risks that are better understood, this decision may be worth revisiting even if there is a substantial cost in switching.

https://doi.org/10.52202/075106-0003

Misunderstood Explanations

Explanations and explainable AI systems are often touted as solutions to many risks inherent in AI systems. On the surface, this thinking might seem reasonable; however, it is important to carefully consider how explanations are used in the context of risk management and if the particular types of explanations offered by an AI system are compatible with the use case. In this context, we consider explanations that provide human users and investigators with accurate information that they can understand and is technically actionable. Indeed, the behavior of any computer program, including any AI system, can be explained at some level. However, for many ML systems, this explanation is not at a level that is understandable or technically actionable. Furthermore, many ML techniques, such as genetic algorithms and deep learning, depend on random number generators for their operation, thus their explanations inevitably have a probabilistic component to them.

There are many ways to divide the space of different types of explanations[19]. Perhaps the most important when it comes to AI systems are explanations that reflect the reason for the underlying decision, as compared to explanations that are consistent with the behavior of the system. The former might be considered "the" explanation for the decision, meaning that it is the singular true explanation. The latter might be regarded as "an" explanation for the decision, being an explanation out of many possible explanations that are consistent.

This distinction is vital when it comes to risk management. An explanation that is based on the underlying decision-making process, expressed in a manner that is understandable to humans at a suitable level of abstraction, allows the human to predict what the system will do and perform root cause analysis of failure.

In contrast, while there is only one explanation that reflects a system's underlying decision-making process, there are an infinite number of explanations that are consistent with any finite set of the system's decisions. While such explanations might be satisfying to human curiosity, it is dangerous to rely on them to predict the behavior of the system. Such explanations also do not allow investigators to perform root cause analysis (beyond a statistical analysis) or prevent future failures. Unfortunately, many AI systems that are sold on the promise of having "Explainable AI," particularly those that make use of gradient methods like Locally Interpretable Model-agnostic Explanations (LIME)[20] and Shapley Additive Explanations (SHAP)[21], produce "an" explanation and are incapable of producing "the" explanation.

The obfuscation of the distinction between "the" explanation and "an" explanation poses a considerable risk because, from the perspective of a human observer, it is impossible to determine if a given explanation is "the" explanation or "an" explanation. Risk management controls that are developed based on "an" explanation for how the system works run the risk of both unnecessarily restricting the behavior of the system and being blind to real risks. For example, an autonomous vehicle might have the ability to detect stop signs. There are many aspects of the sign that the system might detect, such as its shape, color, and markings. If the system fails to see a particular

---

[19] Sheh, R., and Isaac, M. (2018) Defining Explainable AI for Requirements Analysis. In KI-Künstliche Intelligenz, V32, PP 261-266, Springer Berlin Heidelberg

[20] Ribeiro, M. T., Singh, S., Guestrin, C., ""Why Should I Trust You?": Explaining the Predictions of Any Classifier," arXiv, revised 2016-08-09, https://arxiv.org/abs/1602.04938

[21] Lundberg, S., Lee, S., "A Unified Approach to Interpreting Model Predictions," arXiv, revised 2017-11-25, https://arxiv.org/abs/1705.07874

stop sign, it is vital to know exactly which aspect of the sign was missed. An explanation that happens to be consistent with the observed failures is not sufficient.

Human equivalence is sometimes cited as a reason for "the" explanation being too high a bar to expect AI systems to clear. It is true that humans are sometimes unable, or unwilling, to provide "the" explanation for why they do something. However, a primary control that society uses to manage the risk of not being able to perform root cause analysis on human behavior is to remove the human's ability to perform the task. For example, a driver who crashes often enough will lose the ability to drive, be it by having their license suspended or losing their insurance coverage. This control works because removing the ability of the singular human to drive does not prevent other humans from driving. However, for an organization that has deployed AI on a fleet of uncrewed systems that effectively have the same AI "driver," the equivalent would be to "ground" the entire fleet. The grounding of a UAS fleet is not a viable risk control to impose.

Risks of AI System End of Life

When commissioning any mission-critical system, one must consider managing organizational process continuity when the system reaches end of life, both planned and unplanned. Work must continue as the organization transitions to a new, future system, which might include migrating data and configuration, as well as training personnel. Measures often taken to ensure a smooth transition include having configuration and data documented and stored, ready for the new systems, and having organizational processes in place to develop and deploy the new systems. Local backups of critical information help to protect against disruption caused by unexpected end of life of the system, such as failure of the computer that it is running on or a cloud provider that is critical to the operation of the system going out of business.

AI systems that include ML components pose unique challenges. An AI system that has become an important part of the organization has, by definition, incorporated organization-critical information into its learned model. However, ML models are highly specific to a particular learning technique and often even to a specific system. If the system experiences a failure, there can be a significant risk that the learned model will not be useful for a replacement system.

In the context of autonomous systems, this risk can appear at any level of abstraction. For example, an autonomous vehicle may include a management computer that adjusts various control parameters as the vehicle ages and its underlying behavior changes. The computer compensates for these changes and makes the vehicle behave as close as possible to a newer vehicle. If the management computer fails, the replacement computer will not have this learned information unless the old model was backed up prior to failure and is compatible with the replacement computer. If restoration is not possible, there is a risk that the vehicle may behave in an unexpected manner until it has had a chance to relearn its various parameters. This risk is particularly pronounced for situations where the learning controller is used to address relatively unusual situations, such as operating close to mechanical limits, that may not be encountered in post-repair testing.

In the ideal case, this AI risk might be controlled by dictating that equipment that contains ML models must have procedures for such learned models to be migrated to replacement equipment. As a fallback, visibility into this risk means that procedural controls can be put in place, such as ensuring that systems being brought back online after repair also have appropriate retraining. It also becomes especially important for post-repair testing to fully exercise the system. This can take considerable effort, particularly for systems that rely on ML models to address highly complex environments and tasks.

https://doi.org/10.52202/075106-0003

Another example is an autonomous vehicle that relies on navigational path planning from a cloud provider which routes the vehicle according to weather and traffic. The cloud-hosted path planner might also learn optimal paths based on organizational usage patterns. If the cloud provider experiences a disruption or goes out of business, it is unlikely that the learned data can be migrated to a new service. At the very least, comprehensive requirements for the cloud-hosted service would be needed to find, develop, and/or configure a replacement service. There may also be a significant period of degraded performance as the new service relearns the usage patterns that may also pose risks to the organization.

Greater than the Sum of their Parts

Thus far, we have considered cybersecurity and AI risks individually. It is also essential to consider them together, including how AI can change the management of cybersecurity risks and vice versa. This field is still nascent with little existing guidance considering both factors. In this section, we highlight a few examples.

The Dangers of Inexplicability. In today's world of complex IT systems, it is almost expected that IT systems will behave in ways that users do not expect or understand. With the increasing adoption of AI systems, particularly opaque AI, not only can the system behave in a way that is unexpected, but it can also be impossible to determine why the system behaved in that manner. This property is sometimes referred to as inexplicability.

This presents an AI risk in its own right, for it becomes impossible to perform root cause analysis on the failure or determine how to avoid it in the future. For instance, an uncrewed system that is performing surveillance might consistently miss unusual events in a particular sector. In some cases, it may even be impossible to determine if the system failed or if it simply had more information than the user. For example, an uncrewed vehicle might autonomously choose to take a route with which the operator disagrees.

It also presents a cybersecurity risk as detection of malicious activity is a cornerstone of cybersecurity risk management. If the system is known to behave in an unexpected manner and, in some cases, changes its behavior over time as it learns its environment and task, it becomes much more difficult to distinguish legitimate activity from malicious activity.

Worse yet, the presence of systems that behave in inexplicable ways raises the tolerance of people in the organization to inexplicable and unusual behavior more generally. An attacker who is aware of this, particularly on the inside, can more easily deflect questions if unusual activities are detected by blaming the behavior "on the AI," potentially even for systems that have nothing to do with AI.

Attacking the AI. The underlying algorithms of the AI systems present a wide range of attack surfaces that have only recently been studied. These attacks come about largely because AI systems understand the world in a way that is quite different from the way most humans do. This difference in understanding means that an attacker can create situations that seem normal to a human observer and yet can trick an AI into making the decision that the attacker wants it to make.

Adversarial Machine Learning[22] refers to the field of study concerning tricking AI systems containing ML components, by influencing the inputs into the system, so that it makes a decision

---

[22] Vassilev, A., Opera, A., Fordyce, A., Anderson, H., "NIST Trustworthy and Responsible AI, NIST AI 100-2e2023, Adversarial Machine Learning, A Taxonomy and Terminology of Attacks and Mitigations," National Institute of Standards and Technology (NIST), 2024-01-02, https://nvlpubs.nist.gov/nistpubs/ai/NIST.AI.100-2e2023.pdf

that it should not make. For example, uncrewed ground vehicles on public roads must read a wide variety of street signs. In modern systems, this is often accomplished through deep learning, which learns patterns that correspond to symbolic concepts such as stop signs. Crucially however, the learning takes place in a way that cannot be comprehensively verified. It has been shown such systems can be tricked into erroneously identifying one sign as another, with simple, unobtrusive modifications to the sign, such as applying small stickers. A human observer would not interpret the stickers as a change in the meaning of the sign; however, an ML system may misread the image.

Certainly, this is not a property that is unique to ML systems or AI systems in general. Humans are also susceptible to inputs that have been manipulated to make us think or see something that is different from what is there. In the human field, these are known as illusions. Adversarial Machine Learning could be considered the field of study concerning finding illusions that ML systems suffer from and particularly ones that are not immediately obvious to humans who might be monitoring the system.

This parallel with human illusions also suggests a potential preemptive risk management control. For safety-critical tasks, such as humans operating motor vehicles, the environment is designed to minimize the probability and impact of illusions. For example, traffic cones are a bright color and tend to not resemble anything else that one expects to see and might find confusing. Even if someone paints something on the traffic cone, a human is unlikely to confuse it for something else, and even if they did, there are generally many cones marking an area. For ML systems, the equivalent would be determining where such "illusions" might occur and structuring the environment accordingly. This can be difficult as different ML systems can suffer from very different illusions. More transparent AI systems, while still likely to suffer from illusions, may permit greater analysis as to why the illusions happen and how they might be proactively prevented.

There are also attacks on AI that do not involve machine learning but can be equally impactful. One well-known example relevant to both crewed and uncrewed systems was demonstrated in 2020 when an artist dragged 99 mobile devices in a wagon at slow speed across a bridge, tricking Google Maps into reporting a traffic jam on the bridge[23]. While humans would understand a traffic jam to be numerous cars moving slowly, Google Maps can only detect the position of users' mobile devices and makes the assumption that their motion near roads corresponds to the motion of the cars. Crucially, this difference is relatively unknown to users of Google Maps, making it possible to then trick users, including any autonomous uncrewed system that incorporates traffic data from Google Maps, into thinking that there is a traffic jam. An attacker, knowing this, could fake such traffic jams to discourage vehicles from passing through a particular area.

Governance, Social and Regulatory Consequences

Risk management governance is a control that ties organizational objectives to processes, protective assets, and configurations. As such, it must consider and manage the specific technologies used within an organization in the context of that organization itself. Industry best practices and frameworks available, including those of NIST, are excellent resources so that organizations have a known baseline. However, appropriate application is still required for the individual context, particularly when it comes to these new technologies.

In the context of uncrewed systems, there are wide-ranging and novel applications. The challenges already discussed with the application of an existing framework, toolset, or otherwise to a new and novel application also exist in the domain of governance. The governance flavor to

---

[23] Barrett, B., "An Artist Used 99 Phones to Fake a Google Maps Traffic Jam," Wired, 2020-02-03, https://www.wired.com/story/99-phones-fake-google-maps-traffic-jam/

https://doi.org/10.52202/075106-0003

these includes different societal expectations of new technologies, which flow into additional expectations on an organization using technologies (e.g., AI manipulation of images). New technology also increases the scope of impact of a consequence far beyond the traditionally expected bounds of an organization (e.g., accidental poisoning of a shared AI data source).

Without governance review cycles factoring in technologies such as AI and automation, there is also the risk that processes and procedures designed to allow for sound procurement decisions, consistent technology use, incident response and review, and other standard cyber practices are no longer possible because the technology is inherently different 'under the hood.' At the operational level, processes, procedures, and plans need to factor in the capabilities of the new technologies and changes to the operating environment to allow the technologies to function fully.

Using the previous example of allowing additional public network connectivity from a UAV, it may be necessary to change monitoring processes for network intrusions and review risk registers, metrics, and reporting elsewhere in the organization. These actions allow for the application of risk management controls that factor in the new and expanded threat surface related to external connectivity and the possible malicious actions that an adversary could perform, throughout the organization, should they compromise the UAS. Governance controls related to continuity and disaster recovery would also need to be reviewed for allowing connectivity to public networks.

A further aspect that governance needs to consider is the potential capabilities of the new technology in the context of social and regulatory consequences. Just because an organization elects not to use the full capabilities of a technology does not mean it is protected from public scrutiny with potential reputational damage. Nor does it protect the organization from any regulatory controls related to that capability, even if not in use. Organizations must also consider these environmental factors when reviewing governance when integrating new or novel systems into their technology ecosystems.

On the bright side, this review of governance is also beneficial for an organization in achieving improvements in processes and procedures related to the use of the new technology. The process of reviewing the capabilities of a new technology (or novel application) against an organization's objectives for the purpose of identifying and managing risks also allows an organization to develop the most effective and efficient processes for operating and utilizing available capabilities. This improves efficiencies and return on investment (or confirms that the product is not suitable).

Immediately Actionable Risk Management

Cybersecurity

The CSF 2.0 and associated cybersecurity guidelines, such as SP 800-53, help vendors and end users to identify systematic risks in uncrewed systems. It helps users and vendors to divide the system up and, in the context of the six CSF 2.0 functions, determine at a general level where risks lie and how risks might be addressed. This helps to create a healthy cybersecurity ecosystem. However, the guidelines only help if organizations reevaluate their attack surfaces in the presence of these new technologies.

While more specific guidance is being developed, users can employ general NIST guidelines, such as the CSF 2.0, to get a head-start on identifying cybersecurity implications that they may encounter when adopting new technologies. End users with existing or new information technology infrastructure can systematically identify each of the NIST SP 500-53 controls and assign them to responsible individuals.

Organizations can also use this as an opportunity to take a broader view of their cybersecurity policies and procedures to ensure that they are harmonized or, at the very least, non-contradictory.

Organizations can delineate clearer boundaries between their different policies to avoid confusion as to which policy might apply to the "gray zone" that often exists with these increasingly connected devices.

Artificial Intelligence

Organizations can similarly begin to evaluate their AI risk exposure using the AI RMF, even in the absence of uncrewed systems-specific guidance. Until the supply chains for data and software become more accustomed to rigorous AI risk management, there will be a high degree of uncertainty around which particular systems expose an organization to risk and how this risk is manifested. Agencies like NIST are working to provide additional resources and guidance to all stakeholders, with the goal of providing end users with more detailed information to make risk management easier.

In the meantime, organizations will need to do their own research and investigations to determine how to apply these general guidelines to their uncrewed systems. Later in this section, we present a list of questions derived from a workshop of uncrewed systems stakeholders, held in February 2024. Organizations may use these questions to assess their AI risks and tolerances, as well as to educate supply chain players on the need for greater transparency on the underlying AI systems. This increased transparency better informs the end users' risk management decisions.

Organizational Governance

Organizations may leverage existing governance to improve risk management visibility. and therefore, management for UAS.

Organizations, with or without an uncrewed systems cybersecurity or AI subject matter expert, can perform a minimal assessment of this technology through any technology risk management governance process. These processes include third-party risk management governance, appropriate procurement governance, appropriate change management governance, access control and authorization (including non-humans, such as the Uncrewed System itself), and other applicable governance processes. Using existing governance processes would go a long way to improving the visibility of AI risks while waiting for more specific guidance.

Another way to consider uncrewed systems in the context of existing governance processes is to view the AI system as an "unknown" human performing those tasks. While an imperfect analogy, it would be a useful process to run any known human factor risk governance process over the use case of an untrained human performing those tasks to identify and then model potential risks that may require additional controls or management.

Identifying Risks

During the Workshop on Cybersecurity and AI Risk Management for Uncrewed Aircraft Systems in Public Safety, held in February 2024, an interactive session was held to solicit the top concerns that the attendees had regarding cybersecurity and AI risk. Attendees were first invited to submit suggestions for questions that they felt most important. These were collated and then, in a second round, the attendees were asked to vote for their top questions. Most of the attendees were public safety end users, followed by government representatives and researchers. The resulting list of questions is presented below. We stress that this list is not comprehensive. It represents areas of greatest concern to the attendees, however it also highlights possible blind spots in the end users' understanding of these risks.

This list helps to identify the top concerns these stakeholders have in managing cybersecurity and AI risk. The results represent some of the top questions that those who are managing or procuring uncrewed systems should consider when analyzing their risk exposure, including how

https://doi.org/10.52202/075106-0003

the Uncrewed System might present risks that differ from those of more conventional IT systems. It also represents aspects of the cybersecurity and AI risk management challenge for which there is already some base level of understanding and the greatest likelihood for traction in the immediate term. Knowing the answers to these questions, or knowing that the questions could not be answered, and having discussions inspired by the asking of these questions are important first steps to applying the aforementioned resources, such as the CSF 2.0 and AI RMF.

The top 10 questions are presented below, adjusted slightly for relevance to the broader uncrewed systems domain.

1. How secure is the system? What is the attack surface? How, and in what timeframe, will we be notified of breaches of the various severity levels? What cybersecurity stress testing or "Red Teaming" has been performed, on both the system and systems on which it depends?
2. How and where is the data stored? What is the level of encryption and other security methods for the data and its derivatives, such as metadata, error logs, vendor telemetry, and temporary files, in transit and at rest? Who can decrypt or modify the data?
3. How can we determine if the AI is giving incorrect information? What are the possible mission-critical failure modes of the system? Do we have enough access to the system to determine if a decision that we do not agree with is due to an error on the part of the AI system or to otherwise perform root-cause analysis? How can we correct the behavior of the system?
4. What measures are taken to ensure an appropriate level of data privacy? What are the levels of data privacy that have been chosen and how were those decisions made?
5. In what physical locations is the AI processing being done? For instance, is the AI processing occurring on the Uncrewed System, on the controller, on our own servers, or in a cloud data center in-state, in-country, or elsewhere?
6. What AI and cybersecurity challenges have other users run into with this system and what steps have they taken to mitigate risk?
7. Is information about the confidence that the AI system has in its decision available to the operator and/or investigator? What information is available to assist in interpreting the confidence level of this decision?
8. What will the system do if it encounters a situation that is unusual or unexpected? Is the system capable of detecting if it is operating in a situation that was not represented during its development, training, or testing?
9. Where does additional input data to the system, such as maps and positioning, come from; how is it updated and authenticated; and what measures are in place to address intentional and accidental interference, corruption, jamming, spoofing, or similar?
10. What measures are taken to ensure continuity of critical operations in the event that the system undergoes planned or unplanned downtime or end of life? How likely is it that the information and systems required for recovery are affected by the same cyber-attack as the main system? If the AI system suffers a failure, such as effects from a bad model or input data, what plans exist to roll back to a good model?

Summary and Future Work

Cybersecurity. Artificial intelligence. Uncrewed systems. All three of these represent concepts and technologies that have individually resulted in major disruptions to the field of risk management and governance at different points in very recent history. The risks, particularly in less

https://doi.org/10.52202/075106-0003

common use cases, are poorly understood. Very little guidance or precedent exists for the management of risks when these three come together outside of specific, well-resourced sectors.

We provided an overview of some key definitions, concepts, and resources in these rapidly changing domains. Our coverage of these topics is not comprehensive, as the field is too broad for the scope of this paper. Instead, as a starting point, we described a few key, less obvious risks in more detail. We also provided some actionable steps, mostly around gathering information by asking key questions and initiating discussions with other stakeholders, particularly within the software and data supply chains.

The NIST PSCR Working Group on Cybersecurity and AI Risk Management for Uncrewed Aircraft Systems in Public Safety is working to identify areas of most need for resources. The goal is to provide targeted guidance to bridge the gap between conventional risk management and governance and the needs of the Uncrewed Aircraft Systems in Public Safety application. These resources will also have relevance to uncrewed systems and other applications, particularly where there is already a strong risk management culture.

A high priority is the further development of the list of questions that those in charge of managing the procurement and use of uncrewed systems can ask to better understand risks. Associated with this will be the development of guidance that explains each of the questions in greater detail, along with commonly expected answers and the implications of the various answers for the organization's risk management and governance.

Parallel work will be to continue consultations with a broad range of stakeholders, including users, software and supply chain entities, and those representing societal concerns, to determine what other questions have the most impact on improving the state of risk management for uncrewed systems. This effort will also seek to raise the general level of awareness of these novel and less well-understood risks and to identify additional resources that may be needed. We invite all stakeholders to join us in this work.

These efforts will also feed into the development of joint profiles for both the NIST AI RMF and CSF 2.0. While these efforts focus on the Uncrewed Aircraft Systems in Public Safety application, the guidance will also be relevant to broader classes of uncrewed systems and applications beyond public safety.

https://doi.org/10.52202/075106-0003

COMPARATIVE STUDY OF VISION-BASED METHODS FOR REAL-TIME TRAFFIC MONITORING

Diksha Aggarwal,[*] Surafel T. Anshebo,[*] Kevin Kochersberger,[†] and A. Lynn Abbott[‡]

Uncrewed Aerial Vehicles (UAV) technology is advancing dramatically with applications ranging from surveillance to disaster response. One such area is real-time traffic monitoring where UAVs with their vision-based methods can play a significant role in streamlining traffic flow, mitigating congestion, and facilitating quick emergency response in accidents. However, such an implementation is accompanied by challenges related to accuracy and computational overload. Therefore, a comparison of available methodologies in detection and tracking using computer vision is presented in this paper over different traffic scenarios.

Keywords: *Uncrewed Aerial Vehicles (UAVs), Multiple Object Tracking, Accuracy, Single Object Tracking, Computer Vision*

INTRODUCTION

Vision-based surveillance systems are ubiquitous in urban areas. The role played by cameras in providing data for crowd counting, urban planning, traffic management studies as well as disaster response management is extremely valuable[1]. Small Uncrewed Aerial Vehicles (sUAVs) are emerging as viable platforms for surveillance applications, given their ability to maneuver in three dimensions and fly through narrow passages. Furthermore, with the advancement of the capabilities of vision algorithms, equipping sUAVs with vision sensors makes them a suitable technology for aiding in ad-hoc scenarios in traffic safety, streamlining traffic flow, and most importantly, improving responsiveness to emergencies. All these applications are predicated on pattern recognition, object detection, and tracking, which are essentially techniques within the framework of computer vision.

The state-of-the-art techniques in vision-based algorithms can be classified into traditional computer vision algorithms and deep learning-based algorithms. While the former boasts computational efficiency, the latter exhibits robustness in terms of solutions and remains less affected by scaling and occlusion. Both these categories of algorithms are widely studied for traffic applications[2]. While they are distinct in terms of their solution approaches, a common challenge confronted by both approaches is their ability to respond to different lighting conditions and occlusions. Given that the current research aims to deploy UAV-based vision systems for traffic safety, it is imperative that the deployed algorithms can overcome such challenges. Hence, a comprehensive understanding of the performance of each category of algorithms under such challenging conditions becomes crucial.

[*] Graduate Student, Department of Mechanical Engineering, Virginia Tech
[†] Professor, Department of Mechanical Engineering, Virginia Tech
[‡] Professor, Bradley Department of Electrical and Computer Engineering, Virginia Tech

https://doi.org/10.52202/075106-0004

This paper thus proposes to evaluate and compare the performance of the traditional computer vision algorithms and deep learning algorithms, by applying them to vehicle traffic monitoring. This entails both detection as well as tracking of vehicles. Specifically, the Haar Cascade[3], Lucas-Kanade optical flow[4], and SORT[5] algorithms are explored in terms of traditional algorithms, whereas YOLOv5[6] and Deep SORT[7] are the deep learning algorithms that are investigated. Moreover, this work presents the performance of several combinations of the above algorithms under various lighting conditions and views, quantified via state-of-the-art metrics. Since the overarching goal is to understand the performance of the algorithms under challenging conditions like lighting and occlusion, the VisDrone dataset appears to be the most suitable, as it offers the expected challenging data to test the algorithms. The remainder of the article is organized as follows: The next section discusses the motivation for this research and reviews the related work. Following that we provide a brief overview of the dataset and algorithms evaluated. We then present the evaluation metrics with results and finally, we conclude the article with a summary and conclusion.

RELATED WORK

Traditional computer vision methods leverage inherent image attributes such as texture, color, and shape and have been the mainstay for years. These approaches include template matching[3,8] and feature extraction methods like HOG, SIFT, and ORB. Researchers have combined these extracted features with machine learning-based classifiers like Haar Cascade, KNN[9], and SVM[10] to detect vehicles. These detectors perform well but are sensitive to scale variations and less effective in cluttered environments[11]. Vehicle detection has also been explored using deep learning-based methods. These include two-staged approaches like Fast RCNN[12], Faster RCNN[13], and Mask RCNN[14], and one-staged methods like SSD[15], and YOLO[16]. These advanced methods offer improved performance and accuracy in vehicle detection tasks.

Detected objects can be tracked using methods such as Lucas-Kanade optical flow[17], mean-shift tracking, and Kalman filter-based SORT[7]. These tracking approaches are mostly based on predicting the next probable position based on tracking features. However, they can encounter issues related to scaling and occlusion which can affect their performance. These methods have been used in both single and multi-object tracking. In single object tracking (SOT), a vehicle is detected in the first frame, or its position is known a priori. As a result, the performance entirely depends on the tracking algorithm used. However, in multi-object tracking (MOT), a detection method is required at each frame to identify the targets that are leaving or entering the frame. Each detected object is assigned a unique ID and is tracked across frames. The IDs are assigned based on the matching detections with existing tracks based on criteria such as distance, appearance, and motion patterns. As a result, the accuracy of the tracker largely depends on the accuracy and precision of the detection algorithm. Therefore, researchers have been integrating deep learning-based object detection methods with traditional trackers to make them robust. For example, Gomaa et al.[18] combined optical flow with CNN and K-means clustering to remove background and generate features on the segmented vehicles to track. Similarly, other researchers have leveraged deep learning to track the objects for a longer time and reduce the number of ID switches when occlusion occurs. For example, Wojke et al.[7] introduced Deep SORT, an extension of the SORT algorithm with the integration of appearance information by using a re-identification CNN model.

While working with deep learning-based algorithms, one of the major challenges is obtaining a high-quality dataset that provides sufficient labeled images in different conditions. In recent years, several challenges have been organized worldwide to promote research and development in object detection and tracking algorithms like Computer Vision and Pattern Recognition (CVPR) challenges, MOT challenges, and workshops. As a result, datasets such as KITTI[19] for autonomous

 https://doi.org/10.52202/075106-0004

vehicles in cities, and MOT15[20] and MOT16[21] for tracking pedestrians in urban environments are developed. These datasets typically focus on the images taken from a ground vehicle. To perform operations safety-critical operations using sUAVs, an aerial dataset is necessary. The underlying challenge with the drone dataset is that it can produce blurry images due to the camera motion. The scale of the objects also varies with the altitude of the drone. Small angular variations can produce large changes in the camera view which can make it difficult for the algorithm to keep track of the objects. Hsieh et al.[15] presented the first paper with an aerial dataset for counting cars in a parking lot. Du et al.[22] developed UAVDT aerial manually annotated datasets for vehicles for single and multiple object tracking while considering different weather conditions and camera angles. Zhu et al.[23] created the VisDrone dataset with 10 object categories including people, bicycles, cars, vans, trucks, etc. in different traffic scenarios from various angles.

OVERVIEW: DATASET AND ALGORITHMS

The VisDrone2019 dataset[23], developed by the AISKYEYE team at Tianjin University's Lab of Machine Learning and Data Mining in China is used in this research. The dataset comprises 288 video clips, 10,209 static images, and 261,908 frames captured by diverse drone-mounted cameras. The dataset spans a broad spectrum, including various geographic locations (drawn from 14 different cities across China, spanning thousands of kilometers), environments (urban and rural), objects (pedestrians, vehicles, bicycles, etc.), as well as scene densities (ranging from sparse to crowded). The dataset features manual annotations, with over 2.6 million bounding boxes delineating targets of frequent interest. In this study, three image sequences have been chosen. For ease of reference, we have used seq1 for uav0000137_00458_v, seq2 for uav0000268_05773_v, and seq3 for uav0000305_00000_v as shown in Figures 1, 2, and 3 respectively.

Figure 1. VisDrone Dataset Seq1 (uav0000137_00458_v)

Figure 2. VisDrone Dataset Seq2 (uav0000268_05773_v)

Figure 3. VisDrone Dataset Seq3 (uav0000305_00000_v)

Under the category of traditional computer vision, three algorithms are employed, namely, the Haar Cascade for vehicle detection, Lucas-Kanade optical flow, and Kalman filtering-based SORT for vehicle tracking. In the deep learning-based approach, this paper uses YOLOv5 for vehicle detection and leverages a Kalman filtering and reidentification-based tracker, Deep SORT to perform tracking of the detected vehicles.

Haar Cascade

Viola and Jones developed the Haar Cascade[3] method which is distinguished by three key contributions- First, the feature selection process uses Haar features, simple patterns derived from the image, to capture changes in texture or edge orientation. Second, it employs integral images for fast feature calculation, and third, the training uses AdaBoost, an algorithm for selecting the most critical features from a large set and combining them into a strong classifier. This classifier is trained using two kinds of images: Positive images that include objects of interest and negative images that do not include the objects. The process builds a cascade of classifiers, each more complex than the previous, designed to eliminate any incorrect detections.

Optical Flow

The Lucas-Kanade optical flow[4,24] method is used for tracking motion between consecutive images or video frames. It calculates the motion in terms of the displacement of pixels from one frame to the next. It is based on the assumption that the brightness of any object remains constant over time. The algorithm analyzes the pattern and magnitude of these motion vectors to estimate the direction and speed of moving objects in the images. This information is then used to predict the location of bounding boxes in the subsequent frames.

SORT

Simple Online and Realtime Tracking (SORT)[5] developed by Bewley et al. is an object detection-based tracker. It uses an object detection algorithm to detect objects in each frame of a video or image sequence and create bounding boxes on the targets. In the first frame, these bounding boxes will initiate the object tracks, and an identification (ID) is assigned to each object. In subsequent frames, the goal is to associate the detected bounding boxes with existing object tracks produced by the Kalman filter. This is achieved by using an association matrix which is based on Intersection Over Union (IOU). This matrix is then used by the Hungarian algorithm to associate the valid tracks. When an object is occluded or moved out of the frame, the tracking algorithm allows trackers to exist for a short time without new matching detections, giving the object a chance to reappear. If a tracker does not receive a matching detection during this time, it is removed.

https://doi.org/10.52202/075106-0004

YOLOv5

YOLOv5 is an advanced version of the YOLO architecture[6]. Unlike traditional two-stage detectors like R-CNN, which first select region proposals and then classify them, YOLOv5 is a single-stage detector that predicts both bounding boxes and class probabilities directly from the image in one evaluation. This makes it exceptionally fast and suitable for real-time applications. YOLOv5 is different from its previous versions with the use of a more efficient neural network EfficientDet[25]. Moreover, YOLOv5 integrates well with modern deep learning libraries such as PyTorch, which makes the development and deployment easier.

Deep SORT

Deep SORT is an extension of the SORT algorithm[7]. While SORT uses Kalman filtering and the Hungarian algorithm for object tracking, Deep SORT also incorporates deep learning features in the association matrix. Deep SORT uses a reidentification-based convolutional neural network (CNN) to extract deep features from detected objects. These features help in recognizing the same object across different frames, even when its appearance changes due to lighting, viewpoint, or other factors. This helps to improve tracking performance, especially in cases of occlusions or interactions between objects.

EVALUATION

The evaluation of the different combinations of algorithms is divided as follows: evaluation of detection algorithms Haar Cascade and YOLOv5 over all frames of video, evaluation of optical flow for SOT, evaluation of SORT and Deep SORT on SOT with Haar Cascade and YOLOv5.

Evaluation Of Detection Algorithms

YOLOv5 is not trained on a vehicle dataset, thus the weights found on the official website of the YOLOv5 repository could not be used effectively for detecting the vehicles. Therefore, a transfer learning approach has been utilized in this work. The pre-trained YOLOv5 model is trained on a new dataset of vehicles. The vehicle dataset is taken from a GitHub repository[23]. The dataset consists of a series of images containing around 1300 images of vehicles. The dataset is divided into 70% training images, 20% validation images, and 10% testing images. The training is done for 50 epochs in the batch of 16 and it took 1.25 hours to train the model. The metrics used to evaluate the training accuracy of YOLOv5 are precision, recall, and Mean Average Precision (mAP). Precision measures the percentage of the model's positive identifications that were correct. Recall estimates how good the model is at detecting positive cases. Finally, mAP is calculated as the mean of average precision for each category in the dataset. The results of the training are shown in Figure 4. The mAP from this training is 0.773.

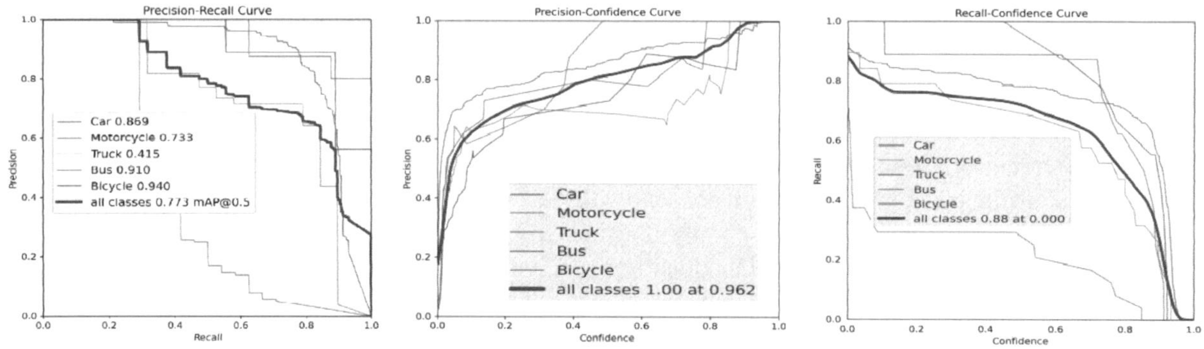

Figure 4. YOLOv5 Training Precision And Recall Curve.

The training weights for the Haar Cascade model are taken from this GitHub repository*. The model was trained using 526 images of cars from the rear (360 x 240 pixels, no scale). The images were extracted from the Car dataset proposed by Brad Philip and Paul Updike taken of the freeways of southern California†.

The detection algorithms are evaluated using recall and precision. The value changes with the tracking algorithms since the tracking algorithms used in this paper generate detection using Kalman filtering. Thus, Haar Cascade and YOLOv5 are implemented on the video and compared with the ground truth data, and recall and precision have been reported as shown in Table 1. It can be that the values are higher in YOLOv5. This is due to the high number of False Positives and the size of the bounding box generated by the Haar Cascade method which highly affects its performance.

Table 1. Performance Metrics For Detection Algorithms Without Tracker On Seq1.

Method	Recall	Precision
Haar Cascade	0.0718	0.0956
YOLOv5	0.5871	0.8999

Evaluation Of MOT Algorithms

For tracking, pre-trained weights of the Deep-SORT model are used and are obtained from the official Deep SORT repository. Evaluation of MOT algorithms is done using the metrics taken from CLEAR MOT[26]. The metrics for identification include ID Precision (IDP), ID Recall (IDR), and IDF1 (ID F1 score). IDF1 is the harmonic mean of IDP and IDR, reflecting the balance between accurately identifying objects and minimizing missed or incorrect identifications. MOT Accuracy (MOTA) consolidates three types of tracking errors into a single performance indicator: the detection errors of False Positives (FPs) and False Negatives (FNs) as well as association errors using ID switches (IDSWs). Since MOTA does not include the localization error, MOTP is defined by CLEAR MOT as the overall tracking precision in terms of average overlap between the detected object and its ground truth. The other trajectory-based metrics[27] used are Mostly Tracked (MT), Partially Tracked (PT), Mostly Lost (ML), and Fragmentations (FM). MT is the number of objects tracked for at least 80% of their span, PT is the number of objects tracked between 20 and 80% of their span, ML is the number of objects tracked for at most 20% of their span and FM is the number of times a track was interrupted by a missing detection.

MOT is performed using three combinations of algorithms: Haar Cascade with SORT, YOLOv5 with SORT, and YOLOv5 with Deep SORT. The results corresponding to each case on seq1, seq2, and seq3, are shown in Tables 2, 3, and 4 respectively. The analysis shows that Haar Cascade with SORT exhibits lower performance as compared to YOLOv5 with SORT. One of the reasons for this observation is the scaling effect and the high number of FNs. The Haar Cascade performs optimally when the objects match the size it was trained on. Another observation is that both trackers demonstrate lower accuracy on seq2 compared to seq1. This may be attributed to the brighter images in seq2, which could have impacted the performance of detection algorithms. Both these findings emphasize that the accuracy of the tracking algorithm depends on the accuracy of

* https://github.com/andrewssobral/vehicle_detection_haarcascades?tab=readme-ov-file

† https://data.caltech.edu/records/dvx6b-vsc46

https://doi.org/10.52202/075106-0004

the detection algorithm. Overall, the evaluation highlights the importance of choosing the right combination of detection and tracking algorithms for optimal performance in MOT.

Table 2. Performance Metrics For MOT On Seq1.

Method	MOTA [%]	MOTP [%]	FM	MT	ML	FP	FN	IDSW	IDF1	IDP	IDR
Haar Cascade + SORT	-2.37	56.7	**12**	0	44	**193**	7023	**11**	0.0044	0.0696	0.0023
YOLOv5 + SORT	**49.6**	**83.83**	62	**18**	14	239	**3261**	26	**0.6244**	**0.8619**	**0.4896**
YOLOv5 + Deep SORT	45.6	79.61	135	**18**	**13**	400	3295	109	0.5438	0.7357	0.4313

Table 3. Performance Metrics For MOT On Seq2.

Method	MOTA [%]	MOTP [%]	FM	MT	ML	FP	FN	IDSW	IDF1	IDP	IDR
Haar Cascade + SORT	-3.94	52.7	98	0	52	712	13350	51	0.0118	0.0915	0.0063
YOLOv5 + SORT	1.39	**84.64**	**19**	0	**48**	615	**12767**	**7**	**0.0993**	0.5224	**0.0549**
YOLOv5 + Deep SORT	**1.75**	70.47	39	0	50	**47**	13286	8	0.0365	**0.7492**	0.0187

Table 4. Performance Metrics For MOT On Seq3.

Method	MOTA [%]	MOTP [%]	FM	MT	ML	FP	FN	IDSW	IDF1	IDP	IDR
Haar Cascade + SORT	-1.8	47.23	**4**	0	69	**131**	6376	**1**	0.0037	0.082	0.0019
YOLOv5 + SORT	**30.09**	**86.05**	21	**13**	**44**	235	**4232**	**1**	**0.4871**	**0.8939**	**0.3348**
YOLOv5 + Deep SORT	13.31	76.71	39	5	53	250	5283	7	0.2697	0.7695	0.1635

In the comparison between SORT and Deep SORT with YOLOv5, the accuracy is nearly consistent for both seq1 and seq2. However, the accuracy of Deep SORT is significantly low on seq3. This can be attributed to the reidentification model being trained on oblique images of cars while seq3 has images taken from a downward-looking camera. This exhibits the challenge of using the same deep learning-based model for different camera views. Moreover, it can be inferred that in MOT, the deep learning-based algorithms are more effective for object detection while traditional algorithms perform better in object tracking. This harmoniously agrees with the evaluation performed by Pereira et al.[29].

Table 5. Performance Metrics For MOT On Seq3 For Yolov5 With Deep SORT With Change In Max_dist.

Parameter	MOTA [%]	MOTP [%]	FM	MT	ML	FP	FN	IDSW	IDF1	IDP	IDR
Max_dist = 0.2	**13.31**	**76.71**	**39**	**5**	**53**	**250**	5283	**7**	0.2697	**0.7695**	0.1635
Max_dist = 0.5	**13.31**	76.6	40	**5**	**53**	255	**5277**	8	**0.2703**	0.7662	**0.1641**

Deep SORT has various parameters that affect its performance like Max_dist, maximum IOU, and minimum confidence. Max_dist is the maximum allowed Mahalanobis distance for matching a new detection to an existing track. These parameters are varied to check the accuracy of MOT. It can be seen that with an increase in Max_dist, its FNs have reduced as shown in Table 5. The Max_dist value increased from 0.2 to 0.7. However, the results didn't show any improvement after 0.5. Varying the minimum confidence parameter of the detection algorithm with Deep SORT did not affect its performance.

Evaluation Of SOT Algorithms

Optical flow is used for performing SOT on a known car location. Its performance is evaluated using the metric given by Soleimanitaleb et al.[28]. The Euclidean distance between the center detected object and its ground truth link as shown in Figure 5 is plotted. This gives a measure similar to MOTP where the overlap between the bounding boxes is measured to find the precision of detection and mAP where the ground truth bounding box is compared with the detected box and returns a score. The other metric to evaluate the performance is using IOU over the frames of the video. IOU metric is used for checking how well our prediction bounding box aligns with our ground truth.

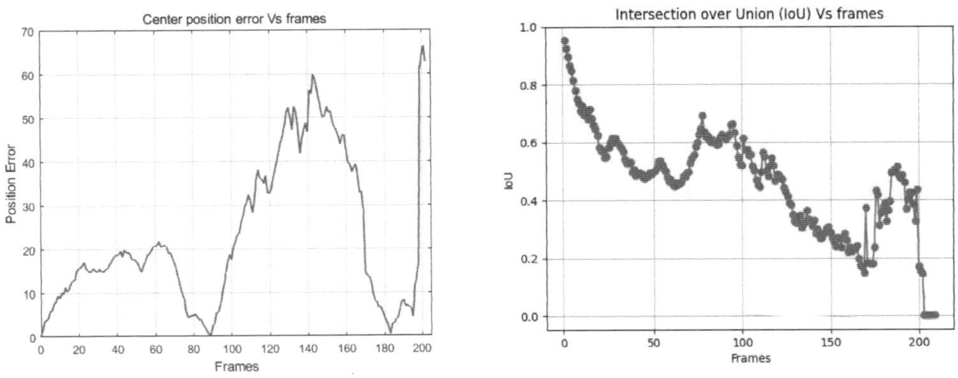

(a) (b)

Figure 5: Using Haar Cascade With Optical Flow (a) Center Position Error, And (b) IOU.

https://doi.org/10.52202/075106-0004

We can observe that as the position error increased after the 90th frame, the corresponding IOU value also decreased. The performance of optical flow in SOT is degrading due to the occlusion and scaling. As the features in the initial bounding box got occluded, the size of the bounding box changed and increased the IOU and hence increased the center position error.

CONCLUSION

We have presented a comprehensive performance evaluation of different detection and tracking algorithms for single object tracking (SOT) and multi-object tracking (MOT) of vehicles. The MOT algorithms used are SORT and Deep SORT. To evaluate their performance, the MOT metrics outlined by CLEAR MOT are used. These metrics give a clear idea about the performance of the two trackers. The algorithms are tested with the datasets of VisDrone with different camera views and lighting conditions. We have observed that in MOT, the deep learning-based algorithms are well suited for object detection while traditional algorithms perform better in tracking. Moving forward, training Deep SORT over a wide range of datasets may enhance its performance. Additionally, more algorithms can be tested for SOT and optical flow can be integrated with other algorithms to improve its performance.

NOTATION

MT	Mostly tracked
PT	Partially tracked
PT	Partially tracked
ML	Mostly lost
FM	Fragmentations
IDSW	ID switches
MOTA	Multiple Object Tracking Accuracy
t	Frame index
GT	Number of ground-truth objects
FN	False negatives
FP	False positives
Ct	Number of matches in frame t

REFERENCES

1. Liu Z, He Z, Wang L, et al. *VisDrone-CC2021: The Vision Meets Drone Crowd Counting Challenge Results*. http://www.aiskyeye.com/.

2. Butilă EV, Boboc RG. Urban Traffic Monitoring and Analysis Using Unmanned Aerial Vehicles (UAVs): A Systematic Literature Review. *Remote Sens (Basel)*. 2022;14(3). doi:10.3390/rs14030620

3. Viola P, Jones M. Rapid object detection using a boosted cascade of simple features. In: *Proceedings of the IEEE Computer Society Conference on Computer Vision and Pattern Recognition*. Vol 1. ; 2001. doi:10.1109/cvpr.2001.990517

4. Lucas BD, Kanade T. ITERATIVE IMAGE REGISTRATION TECHNIQUE WITH AN APPLICATION TO STEREO VISION. In: Vol 2. ; 1981.

5. Bewley A, Ge Z, Ott L, Ramos F, Upcroft B. Simple online and realtime tracking. In: *Proceedings - International Conference on Image Processing, ICIP*. Vol 2016-August. IEEE Computer Society; 2016:3464-3468. doi:10.1109/ICIP.2016.7533003

6. Redmon J, Divvala S, Girshick R, Farhadi A. You only look once: Unified, real-time object detection. In: *Proceedings of the IEEE Computer Society Conference on Computer Vision and Pattern Recognition*. Vol 2016-December. ; 2016. doi:10.1109/CVPR.2016.91

7. Wojke N, Bewley A, Paulus D. Simple online and realtime tracking with a deep association metric. In: *2017 IEEE International Conference on Image Processing (ICIP)*. IEEE; 2017:3645-3649. doi:10.1109/ICIP.2017.8296962

8. Zhuang X, Kang W, Wu Q. Real-time vehicle detection with foreground-based cascade classifier. *IET Image Process*. 2016;10(4). doi:10.1049/iet-ipr.2015.0333

9. Putri DIH, Martin, Riyanto, Machbub C. Object detection and tracking using SIFT-KNN classifier and Yaw-Pitch servo motor control on humanoid robot. In: *2018 International Conference on Signals and Systems, ICSigSys 2018 - Proceedings*. ; 2018. doi:10.1109/ICSIGSYS.2018.8373566

10. Moranduzzo T, Melgani F. A SIFT-SVM method for detecting cars in UAV images. In: *International Geoscience and Remote Sensing Symposium (IGARSS)*. ; 2012. doi:10.1109/IGARSS.2012.6352585

11. Ward IR, Laga H, Bennamoun M. RGB-D Image-Based Object Detection: From Traditional Methods to Deep Learning Techniques. In: *Advances in Computer Vision and Pattern Recognition*. ; 2019. doi:10.1007/978-3-030-28603-3_8

12. Hsu SC, Huang CL, Chuang CH. Vehicle detection using simplified fast R-CNN. In: *2018 International Workshop on Advanced Image Technology, IWAIT 2018*. ; 2018. doi:10.1109/IWAIT.2018.8369767

13. Fan Q, Brown L, Smith J. A closer look at Faster R-CNN for vehicle detection. In: *IEEE Intelligent Vehicles Symposium, Proceedings*. Vol 2016-August. ; 2016. doi:10.1109/IVS.2016.7535375

14. Lou L, Zhang Q, Liu C, Sheng M, Liu J, Song H. Detecting and counting the moving vehicles using mask R-CNN. In: *Proceedings of 2019 IEEE 8th Data Driven Control and Learning Systems Conference, DDCLS 2019*. ; 2019. doi:10.1109/DDCLS.2019.8908877

 https://doi.org/10.52202/075106-0004

15. Chen Z, Guo H, Yang J, et al. Fast vehicle detection algorithm in traffic scene based on improved SSD. *Measurement (Lond)*. 2022;201. doi:10.1016/j.measurement.2022.111655

16. Zhang Y, Guo Z, Wu J, Tian Y, Tang H, Guo X. Real-Time Vehicle Detection Based on Improved YOLO v5. *Sustainability (Switzerland)*. 2022;14(19). doi:10.3390/su141912274

17. Liu Y, Lu Y, Shi Q, Ding J. Optical flow based urban road vehicle tracking. In: *Proceedings - 9th International Conference on Computational Intelligence and Security, CIS 2013*. ; 2013. doi:10.1109/CIS.2013.89

18. Gomaa A, Abdelwahab MM, Abo-Zahhad M, Minematsu T, Taniguchi RI. Robust vehicle detection and counting algorithm employing a convolution neural network and optical flow. *Sensors (Switzerland)*. 2019;19(20). doi:10.3390/s19204588

19. Geiger A, Lenz P, Stiller C, Urtasun R. Vision meets robotics: The KITTI dataset. *International Journal of Robotics Research*. 2013;32(11). doi:10.1177/0278364913491297

20. Yang M, Jia Y. Temporal dynamic appearance modeling for online multi-person tracking. *Computer Vision and Image Understanding*. 2016;153. doi:10.1016/j.cviu.2016.05.003

21. Milan A, Leal-Taixe L, Reid I, Roth S, Schindler K. MOT16: A Benchmark for Multi-Object Tracking. Published online March 2, 2016. http://arxiv.org/abs/1603.00831

22. Hsieh MR, Lin YL, Hsu WH. Drone-based Object Counting by Spatially Regularized Regional Proposal Network. Published online July 19, 2017. http://arxiv.org/abs/1707.05972

23. Zhu P, Wen L, Du D, et al. *Detection and Tracking Meet Drones Challenge*. http://www.votchallenge.net/.

24. Xing X, Yongjie Y, Huang X. Real-Time Object Tracking Based on Optical Flow. In: *2021 International Conference on Computer, Control and Robotics, ICCCR 2021*. ; 2021. doi:10.1109/ICCCR49711.2021.9349376

25. Tan M, Pang R, Le Q V. EfficientDet: Scalable and efficient object detection. In: *Proceedings of the IEEE Computer Society Conference on Computer Vision and Pattern Recognition*. ; 2020. doi:10.1109/CVPR42600.2020.01079

26. Bernardin K, Stiefelhagen R. Evaluating multiple object tracking performance: The CLEAR MOT metrics. *EURASIP J Image Video Process*. 2008;2008. doi:10.1155/2008/246309

27. Wu B, Nevatia R. Tracking of multiple, partially occluded humans based on static body part detection. In: *Proceedings of the IEEE Computer Society Conference on Computer Vision and Pattern Recognition*. Vol 1. ; 2006. doi:10.1109/CVPR.2006.312

28. Soleimanitaleb Z, Keyvanrad MA. *Single Object Tracking: A Survey of Methods, Datasets, and Evaluation Metrics*.

29. Pereira R, Carvalho G, Garrote L, Nunes U. *Sort and Deep-SORT Based Multi-Object Tracking for Mobile Robotics: Evaluation and New Data Association Metrics*

 https://doi.org/10.52202/075106-0004

OPEN TOOLS FOR TRUSTED FLIGHT AND UTM RESILIENCE

Hrishikesh Ballal,* Govind Singh,† and Rhythm Chopra‡

This paper covers two crucial aspects of digital security in the context of drone flights and advanced (e.g. Beyond Visual Line of Sight (BVLOS)) operations. Trusted Flight is a set of software services that provide a security layer to operators and manufacturers to conduct flight operations using the mature and well-tested Public Key Infrastructure and commonly available identity and authentication standards. We will cover the concept of "trusted flights" and its open implementation in Ardupilot and PX4, two popular open-source flight controllers, and the Aerobridge Management server to help with the automation of the security workflow.

In the second part of the paper, we focus on Unmanned Traffic Management (UTM) and the activities within the Global UTM Association (GUTMA) and Dronecode Foundation. Specifically, we introduce the UTM Adapter working group and the contributions to ensure that UTM services are cyber safe from the get-go. Additionally, some of the authors led a taskforce within GUTMA on UTM cybersecurity and resilience to provide an industry perspective for security and open implementations for UTM services. The GUTMA task force aimed to address the cybersecurity aspect within the context of certification / validation of UTM systems. The task force provided a cyber security risk assessment template framework as a starting point for UTM service provider certification efforts in the European Union.

INTRODUCTION

Security in the context of drone flights requires a holistic approach covering cyber-physical systems. We will build a holistic perspective on cybersecurity and the related activities in the context of flight operations and Unmanned Traffic Management (UTM) to enable secure operations and cyber security from the get-go. The common theme between these concepts is the open implementations developed to enable them. Trusted Flights enable drone operators to provide a security layer via digital flight permissions (e.g. one time permission (OTP)) and flight log signing and management within the organization for post-flight audit and security. The trusted flight system allows operators to generate a permission object as a JSON web signature that is downloaded from its digital infrastructure and must be transferred to the drone and the flight controller in order to conduct the mission. The Flight controller is designed so that it can only arm the drone once the

* Founder / Project Maintainer, OpenUTM + Aerobridge Server, Openskies Aerial Technology, Dublin, Ireland
† Principal Engineer, Secure Systems Research Center, Technology and Innovation Institute, Abu Dhabi, United Arab Emirates
‡ Software Engineer, Independent Contributor, Prague, Czech Republic

https://doi.org/10.52202/075106-0005

validity of the token has been verified. In addition, features like public key rotation and flight log signing by the flight controller ensure post-flight checks and verification assurances. We cover the technical mechanism and the benefits of trusted flight management to add security in drone operations in this paper.

On the UTM side, we will cover the activities within the Global UTM Association (GUTMA). From December 2023 till March 2024 GUTMA members co-led by Openskies Aerial Technology and Technology Innovation Institute (TII) (authors of this paper) around cyber-proofing UTM services. Some members of GUTMA are undergoing or considering EU Certification and this task force was aimed at helping the membership build understanding and knowledge base so that the UTM services are cyber-proof from the get-go. We will cover the content of the Technical White Paper that not only delves into the findings, insights, and recommendations derived from the analysis within the task force. This white paper also provides a template to build a threat-based risk assessment and mitigation measures of the identified security challenges in UTM operations. This white paper will serve as a crucial reference for industry stakeholders seeking to enhance the security posture of UTM systems and help with UTM validation / certification efforts. Finally, we share the open-source implementation of the security concepts and verification tools for UTM providers to validate their systems.

Together this builds a comprehensive and holistic framework for flight security and provide opensource tools for verification and deployment within a operators internal environment.

TRUSTED FLIGHT

In the context of this work, trust in drone flight operations focuses on the provision of digital infrastructure and workflows to ensure that only authorized (by an organization) flight plans, vehicles, personnel, and software can conduct a mission. These workflows cryptographically ensure that only the approved flight plan can be executed. The stack ensures that the arming of the vehicle will be disallowed if the cryptographic and other checks fail by building the mechanism in the flight controller itself. This means that UAV hardware manufacturers or operators control what software can and cannot run by refusing to run unsigned / untrusted flight plans. Hardware and software are not only secured for its owner but also secured against its owner. Specifically, this stack covers four specific aspects of drone flight operations:

1. Who is flying the drone? If there is an identity and authentication layer available for the operator across, only personnel and GCS authorized by the company can be allowed to fly.
2. Where is the drone hardware and software made? Who built the hardware and flight controller software / firmware for the aircraft and how to make sure that it is associated with the identity of the company operating the aircraft. This means that the operators can get a remote "kill switch" to prevent any personnel or vehicle from arming the drone.
3. How was the flight plan approved? Is there a mechanism to associate the flight plan with a "permission to fly" for the aircraft, operator, and the vehicle. In other words, without a digital permission linked to the flight plan and the vehicle no flying / arming is allowed.
4. Did the drone fly the approved flight plan? Once the flight plan is validated and executed, the specified drone fly only the specified flight plan.

Together these systems form a holistic trusted flight mechanism where confidence and assurances can be made to ensure security of the drone system. There are other important / major considerations for drone flight security e.g. C2 link encryption and others, but we cover those aspects in another paper. For this work, we will cover the four aspects listed above with a specific focus on

https://doi.org/10.52202/075106-0005

open tools being developed in Ardupilot and PX4, two of the most popular open-source flight controllers for UAVs and other open source projects developed by the authors.

Trusted Flight Components

Figure 1 below shows the main components of this trusted flight system. They operate together, and are linked, however they are not dependent on each. The authors have developed open-source implementations for these concepts; however, these can be swapped as necessary with own proprietary software if they adhere to the openly published API specifications.

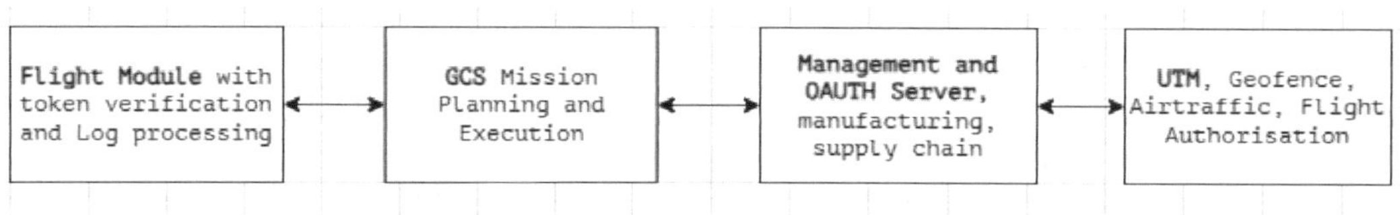

Figure 1. Trusted Flight Components

- *Flight Permission Token:* This is the central component of the trusted flight system; this is a token represented as a JSON Web Token (JWT) issued by the management server once the mission is approved and is transferred to the vehicle. This token has information about the flight plan, the pilot etc. and serves as cryptographic time limited proof of permission to fly / perform the operation.
- *Flight Module:* This is a specialized version / build of a flight controller / auto pilot software that is linked to the company's domain. This flight controller contains the public key of the company's domain and the root certificate. This ensures that only tokens signed by the company's domain are processed and approved. This flight controller uses publicly available Public Key Infrastructure to validate and verify the tokens and subsequently allow arming and other pre-flight checks. This is the most critical cyber-physical system that digitally links the vehicle to the company and its digital infrastructure.
- *Ground Control Station:* The Ground Control Station is a communications mechanism that mainly deals with moving the permission token issued by the management server to the vehicle. We have implementations in QGCS that mainly use Mavlink to send the token to pre-defined locations in the flight controller. In some cases, it will interact with the UTM system to offer UTM services e.g. air-traffic information etc.
- *Management Server:* A optional but a crucial component of automating / performing operations at scale. This component helps in managing the permission artifacts and interactions with the GCS and storing records and keys. While all of this can be done manually, having a management server or extending the fleet management server to accommodate this security layer will ensure at-scale operations.
- *UTM:* For advanced operations, a UTM system must be able to exchange the flight plans and operational intents with peer UTM providers and perform UTM services like strategic deconfliction or remote ID tracking etc.

Trusted Flight Data and Workflow

To understand how the system works together, we consider a simple case of an operator building a flight plan on QGCS and then submitting the plan to the drone to fly. Without the Trusted flight system, this can simply be achieved with the following steps:

https://doi.org/10.52202/075106-0005

1. Connect the drone to the GCS
2. Use the Ground Control system interface to build a flight plan
3. Send the plan to the vehicle
4. Arm the drone and conduct the mission

This workflow has many security-related issues that are introduced at the beginning of the paper. Crucially, the presented above workflow assumes that there is a trust in the pilot and the pilot flies the mission that he is supposed to fly. The trusted flight system provides a security layer for the operators to ensure that only the approved vehicles and flight plans are processed, and appropriate digital logging is provided.

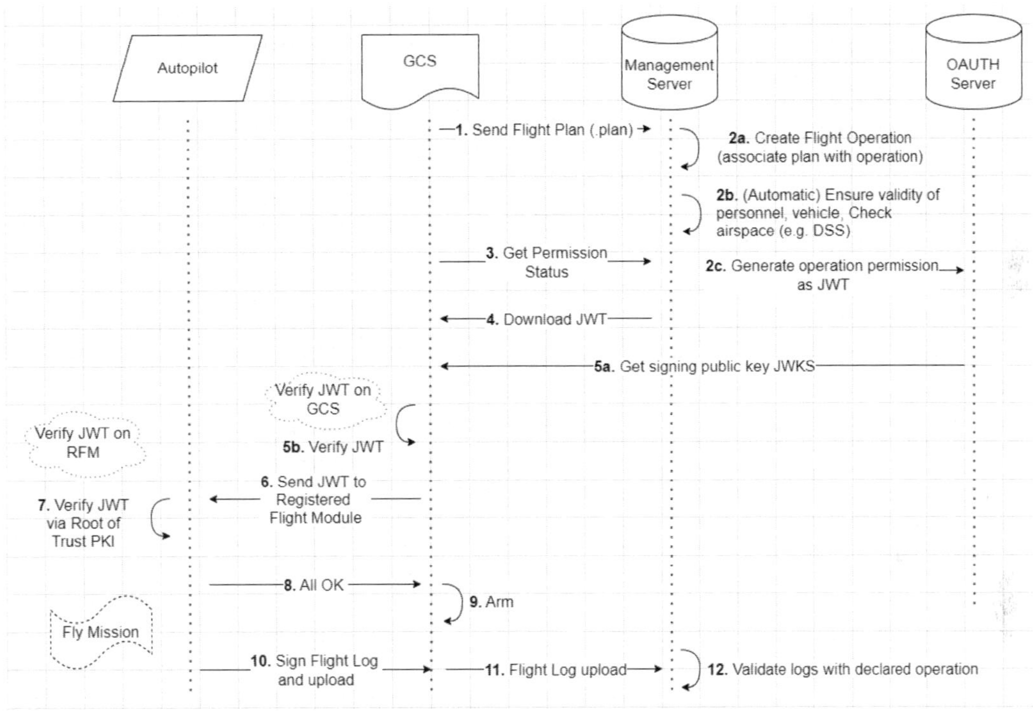

Figure 2. Trusted Flight workflow

To understand how this workflow will work in the context of trusted flights, we use the workflow depicted in Figure 2. A step-by-step depiction of the process is below, these steps will be largely automated in a digital context and assumes integration with the GCS, Management server and the UTM system:

- *Step 0:* Custom firmware is flashed on the vehicle that includes the root certificate and issuer name embedded in the firmware and stored / loaded from read-only memory.
- *Step 1:* The operator builds a flight plan (or loads a flight plan that is already stored in the management server).
- *Step 2a:* The management sever processes the flight plan and creates an operation i.e. associates this plan with a vehicle and person in the organization.
- *Step 2b:* The management server performs additional checks e.g. interaction with the UTM system etc. to ensure de-confliction.

https://doi.org/10.52202/075106-0005

- *Step 2c:* Once the initial checks are complete, the Management server contacts the OAUTH server of the organization to generate a permission artifact as a JWT, see the permission token details for the claims and sample token.
- *Step 3 and 4, 5:* The GCS then downloads the permission token and verifies the artifact.
- *Step 6:* Once the permission token is validated, the token is sent to the vehicle and the mission is also sent to the vehicle.
- *Step 7:* The flight controller verifies the token is valid and checks against the public key in ROMFS and the claims to ensure that the permission is granted for the vehicle, also if a plan file hash is provided, the firmware hashes the loaded file and cross checks against the provided has in the token.
- *Step 8:* The GCS sends the commands to arm if the Step 7 is executed properly
- *Step 10:* Once the flight is completed the flight controller signs the flight log and sends it to the management server, the management server verifies the signature since it has the public key of the flight controller.

This "full cycle" flow ensures that the objectives of the security system are met, and a digital proof and verification system is in place to ensure security audits.

Trusted Flight Permission Token

The permission token is expressed as a JWT and has the following claims, Figure 3 shows a decoded permission token issued by the management server. Using JWT ensures that we utilize existing tools and libraries to validate and parse the data. In addition to the standard claims provided by JWT, we can also add additional claims that can be verified by the flight controller.


```
{
  "iss": "https://id.openskies.sh/",
  "exp": 1640024322,
  "iat": 1640020722,
  "sub":
"g37lXifAQoBfVuQZxT3VJFjXIMgdfXIOpa2LdWBQ@clients",
  "scope": "",
  "typ": "Bearer",
  "flight_plan_id": "12818e87-4c96-4e4c-8c63-
82b8e12c3b73",
  "flight_operation_id": "3408bce9-dbab-4665-abfc-
8ea03b0ad871",
  "plan_file_hash":
"a2a201efa111dde5aaa8f42b7b46d17b595e39d850500dbacd9f79a
ef0bbf68e"
}
```


Figure 3. Flight Permission token claims

We first ensure that the token is decoded and validated correctly and then additionally perform the following computation:

- Utilize the *plan_file_hash* to compute the hash of the flight plan loaded on the flight controller to ensure that the same plan is loaded on the vehicle.
- The *flight_plan_id* and *flight_operation_id* establishes the link between the token and the plane and operation on the management server for auditing.

- The *sub* claim can be used to establish the link between the vehicle / RFM to the flight plan.

Flight Controller Modifications

The core of the trusted flight system consists of modifications to the autopilot to enable additional checks in the pre-arming system.

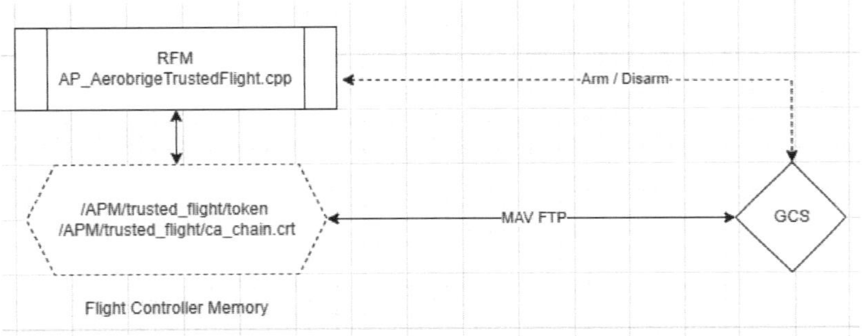

Figure 4. Flight Controller Architecture (Ardupilot implementation)

In the context of Ardupilot, we implemented a new module (AP_AerobridgeTrustedFlight) for handling all the operations and validations related to Trusted Flight. It is wired into the existing pre-arm checks to ensure these validations are performed before arming and cannot be by-passed even with forced arming, given the Trusted Flights feature was enabled during the ardupilot build for RFM. AP_AerobridgeTrustedFlight module expects the certificate chain and JWT token to be available on the filesystem path (/APM/trusted_flight/) as depicted in the Figure 4, failing which the pre-arm checks for the RFM will fail during arming preventing any unauthorized operation to execute.

OPEN IMPLEMENTATION OF TRUSTED FLIGHT

This section provides details about the opensource tools available to enable at-scale automation of the trusted flight system. These contributions are made by the authors to the community as pull requests to opensource projects. And in some cases, this additional security layer provided by these tools will ensure reliable and safe operation of drones in the airspace powered by mature, publicly available cryptography tools and systems. These tools provide an easy on ramp for operators and manufacturers to ensure an elevated level of operational security and data assurance.

 https://doi.org/10.52202/075106-0005

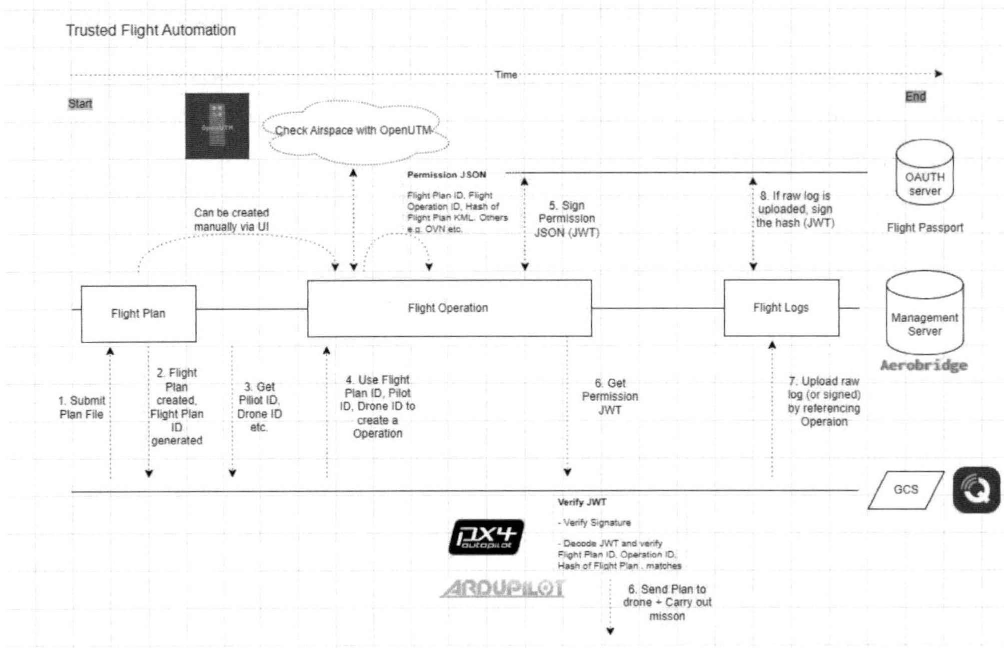

Figure 5. Trusted Flight Components implemented with Management Server and Auto-pilots

The authors have built open-source implementations for this in various projects. The PX4 and Ardupilot software repositories have existing pull requests with the trusted flight module. The Management server is implemented as Aerobridge* management server and can help an organization manage its flight permissions process and issuance of JWT tokens. There are additional features in the Aerobridge management server that ensure the trackability of the components, the provision of signed firmware etc. The QGCS is an opensource ground control station that has integrations and support for the UTM and the management server. Finally, the UTM system is implemented in OpenUTM† to ensure a standard compliant UTM system that utilizes the latest standards e.g. the ASTM standard for strategic de-confliction.

CYBER SECURE UTM

In the rapidly evolving landscape of aviation, the need for robust cybersecurity measures within Unmanned Traffic Management (UTM) systems has never been more critical. As drones become increasingly integrated into airspace operations for various applications, from delivery services to surveillance and beyond, ensuring the integrity, confidentiality, and availability of UTM services is paramount.

At the heart of this necessity lies the imperative to protect against a myriad of cyber threats that could compromise the safety, efficiency, and reliability of unmanned aerial vehicles (UAVs) and

---

* Aerobridge Management Server, available at: https://www.github.com/openskies-sh/aerobridge

† OpenUTM, available at: https://github.com/openutm

the airspace they navigate. From malicious intrusions seeking to disrupt operations to unauthorized access attempts aimed at stealing sensitive data, the potential risks are diverse and ever-present.

To address these challenges, regulatory bodies such as the Federal Aviation Administration (FAA) and the European Union Aviation Safety Agency (EASA) have issued comprehensive recommendations, outlining stringent security requirements for UTM systems[*†]. These guidelines emphasize the importance of implementing robust security controls across all facets of UTM infrastructure, including USSP (U-space Service Provider), CISP (Communication and Information Service Provider), and operator systems. FAA's Near-Term Approval Process (NTAP) and the UTM Pilot Program, along with the EU Regulation 2021/664 on U-Space and accompanying AMC/GM, recommend information security using some generic standards like ISO27001.

All regulatory requirements and recommendations necessitate a comprehensive framework from industry bodies responsible for implementing and deploying UTM services. Recognizing this gap, the authors, co-led a task force[‡] within GUTMA, a UTM industry association, on Secure and Resilient UTM. This task force aimed to explore the implications for UTM service providers regarding security considerations for UTM services deployment. Because of the task force's findings, there emerged a need to create comprehensive, updated tools capable of verifying compliance against a specific class of ever-evolving cybersecurity threats.

UTM Adapter Project

The UTM adapter[§] project seeks to mitigate the fragmentation observed between operators and UTM Service Providers (UTMSPs) in the UAS Traffic Management (UTM) ecosystem. Operators often utilize diverse Ground Control Stations (GCS) and operational procedures to communicate data to the UTM systems. This leads to inconsistencies and interoperability challenges when interfacing with UTMSPs. The UTM adapter project serves as a middleware solution that can be plugged into any GCS, bridging the gap between these heterogeneous operator systems and the standardized interfaces expected by UTMSPs. This project provides a standardized interface for operators to connect with UTMSPs, the adapter eliminates fragmentation and streamlines data exchange, ensuring seamless integration into the broader UTM network. Ultimately, the UTM adapter enhances operational efficiency, promotes interoperability, and facilitates the adoption of UTM services across diverse operator platforms.

Additionally, As depicted in Figure 6, this approach enables harmonized security measures across the UTM ecosystem. By standardizing the interface between operators and UTM Service Providers (UTMSPs) through the UTM adapter, security protocols and mechanisms can be uniformly enforced and monitored. This harmonization ensures consistent implementation of cybersecurity measures, such as authentication, encryption, and access control, across all interactions within the UTM network. As a result, the UTM adapter not only addresses interoperability

* https://www.faa.gov/uas/advanced_operations/traffic_management
† https://www.easa.europa.eu/en/document-library/acceptable-means-of-compliance-and-guidance-materials/amc-and-gm-implementing
‡ GUTMA secure-and-resilient-utm Taskforce, available at: https://www.unmannedairspace.info/latest-news-and-information/gutma-launches-members-task-force-on-secure-and-resilient-utm/

§ Dronecode UTM Adapter, available at: https://github.com/Dronecode/utm-adapter.

https://doi.org/10.52202/075106-0005

challenges but also strengthens the overall security posture of UTM operations, promoting trust, resilience, and confidence in the system.

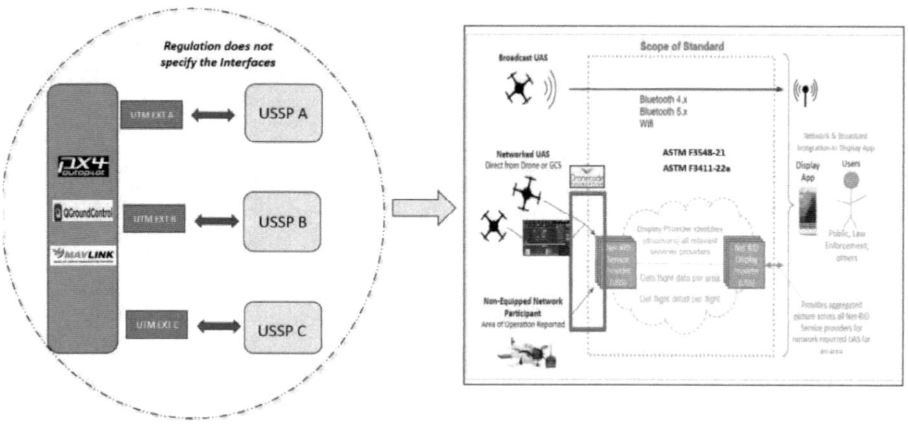

Figure 6. UTM Adapter Conceptual Diagram

UTM Service cyber-security verifier

A new project was created* to address the need for a suite of tools for UTM service providers to verify their compliance against the latest cyber security threats. The goal of the UTM cyber-security verifier project is to provide a toolset that can be run against the UTM services to monitor conformance and response to security threats. This initiative aims to address cybersecurity concerns of regulators associated with certification and validation of UTM systems by establishing a robust security framework grounded in industry standards and developed by a community. The findings and tools developed through this initiative will be published for wider dissemination within the industry.

This projects ultimately aims to bolster UTM security using Confidentiality, Availability, and Integrity cybersecurity principles as shown in Figure 7, they aim to enhance the resilience of UTM services against potential cyber threats and ensure their continued reliability and safety. These tools provided by this project play a vital role in assessing, validating, and continuously monitoring the cybersecurity posture of UTM systems, enabling stakeholders to detect and mitigate vulnerabilities effectively. By aligning with EASA recommendations and leveraging advanced cybersecurity solutions, the aviation industry can confidently navigate the complexities of UTM security, ensuring the seamless and secure integration of drones into our skies.

* UTM Cyber security verifier, available at: https://github.com/tiiuae/UTM-Security-Verifier

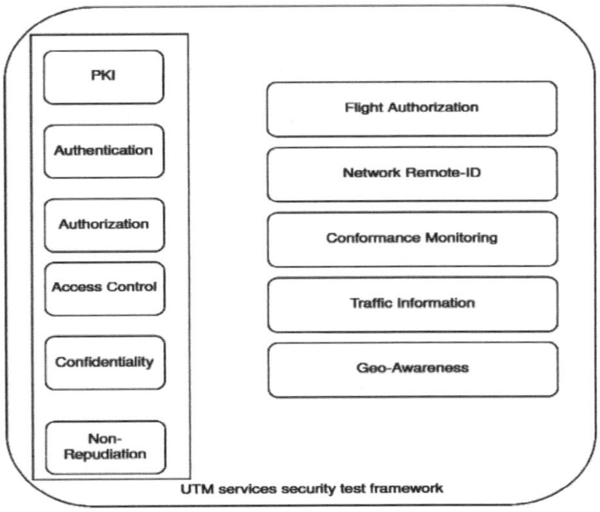

Figure 7. UTM Cybersecurity Attributes.

The UTM cyber security verifier project aims to verify the security features within individual UTM (UAS Traffic Management) services, aiming for a comprehensive assessment of cybersecurity measures across the UTM ecosystem. This initiative focuses on evaluating the effectiveness and resilience of these measures, with the goal of enhancing the overall security posture of UTM systems. The deliverables of this project include a test framework, tailored for examining the functional specifications of UTM adapters to ensure compliance with adapter specifications. Additionally, a security test framework for UTM services is developed, encompassing the creation of test cases as depicted in Figure 8, to validate each security feature against specified requirements and the documentation of verification results, including audit logs. These deliverables are designed to be applicable to any Ground Control Station (GCS) and UTM provider, facilitating the implementation of robust security measures across the UTM landscape.

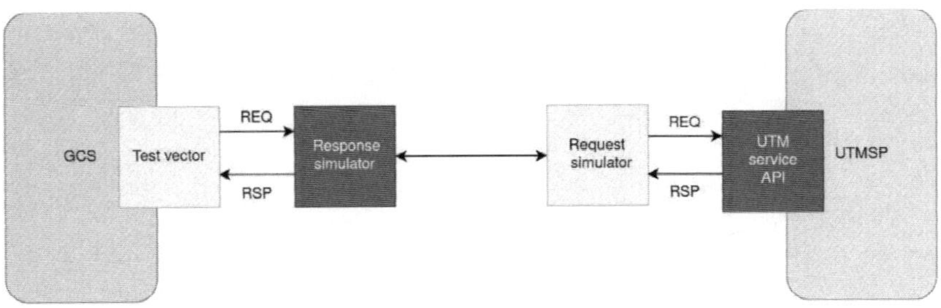

Figure 8. UTM Security Verification Framework

CONCLUSION

In this paper we introduced the concept of trusted flight and shared a series of open tools that are available for the community to ensure safe and secure integration of drones in the airspace.

https://doi.org/10.52202/075106-0005

ACKNOWLEDGMENTS

The authors acknowledge the contributions of the Ardupilot, PX4, OpenUTM, Aerobridge and UTM Adapter community in developing and refining these concepts. In addition, India's No Permission No Takeoff policy inspired early thinking around development of mission security and the operations management server.

ADVANTAGES OF AUTOMATIC TARGET RECOGNITION MULTI-VEHICLE SYNERGY IN UNDERWATER ENVIRONMENTS WITH LIMITED COMMUNICATION

Jordan Liss*, Vincente Nguyen†, and Joseph Mattern‡

Automatic Target Recognition (ATR) is a technology that allows Unmanned Underwater Vehicles (UUVs) to identify, classify, and take actions on specific targets without human intervention. In large areas of operation, single vehicle missions are slow and subject to mission failure if there are any issues with the UUV. To combat this, UUVs equipped with ATR capabilities can be deployed in swarms to achieve tasks with increased efficiency, redundancy, and resilience. However, the introduction of multiple UUVs simultaneously stress the mission due to the limitations of the environment. UUVs tend to suffer from low fidelity data, unreliable acoustic communications, and limited communications bandwidth. The introduction of multiple UUVs in a mission greatly exasperate these issues. We will be discussing how the limitations of underwater environments affect the performance of multi vehicle UUV mission, and we will propose a joint ATR and autonomy solution that limits the impact of underwater conditions on multi-vehicle UUV missions. The ATR solution leverages both low fidelity sonar images and high-fidelity camera images to make detections of various confidences that can be updated by any UUV. The autonomy solution leverages template behaviors to allow each vehicle to safely operate by simulating the behavior of other UUVs. These template behaviors reduce the growth of navigational uncertainty for other vehicles to provide better mission performance.

INTRODUCTION

Problems with Multiple UUVs Mission

The implementation of multiple underwater unmanned vehicles (UUVs) for a single mission presents a myriad of challenges across various environments. These challenges encompass technical, logistical, and operational aspects, stemming from issues such as communication (mainly acoustic) and coordination between UUVs, resource allocation, mission planning complexity, environmental uncertainties, and the need for robust autonomous decision-making systems. Furthermore, integrating multiple UUVs into a cohesive operational framework demands addressing concerns regarding power management, sensor fusion, navigation accuracy, and collision avoidance. Despite the potential advantages of deploying multiple UUVs for enhanced mission capabilities, the practical realization of such endeavors remains hindered by the intricate interplay of these multifaceted challenges.

* Software Engineer, Engineering Department, SeeByte, Inc. Jordan.liss@SeeByte.us
† Product Manager, Engineering Department, SeeByte, Inc. Vincente.nguyen@SeeByte.us
‡ Software Engineer, Engineering Department, SeeByte, Inc. Joseph.mattern@SeeByte.us

https://doi.org/10.52202/075106-0006

Challenges in Underwater Communications

The nature of underwater acoustic communications is inherently fraught with challenges, making it a uniquely difficult, low-fidelity, and error-prone method of data transmission. These challenges arise from the physical properties of water and the way sound waves propagate through it (Reference 1). Unlike air, water is a denser fluid medium with vastly different sound propagation physics, which affects the speed, range, and fidelity of acoustic signals. Furthermore, the underwater environment is cluttered often not accurately mapped in real-time with various sources of propagation noise, such as marine life, underwater terrain, and human-made structures, that can reflect, absorb, or scatter sound waves, leading to signal degradation and loss.

The variability in water temperature, salinity, and pressure at different depths also impacts the sound speed profile, causing refraction of sound waves. This variability can lead to multi-path propagation, where signals reach the receiver at different times and angles, causing phase shifts and temporal spreading. The phenomenon significantly complicates signal interpretation, leading to delays, distortion, and decreased data rates. Additionally, the presence of background noise from natural sources like marine life and wave action and anthropogenic sources such as ship traffic and industrial activities further obscures the signal, increasing the error rate and reducing the reliability of data transmission.

Bandwidth limitations are another critical issue in underwater acoustic communications. The usable frequency range for underwater communications is much narrower than that in above surface communication systems, like radio waves, constraining the amount of data that can be transmitted over a given time period. Higher frequencies, while offering better bandwidth and hence higher data rates, are absorbed more quickly in water, limiting their range. Conversely, lower frequencies can travel longer distances but offer limited bandwidth, leading to lower data transmission rates. This trade-off between range and bandwidth necessitates a delicate balance, often resulting in compromises on data rates, fidelity, and error rates in underwater acoustic communications.

Neptune Autonomy Solution

Autonomous underwater vehicles (AUV), and more recently unmanned surface vehicles (USV), have grown in acceptance to the point where their relevance is unquestioned. The usage of AUVs in oceanography or mine-counter measures (MCM) applications is nowadays routine with dedicated teams trained to operate unmanned autonomous assets in day-to-day missions. The number of navies operating unmanned autonomous assets continues to increase, and unmanned maritime systems (UMS) are also starting to prove capable in other applications such as Anti-Submarine Warfare and intelligence gathering (Reference 2). Neptune autonomy software solution tackles these varying multi-vehicle UUV and provides a unified environment where collaborative and adaptive missions can be planned in an effective way. Notably, Neptune autonomy solution can efficiently manage the water space, handling dynamic mission objectives, asserting safety margins/collision avoidance between multi-vehicles, while empowering the operators.

Neptune provides efficient management of water space as a function of time, space, and energy by achieving the necessary shared objectives and avoid repeated objectives. The autonomy embedded software running on all the vehicles in the mission setups the most optimal route to accomplish a survey mission with input from the operator during the pre-mission stage. For example, an operator is aided setting up intelligent launch and recovery offsets during the pre-mission stage. The Neptune topside software during the upload process can automatically split vehicles in a squad to their respective launch and recovery (L&R) locations and set offsets. The L & R objectives are separated in depth and altitude following safe transit depth plane tolerances automatically

https://doi.org/10.52202/075106-0006

determined by the software. The topside master mission file, which is disseminated to the all the vehicles before mission execution encodes all the vehicle safe transit planes, mission objectives, and respective vehicle surveys. All vehicles have a copy of the missions all the other vehicles are going to execute, in addition to their own.

Neptune can generate new dynamic objectives to be shared with the squad (i.e. re-tasking if a vehicle goes down) using only acoustic communication for information dissemination. Acoustic vehicle-to-vehicle (V2V) received messages include a vehicle's current objective and progress through the set of objectives. In order to avoid latencies in communication, the Neptune acoustic message schema does not require message acknowledgments instead each vehicle uses set timeslots to broadcast communication to vehicles within listening range. In addition, in the event a single vehicle has limited communication to the dry-side command and control (C2) station, but individual UUV communication is reliable. Vehicle message statuses can use the mesh network of vehicles to hop information to the topside C2 station. When communication is reliable between n number of nearby vehicles, the sub-squad of vehicles performs dynamic deconfliction of each other. The hive-mind knowledge of all the vehicles especially one's attempting to perform the same objective do the following: 1) Vehicle will not attempt to start an objective that it knows has been completed or is already being performed by another 2) When an acoustic message is received from another vehicle, current objectives are checked and ensured that the vehicles are not attempting to perform the same objective. Most importantly, algorithm logic focuses on vehicle safety is used to determine which of the vehicles should abort any superfluous objectives and which will continue.

Neptune most importantly empowers the operators to tailor mission for their squad and implement unique strategies based on preferred assets, and hierarchy to all assets in question. Once the mission is underway, the actual ordering of objectives is affected by generation of unplanned objectives (i.e. ATR hits), conflicting dynamic exclusion zones, time changes between GPS fixes, environmental anomalies, and unique behaviors for planned and unplanned objectives. Neptune hive-mind makes efficient use of all the vehicles in the squad by keeping them all busy performing mission objectives and when any vehicle is not engaged in an objective, Neptune searches for new objectives for them to perform (See Appendix 1 for more detail).

ADVANCED MULTI-MODAL ATR

Advancements in computer vision, machine learning and deep learning have evolved to a state where real-time detection and performance exceeds human abilities. SeeByte's approach to automatic target recognition in forward looking sonar (FLS) and electro-optical camera (EO) utilizes a convolutional neural network (CNN) and single-shot detector (SSD) backbone. SeeByte automatic target recognition (ATR) algorithm solution has been showcased in various underwater environments on various underwater vehicles using various combinations of camera and sonar fusion solutions.

SeeByte ATR system is a deep neutral network (DNN) framework that can rapidly evaluate multiple algorithms for a given application. These internal tools consist of open-source tools, custom built tools, as well as closed source tools from third parties. Many of the software solutions and processes were evaluated for this application, and a CNN with a backbone was selected after rigorous testing. The convergence of using a CNN with an SSD as SeeByte's chosen methodology provides the best balance between positive detections and inference time. Most notably, SeeByte ATR intended deployment is real-time on edge devices on underwater vehicles where it was optimized for minimal power consumption and controlled heat generation/dissipation. The SeeByte ATR solution model dataset development was generated by manually tagging footage stream, frame by frame, by subject-matter experts with great deliberation and trained-eye accuracy. This

https://doi.org/10.52202/075106-0006

method resulted in a highly robust CNN that is vehicle agnostic, EO device agnostic, and viewing angle agnostic (with limits) that has been tested and verified to be able to detect an MLO target before an expert operator can.

Localization and Navigation Correction

The deployment of multiple underwater unmanned vehicles (UUVs) in a coordinated squad introduces a powerful dimension to enhancing localization and navigation correction capabilities far beyond the reach of single-vehicle operations. This collaborative approach leverages the strengths and capabilities of individual UUVs to achieve a synergistic effect, significantly improving the accuracy and reliability of underwater missions. For instance, when one UUV is equipped with advanced ATR capabilities, it can act as a beacon or reference point for other vehicles in the vicinity, facilitating a dynamic, real-time enhancement of navigational precision and spatial awareness among the fleet.

This collaborative framework operates on the principle of shared situational awareness. The ATR-equipped UUV can detect, identify, and relay information about underwater landmarks or objects of interest to its counterparts. The critical contact information, once shared, enables other UUVs to cross-reference their own positional data against the identified landmarks, allowing for immediate correction of navigational errors and drifts. Such an approach is particularly beneficial in GPS-denied environments, where traditional navigation systems are prone to cumulative errors over time.

The utilization of multiple UUVs in this manner enables the distribution of different sensing and processing tasks among vehicles, optimizing the use of available resources. For example, a UUV with enhanced ATR capabilities can focus on detailed object detection and identification, while others concentrate on mapping, environmental monitoring, or direct intervention tasks. Through inter-vehicle communication protocols, these UUVs can share critical data and updates, enabling the fleet to operate as a cohesive unit with enhanced spatial intelligence. This not only increases the operational efficiency and coverage area, but also significantly reduces the likelihood of navigational errors, as multiple data points and observations contribute to a more accurate and robust understanding of the underwater environment. Neptune autonomy solution is the ideal framework for solving these many challenges in an optimized, operator friendly, third-party compatible, and reliable/repeatable way.

Camera and Sonar Fusion ATR Single Vehicle Use-Case(s)

Recently SeeByte, Inc in collaboration of other ocean vendors have proved on-vehicle the capability of using SeeByte ATR for detection, localization, navigation, inspection, identification, and neutralization of mine-like objects (MLOs) on the seafloor, in the water column, and at the water surface. The project funded by Defense Innovation Unit (DIU) to develop and demonstrate an untethered prototype Autonomous EOD Maritime Response Vehicles (AEMRV) system that can locate, identify, and interrogate a subsea threat. The subsea robotics vehicle prototype derived from VideoRay, Inc.'s EOD Maritime Expeditionary Standoff Response (MESR) ROV Defender platform (includes a HD 1920 x 1080, 25 FPS camera, Oculus Binocular camera and Blueprint Subsea Oculus Dual Frequency Sonar). The ROV prototype was modified with additional payloads and systems, including the Greensea IQ, Inc OPENSEA Edge platform, Oceancomm M2 acoustic modems, Seatrac Ultra Short Baseline (USBL), and VideoRay lithium battery pods to meet AEMRV mission requirements.

The SeeByte ATR AEMRV technological demonstration in September 2023 in San Diego, CA, USA showcased accurate, reliable and scalable MLO/non-MLO detection in ranges of up-to 40 m from the vehicle, in addition to close ranges of roughly 1 meter from the vehicle. Examples of SeeByte ATR detection on real-time data streams are seen in the Figure 1 and Figure 2 below:

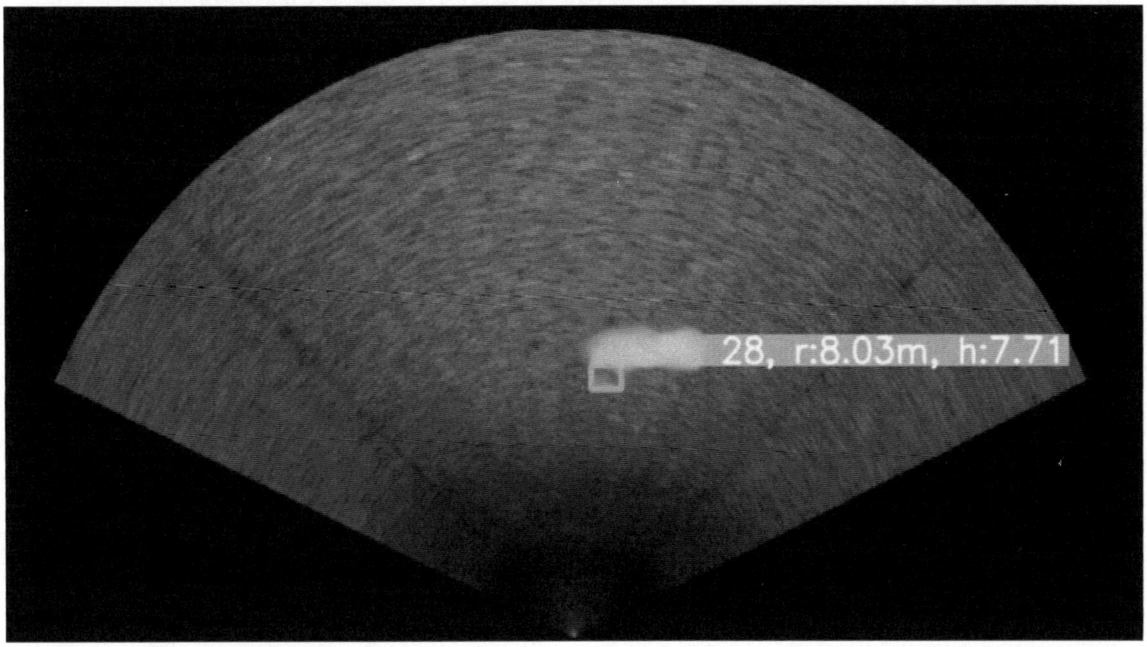

Figure 1. Sonar Frame Capture of Detected MLO with ATR Detection Overlay.

Vehicle range = 8.03 m and detected height of 7.71 m (Image Intentionally Blurred)

https://doi.org/10.52202/075106-0006

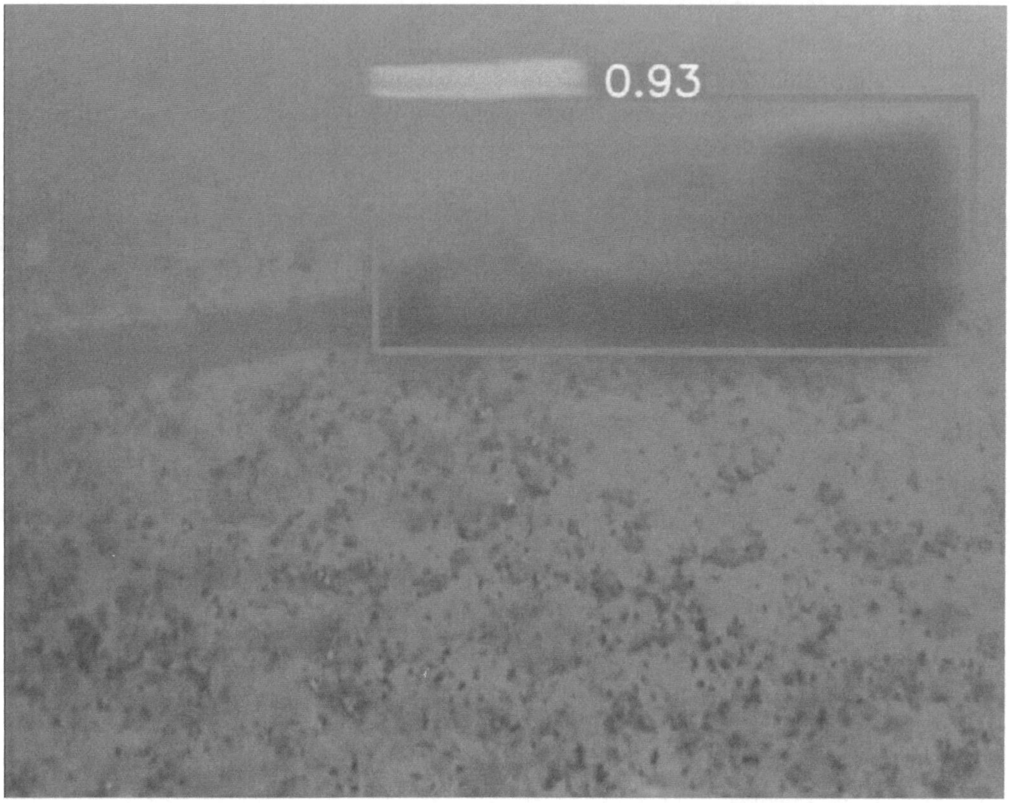

Figure 2. Close Range Image of MLO Detection and ATR Overlay (Image Intentionally Blurred)

The most important takeaways from the demonstration was the ability to show accurate and reliable fused ATR solutions from streaming both the FLS data and camera stream from the Videoray HD camera/Numurus Stereo camera. The object of interest (MLO or non-MLO) in the sonar and camera stream was converted into a single landmark object in the vehicle referenced map and localization of the landmark was improved iteratively (based on spatial and temporal confidences of each ATR detection). The fused ATR solution allowed the AEMRV vehicle to autonomously navigate to new objectives in a mission to detect, track, and inspect an object-of-interest in real-time as seen in Figure 3.

https://doi.org/10.52202/075106-0006

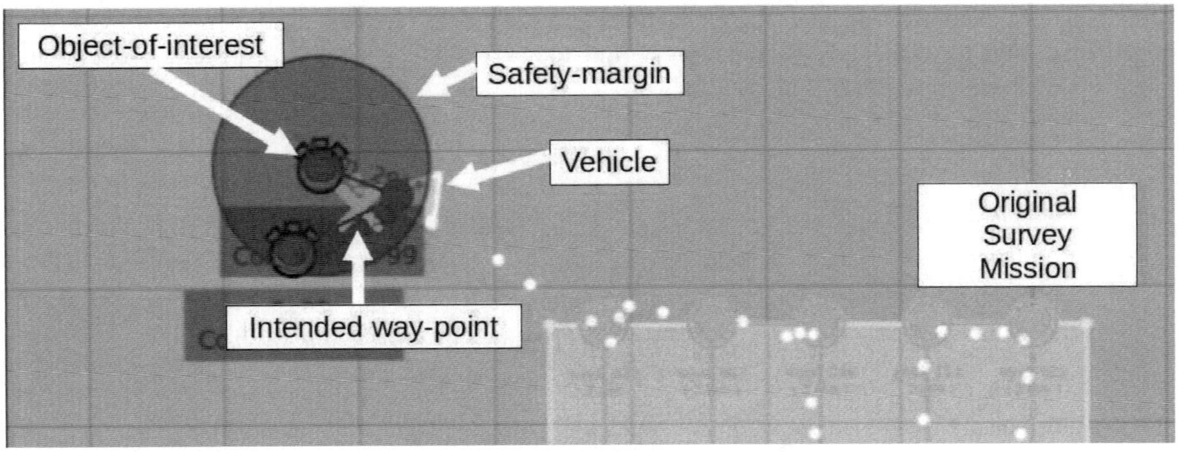

Figure 3. Screenshot of C2 Showcasing the Vehicle Detecting, Locating, and Tracking MLO within 5 m of vehicle.

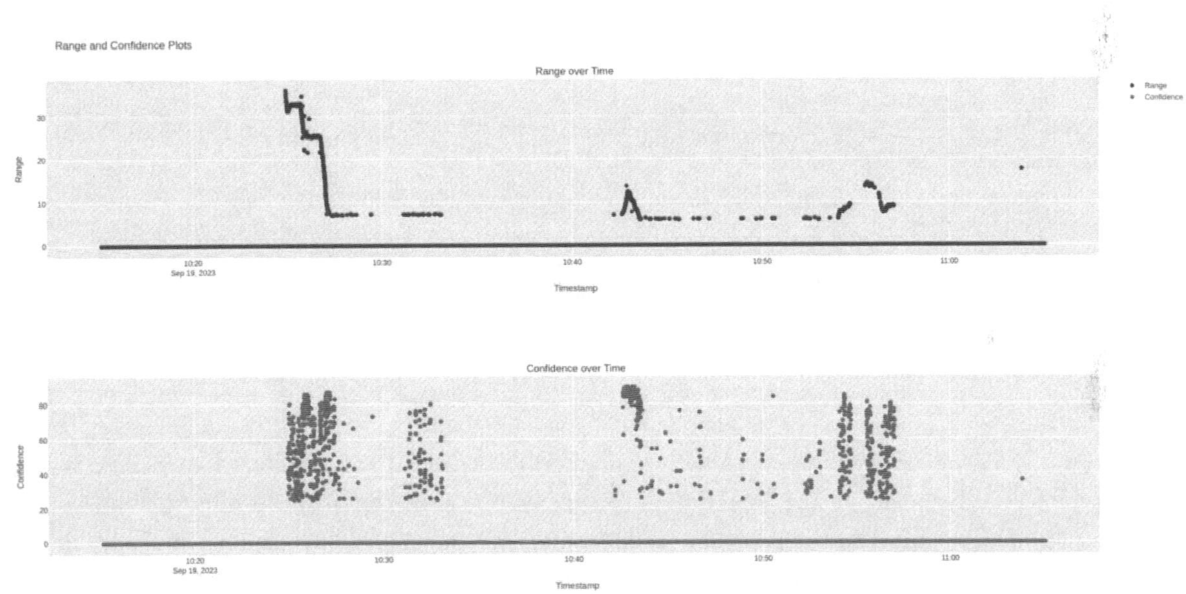

Figure 4. SeeByte ATR performance of estimated object range (on the top plot in blue) and detection confidences (on the bottom plot in red)

The SeeByte ATR multi-modal ATR tool shows detection confidence and classification performance immunity between 1m to 40 m. The detection confidence behavior is not biased or skewed as a function of range of the vehicle from the object of interest as seen in the bottom plot of Figure 4. The high confidences detection occurred at both 1 m away and 40 m from the vehicle. Front-seat end-users of the data stream were able to filter the data based on classification tag, detection confidence, distance, time consistency, and spatial consistency. In Figure 5, is an example

https://doi.org/10.52202/075106-0006

of the capability provided by leveraging the SeeByte ATR data stream. End-users of the AEMRV dataset were able to accurately detect, track, and localize the location of a potential MLO in a survey mission with no a priori knowledge.

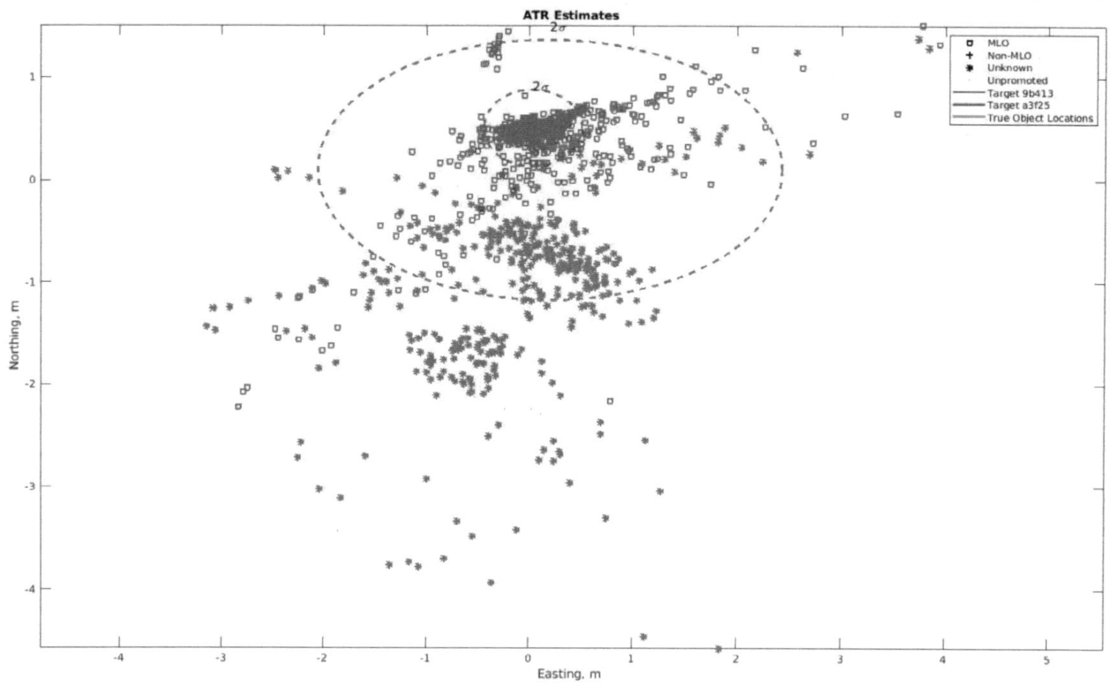

Figure 5. ATR performance showing high clustering of target observations are within +/- 1.5 m in Easting and +/- 0.5 m in Northing for the target a3f25, as seen by the green probability dashed circle

The determined object-of-interest (which happened to be a real MLO detected off the coast of San Diego, CA, US) based on the SeeByte ATR data stream and real-time temporal and spatial confidence modelling allowed the AEMRV untethered and fully autonomous accurately track and localize the object to within +/- 1 m of the actual position. Most importantly, the ATR data allowed the vehicle to identify, station keep within close range, and to deploy a neutralization charge (again all autonomously).

On a side note, the success of the fused SeeByte ATR solution proven on the AEMRV ROV vehicle, paved way for a similar application on an amphibious underwater crawler vehicle called the A2RV funded by DIU. The same SeeByte fused sonar and camera ATR solution was integrated into a the A2RV with no changes and despite the sonar viewing angle of MLOs being closer to 5 degrees with respect to the seafloor, instead of 25-35 degrees in the AEMRV ROV mission. The ATR accuracy and performance on the A2RV vehicle was equivalent to the original model set use-case being only ROV sonar and camera imagery training data. Shows the robustness of the SeeByte ATR CNN model.

The single vehicle use-case (on ROV or crawler) shows the power and capability of the SeeByte ATR software solution for MCM applications for all types of vehicles, especially maritime vehicles. The lessons learned from the vehicle single use-case set the foundation for multi-vehicle ATR capability and with the Neptune autonomy software solution the value add for MCM is exponential.

SYNERGY OF MULTI-VEHICLE AUTONOMY WITH MULTI-MODAL ATR

The synergy of multi-vehicle autonomy combined with multi-modal ATR significantly amplifies the operational resilience and efficiency of underwater missions. The Neptune approach not only maximizes the use of available resources but also introduces a layer of redundancy that is critical for mission success in the challenging and unpredictable underwater environment. By distributing ATR capabilities across multiple vehicles, a joint ATR multi-vehicle solution ensures that even in the event of a vehicle failure, the mission objectives can still be met without substantial disruption. The section delves deeper into the advantages and mechanisms behind this innovative approach.

Hardware Redundancy and Mission Continuity

One of the most significant advantages of a Neptune multi-vehicle strategy is the inherent hardware redundancy software framework flexible architecture it provides. In a scenario where one vehicle experiences a failure, whether due to technical malfunctions, energy depletion, or environmental hazards, the remaining vehicles in the fleet can reassess the mission plan and redistribute the tasks among themselves. Figure 6 shows an example of the multi-vehicle coordination mission running within the SeeTrack C2 software dryside. This adaptive reassignment of objectives managed by Neptune allows the squad to facilitate shared ATR responsibilities/capabilities, where each vehicle is equipped with sensors and processing units capable of performing target detection and identification. Therefore, the loss of a single unit does not cripple the mission, as the collective capabilities of the fleet ensure the continuity and completion of objectives.

Distributed Sensing and Enhanced Coverage

The deployment of multiple vehicles, each with multi-modal ATR capabilities, also enhances the spatial coverage and sensing diversity of the mission. Different vehicles can be equipped with varied sensor modalities, such as sonar, optical imaging, and chemical sensors, enabling a comprehensive assessment of the underwater environment from multiple perspectives. The multi-modal approach to ATR not only improves the accuracy of target identification but also reduces the likelihood of false positives, as objects of interest can be verified through cross-modal validation.

Collective Data Fusion and Decision-Making

A joint ATR multi-vehicle solution benefits from the collective processing power and data fusion capabilities of the fleet. By aggregating and analyzing data from multiple sources in real-time, the vehicles can make more informed decisions about navigation, object identification, and mission strategy. The collaborative decision-making process is particularly valuable in complex or

https://doi.org/10.52202/075106-0006

rapidly changing environments, where the synthesis of diverse data streams can reveal insights that would not be apparent from a single source.

Scalability and Flexibility

A multi-vehicle approach with shared ATR capabilities offers remarkable scalability and flexibility in mission planning and execution. Depending on the objectives and available resources, the number of vehicles and their sensor configurations can be adjusted to optimize performance. The scalability ensures that missions can be tailored to specific requirements, whether that involves covering a large area in a search-and-identify operation or focusing multiple vehicles on a single target for detailed analysis.

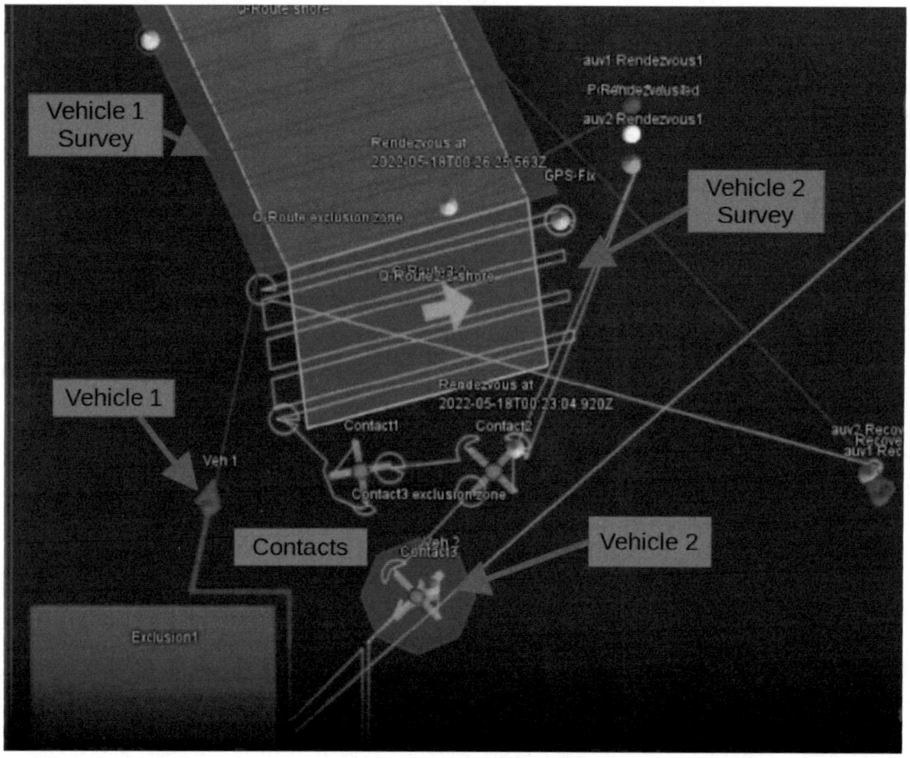

Figure 6. SeeTrack C2 Software with Multi-Vehicle Coordination during a Survey Mission of Potential MLO Threats

CONCLUSION

The implementation of multiple UUVs for a single mission in limited acoustic communication environment can be improved by synergizing the different multi-model with Neptune multi-vehicle autonomy solution. The demonstration of using multi-modal SeeByte ATR in a single use-

https://doi.org/10.52202/075106-0006

case on an ROV with a FLS showed very fruitful potential for vehicle reliance, and reliability without operator-in-the-loop. The benefits are scaled exponentially when deployed on multi-vehicles via Neptune multi-vehicle autonomy solution. The SeeByte ATR has shown great potential on various vehicle hardware configurations in automating MCM and empower third-party hardware vendors and provide great value to the operator with minimal to non-existent interfacing. When deploying SeeByte ATR at scale for multi-vehicle deployment via Neptune's autonomy engine it strengthens the capabilities through hardware redundancy and mission continuity, distributed sensing and enhanced coverage, collective data fusion and decision-making, and mission scalability and flexibility. Future synergy of multi-vehicle autonomy with multi-model ATR will include deploying the Neptune and SeeByte ATR software on various types of autonomous multi-vehicles (including holonomic vehicles (i.e. ROVs), amphibious vehicles, unmanned surface vehicles [USVs]). The Neptune autonomy solution is not constrained by vehicle type and the SeeByte multi-modal ATR CNN model is flexible enough to be used on various types of sonars, cameras, magnetometer, Echoscope (Reference 4), depth-sensing cameras, and Li-DAR hardware solutions.

ACKNOWLEDGMENTS

Some images (i.e. Figure 3 and Figure 5) were provided by Greensea IQ. Figure 3 is a cropped screenshot of C2 front-seat software EOD Workspace.

APPENDIX:

1. Neptune Multi-Vehicle Implementation (Reference 2)

- Neptune constructs a list of objectives in from the mission file. The list is ordered the same as objectives in the Neptune mission file.
- Neptune removes from the list, objectives that are not available. An objective might be not available for the following reasons:
    - It has already been completed, by any vehicle in the squad.
    - It is current being performed by another vehicle.
    - It was previously attempted by this vehicle but failed. It will not be attempted again by this vehicle but may be attempted by another vehicle.
    - This vehicle does not have the capability to complete the objective. For example, reacquire objectives cannot be attempted by a vehicle that does not have a reacquire capable behavior.
    - It may be de-conflicted as described in section Implementation.
- Once these objectives have been removed from the list, objectives that remain in the list are available for the vehicle to perform.
- Neptune will tell the vehicle to attempt the first available objective in the list.

REFERENCES:

1 LURTON, XAVIER. Introduction to Underwater Acoustics: Principles and Applications. SPRINGER-VERLAG BERLIN AN, 2016.

2 https://www.SeeByte.com/products/neptune/

3 https://www.SeeByte.com/products/atr-system/

4 https://www.codaoctopus.com/products/3d/echoscope-sonar-gallery

https://doi.org/10.52202/075106-0006

SIMPLIFIED VEHICLE OPERATIONS: PATHWAY TO AUTONOMY

Francis X Govers III

Simplified Vehicle Operations (SVO) is the concept of increasing safety and utility of aircraft or helicopters by eliminating the physical skill requirements of aircraft operations. Loss of Control of aircraft contributes to 70% of GA accidents. Can we reduce this number by making aircraft dramatically easier to fly? What sorts of things could be automated, and how would this affect aviation as we know it?

INTRODUCTION

There are a number of "revolutions" taking place almost simultaneously across the aviation marketplace. First of all is a revolution in small, high reliability, high performance computers. Second is a revolution in sensors and sensor processing, with both getting dramatically smaller and easier to use. Third is the interest in eVTOL, UAV, and other applications of multirotor and tiltrotor aircraft that do not use traditional control and surfaces to maneuver the aircraft, and instead rely on advanced FCC (flight control computers) to manage multiple control loops to engineer stability on an unstable system. Manual control (individual throttle control) of such an aircraft would be very difficult to impossible, making computer assistance a requirement. Finally, we have a revolution in artificial intelligence, allowing new ways of creating interfaces to vehicles and information

WHY SVO?

According to the FAA, 70% of General Aviation accidents (across aircraft, rotorcraft, experimental and ultralights) involve loss of control. The US Helicopter Safety Team (USHST) reports at 41% of helicopter accidents involve loss of control. Loss of control is a situation where the pilot of the aircraft is no longer controlling the flight path of the vehicle. This includes stall/spin accidents in fixed-wing aircraft, overspeed or overstress resulting in structural failure, loss of directional control on the runway or during taxi, or minimum controllable airspeed in a twin engine plane. In helicopters, which are far more difficult physically to control, the pilot may lose control in a hover, have a "settling with power" or vortex ring state uncontrollable descent, or face an LTE (Loss of Tail Rotor effectiveness) accident. Even a simple hard pull on the collective in a turbine helicopter may over torque the rotor system and do a million dollars of damage to a gearbox.

The premise put before you is to question if the addition of automation and/or autonomy in the aircraft could reduce the amount of risk in aircraft by adding an additional safety barrier to loss of control accidents by reducing the amount of physical skill involved in piloting, and creating an aircraft that "refuses to crash".

IMAGINE A WORLD – SVO EXAMPLE

Let's image one version of SVO. The objective in this thought experiment is to imagine if flying was as simple as driving from a physical standpoint. To start an airplane, we get in the plane and turn the key. The engine starts. No mixture, no prop, no priming. All the electrical systems come

https://doi.org/10.52202/075106-0007

on and all the avionics start up. The radio tunes itself to the ground control frequency, and the ADIS information is displayed on the display in front of you. Once you get clearance, you drive the plane to the runway, using the yoke to point the plane down the taxiway and a brake pedal to adjust speed. During the taxi, the plane performs the runup by itself. By the time you arrive at the departure end, the plane is ready to go. The radio changes to tower frequency when the ground controller says "contact tower on 122.3". You have already downloaded your flight plan from your tablet to the aircraft's nav system. Then you get clearance and take the runway. You push the throttle to the stop and pull full back on the control yoke. The aircraft flies off the runway. You hold this configuration until reaching cruise altitude, then release the yoke and the aircraft levels off and turns on course for your destination. For landing, you put the throttle in "descend" detent, and point the aircraft at the runway with the yoke. The aircraft flies at the correct speed and you steer to the runway, pointing the nose of the aircraft at the numbers. The aircraft flares by itself, and touches down. You drive off the runway and to parking. A no point did you worry about speeds, angle of bank, coordination, or keeping a heading.

This would be a simplified control based paradigm for SVI. It still uses the controls but in a simplified manner. The next paradigm is very different.

BUTTON CONTROL – SVO PARADIGM

Another type of control technique for SVO we might call the "button" paradigm. For this mode, the pilot never touches the control yoke or stick, and instead uses a touch screen interface with buttons and sliding indicators. The startup and preflight are as above. For taxi, the operator selects the runway (and taxiway path) and hits "taxi" button. The aircraft moves by itself to the proper runway, avoiding other aircraft and other obstacles as necessary. The aircraft performs the runup by itself, and announced "ready for takeoff". The pilot pushes a button for "take runway for takeoff" and once the aircraft is in position, and clearance received, pushes a "takeoff" button. The aircraft rolls and takes off, assuming a best rate of climb. Now the flight plan is engaged. At any point, the pilot can use "knobs" to modify the flight, turning the nose right or left, or climbing or descending by adjusting the altitude setpoint. The pilot selects the proper runway for landing, and the aircraft lines itself up for landing. Let's say another aircraft cuts in front of you on downwind. You push a "circle" button to do a quick 360 to give spacing.

The entire flight is made without touching any controls. For helicopters, SVO reduces workload further by providing position hold hovering, vector controlled hover taxi, auto climbout, level flight, descent and hover at the other end. All coordination is controlled by the computer, which manages yaw and balances collective, cyclic, and anti-torque inputs.

WHAT ABOUT FAILURES

SVO does not mean that the aircraft reverts to manual control after a systems failure. Indeed, for eVTOL multicopter and other complex aircraft that may not be possible. The control system has to be designed to "fail safe" and to drop to a safe degraded mode. SVO systems may have a "pilot-on-the-loop" mode where the pilot can put in "knobs" commands – airspeed, altitude, heading, rate of climb/descent, airspeed, and have control of the flight path while still having the safety of the control system.

Another mode may include a joystick to "point" the aircraft – climb up, descend down, turn right, left.

SVO design must provide for fallback modes that still control the aircraft and provide safe operations.

WORRY FREE CONTROL

https://doi.org/10.52202/075106-0007

A side benefit of SVO can be the concept of "worry free control", which can also be called "Care-free maneuvering". In this mode, the full operational envelope of the aircraft can be exploited, but with limitation to prevent loss of control. This may include g-limits, angle-of-attack limiters, bank angle limits. Other monitors may provide envelopes around power settings, torque limits, or propellor rpm. The concept is to allow the pilot to move the controls at any rate, to the stops, without exceeding the performance envelope of the aircraft. This may require auto leveling, coordination, or changing trim levels and force feedback detents.

NEXT STEPS

The industry has been working to define the rules around SVO. GAMA, the General Aviation Manufacturer's Association, has a SVO committee as does the NBAA. They are working with the FAA to determine the limitations around SVO, and to define if a new rating for SVO aircraft will be necessary.

Regardless, the safety benefits of SVO to reduce accidents must be investigated and developed.

REFERENCES:

1. Lombaerts, Thomas, John Kaneshige, and Michael Feary. "*Control Concepts for Simplified Vehicle Operations of a Quadrotor eVTOL Vehicle*." AIAA Aviation Forum, 15-19 June 2020, virtual conference. NASA Technical Reports Server, https://ntrs.nasa.gov/citations/20205002256.

2. GAMA EPIC SVO Subcommittee. "A Rational Construct for Simplified Vehicle Operations" 20 May 2019.

3. Allison Duquette. "FAA Fact Sheet – General Aviation Safety" June 19, 2012. Federal Aviation Administration.

TOP DRONE APP RISKS

Godfrey Nolan, Rudy Mendoza, David Kovar, Bobby Sakaki, Benjamin Sugg

This paper seeks to establish a foundational security risks checklist, capturing the breadth of vulnerabilities from ground control to cloud storage. We have laid out a structured pathway for identifying these risks, understanding their potential impact, and implementing strategies to mitigate them. The checklist encompasses various aspects of drone security, ranging from communication link protection to defense against OSINT exploitation. Our intention is to spark a conversation among experts, enthusiasts, and professionals in the drone community, soliciting contributions that will enrich the content and utility of this document. We encourage anyone with expertise, feedback, or novel ideas to engage with us in refining this living document. Upon integrating the community's insights, we will proceed to formalize a first draft. This comprehensive version, bolstered by collaborative wisdom, is set to be presented to OWASP, marking hopefully a significant step towards fortifying the integrity of drone operations against cyber threats.

INTRODUCTION

Drones, with their increasing integration into commercial, public safety, and recreational sectors, present a complex matrix of security challenges that must be rigorously managed to safeguard sensitive data and ensure operational integrity. As these unmanned aerial systems (UAS) become more sophisticated, so too do the potential cybersecurity threats they face. From the interception of transmitted data over insecure communication links to the exploitation of publicly available information to breach development systems, the risks are diverse and multifaceted. Ensuring the security of these systems involves addressing vulnerabilities in software and firmware, securing data both in transit and at rest, and protecting against physical tampering. Moreover, the rise of cloud computing and extensive use of open-source intelligence (OSINT) by malicious actors add layers of complexity to securing drone operations. This expanded threat landscape requires a comprehensive and proactive approach to drone security, incorporating robust technological safeguards, strict operational protocols, and continuous vigilance against emerging threats.

This document represents an initial effort to outline the fundamental security risks associated with drone operations. It serves as the foundation for creating an OWASP-style (Open Web Application Security Project) checklist, intended to evolve into a living document on OWASP.org that will be updated and refined over time. Additionally, this initiative seeks to engage the broader drone community, inviting feedback and contributions to the first draft of the checklist. By fostering collaborative input, we aim to ensure that the checklist remains relevant, comprehensive, and effective in addressing the dynamic security challenges faced by drone technology in various sectors.

https://doi.org/10.52202/075106-0008

CONNECTED COMPONENTS

Figure 1 illustrates the interconnected components of a modern drone that collaborate to gather and communicate information, highlighting their potential vulnerabilities to exploitation.

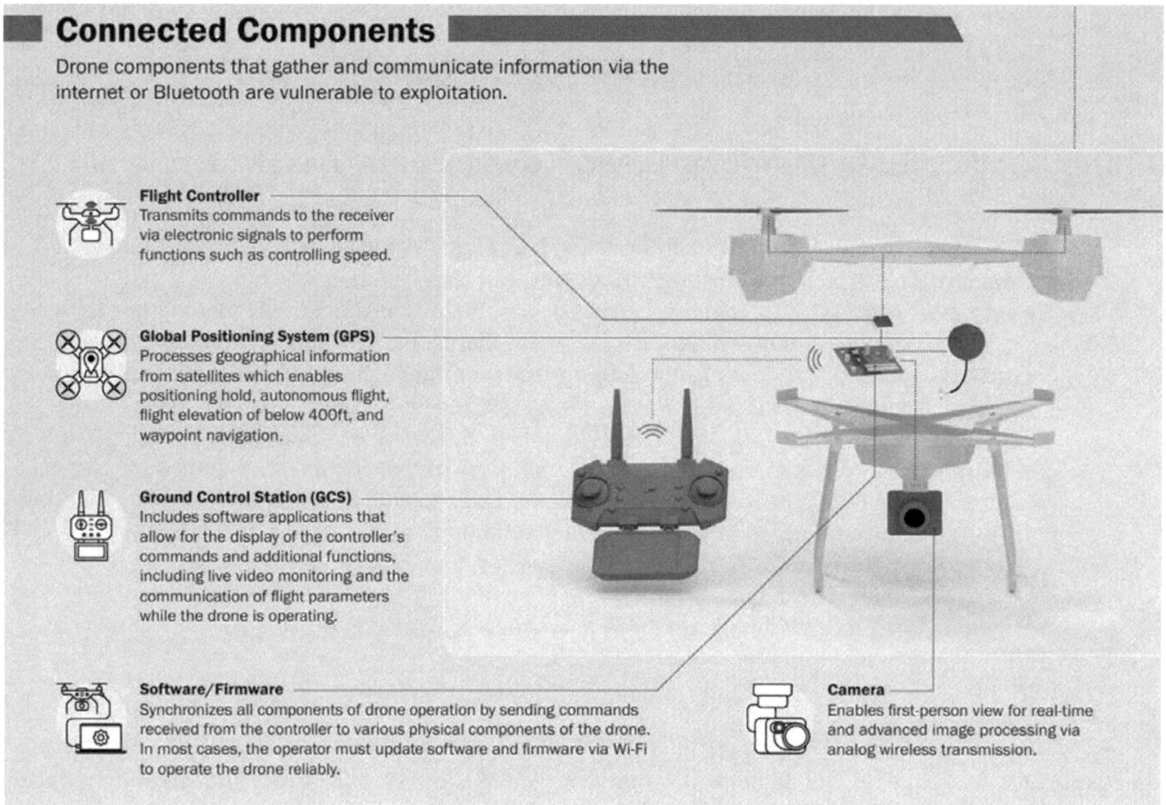

Figure 1. Connected components on a drone from CISA [1].

The flight controller is the brain of the drone, transmitting commands to various parts to perform critical functions like controlling speed. The Global Positioning System (GPS) processes geographical data from satellites, facilitating capabilities such as positioning hold, autonomous flight, and precise waypoint navigation. The Ground Control Station (GCS) provides the user interface for the pilot, displaying the controller's commands and facilitating additional functionalities such as live video monitoring and in-flight parameter adjustments. The software and firmware component acts as the orchestrator, synchronizing the operation of the drone by relaying commands from the controller to the physical components and managing essential updates via Wi-Fi connections. Lastly, the camera serves as the drone's eye, providing first-person views for real-time monitoring and enabling advanced image processing through analog wireless transmission. Each of these connected components, while critical for the drone's operation, also presents a vector for potential security risks if not adequately protected.

OWASP DRONE SECURITY CHECKLIST

This document aims to catalog the potential points of attack on a drone, as depicted in Figure 1. Our endeavor is to document each area of vulnerability, articulate the associated risks, and delineate the corresponding mitigation strategies. By providing this overview, we seek to equip stakeholders with the knowledge to reinforce the defenses of their unmanned systems against a spectrum of

https://doi.org/10.52202/075106-0008

cyber threats. We look forward to any input from the community on any changes, updates or omissions in this list.

1. Insecure Communication Links

Risk: Data transmitted over the drone's communication links (RF, Wi-Fi) can be intercepted or tampered with due to weak encryption or lack of secure protocols.

Mitigation: Use strong encryption standards and secure communication protocols. Validate the security of RF links and implement measures against eavesdropping and data manipulation.

2. Weak Authentication Mechanisms

Risk: Default or weak credentials in drone systems, including UAVs and GCS, can allow unauthorized access.

Mitigation: Implement robust authentication mechanisms, enforce strong password policies, and use multi-factor authentication where possible.

3. Firmware and Software Vulnerabilities

Risk: Unencrypted firmware and vulnerable bootloaders can lead to unauthorized modifications and exploitation.

Mitigation: Secure the firmware update process, use encryption, and perform regular security audits of the software and firmware.

4. Insufficient Physical Security

Risk: Physical access to drones or their components can lead to unauthorized use, data theft, or hardware tampering.

Mitigation: Implement tamper detection mechanisms, secure physical interfaces like USB ports, and restrict physical access to critical components.

5. Insecure Supply Chain

Risk: Compromised components from suppliers can introduce hidden vulnerabilities.

Mitigation: Perform thorough security vetting of suppliers, use trusted sources for hardware and software components, and monitor the integrity of the supply chain.

6. Inadequate Privacy Protection

Risk: Drones collecting sensitive data without proper safeguards can lead to privacy breaches.

Mitigation: Ensure compliance with privacy laws, implement data minimization practices, and secure any collected data both in transit and at rest.

7. Network Service Vulnerabilities

Risk: Network services on drones and ground control systems can have vulnerabilities that allow unauthorized access or denial of service attacks.

Mitigation: Harden network services, perform regular vulnerability scanning, and implement network segmentation and firewalls.

8. Unsecured Third-party Components

Risk: Using vulnerable third-party software libraries and components can compromise drone security.

https://doi.org/10.52202/075106-0008

Mitigation: Keep third-party components up-to-date, and conduct security reviews of all third-party code used in drone operations.

9. Lack of Secure Update Mechanism

Risk: The absence of a secure mechanism for updating drone software and firmware can allow attackers to install malicious updates.

Mitigation: Use cryptographic signatures for updates, ensure secure delivery of updates, and verify integrity before installation.

10. Inadequate Logging and Monitoring

Risk: Insufficient monitoring of drone operations can delay the detection of security breaches or operational anomalies.

Mitigation: Implement comprehensive logging and monitoring systems, and use real-time alerting for suspicious activities.

11. Insecure Data Storage

Risk: Sensitive data stored on drones or in GCS can be accessed if data storage is not secure.

Mitigation: Encrypt stored data and implement strong access controls.

12. Spoofing and Replay Attacks

Risk: Risks of GPS spoofing, ADS-B data spoofing, or command replay attacks that could mislead or take control of the drone.

Mitigation: Implement anomaly detection, secure GPS protocols, and message authentication.

13. Radio Frequency (RF) Interference and Jamming

Risk: Drones can be disrupted or controlled through intentional RF interference or jamming.

Mitigation: Use frequency hopping, spread spectrum technologies, and jamming detection systems.

14. Sensor Vulnerabilities

Risk: Sensors like GPS, ADS-B, and others can be exploited to feed incorrect data to the drone systems.

Mitigation: Validate sensor inputs, use redundant sensor systems to verify data accuracy.

15. Inadequate Incident Response

Risk: Lack of preparedness for handling security incidents affecting drones.

Mitigation: Develop and regularly update an incident response plan specific to drone operations.

16. Cloud Storage and Data Management Vulnerabilities

Risk: Inadequate security controls for drone data stored in the cloud (e.g., videos, logs, images) can lead to unauthorized access and data breaches.

Mitigation: Use encrypted storage solutions, enforce strong access controls, and apply best practices for cloud security. Regularly audit cloud storage and data management practices.

17. Exploitation of Publicly Available Information (OSINT)

Risk: Hackers using publicly available data or OSINT (open source intelligence data) from platforms like LinkedIn, Twitter, and GitHub to gather information on drone development teams, access codes, or deployment practices.

Mitigation: Implement strict policies on what can be shared publicly by development teams, monitor public data for potential leaks, and educate teams on the risks of sharing sensitive information.

18. End-of-Life and Decommissioning Risks

Risk: Inadequately secured decommissioning processes can leave residual data accessible, or hardware could be reused maliciously.

Mitigation: Define and enforce clear procedures for data sanitization and hardware disposal. Ensure all sensitive data is irrevocably destroyed or securely archived.

19. Interoperability and Integration Issues

Risk: Integrating various systems and technologies without a cohesive security strategy can introduce vulnerabilities.

Mitigation: Design and implement a secure integration framework; conduct security assessments of all integrated systems.

20. Third-party Services and API Security

Risk: External APIs used by drones or GCS might be insecure, providing a pathway for attacks.

Mitigation: Securely configure and regularly monitor all external APIs. Ensure API security practices are in place, including authentication, throttling, and encryption.

21. User Error and Misconfiguration

Risk: Incorrect configuration of drone systems by users can expose them to risks of unauthorized access or malfunction.

Mitigation: Provide comprehensive training for users, implement configuration management tools, and conduct regular audits of system configurations.

22. Environmental and Operational Hazards

Risk: Drones operating in complex environments may encounter situations that can be exploited to cause physical or functional failures.

Mitigation: Use environmental sensing technologies to detect and respond to hazards. Implement robust fail-safe and recovery procedures.

23. AI and Autonomous Decision-Making Vulnerabilities

Risk: AI-driven components of drones can be susceptible to attacks such as data poisoning or adversarial attacks, leading to flawed autonomous decisions.

Mitigation: Validate and secure AI training data, implement layers of decision validation before autonomous actions are taken, and monitor AI systems for signs of manipulation.

CONCLUSION

In conclusion, this paper has presented an initial checklist detailing the multifaceted security risks associated with drone operations. Our goal is to foster an ongoing dialogue on these issues, and we warmly invite input from the community. If you have insights or additional considerations,

https://doi.org/10.52202/075106-0008

please do not hesitate to reach out to the authors. Following a thorough review of the feedback and any subsequent updates, our next step will be to assemble the first draft of the checklist. This draft will then be submitted to OWASP for consideration and inclusion, contributing to the collective effort of safeguarding the increasingly pivotal realm of drone technology.

[1] Secure your Drone – Privacy and Data Protection Guide

https://www.cisa.gov/resources-tools/resources/secure-your-drone-privacy-and-data-protection-guidance

CHALLENGES AND PROGRESS IN PREDICTIVE MAINTENANCE OF LONG-ENDURANCE & LONG-RANGE UNCREWED PLATFORMS

Kenneth Lu[*], Sanja Cvijic[†], Arjuna Balasuriya[‡]

Today's uncrewed platforms are typically operated by humans using remote control to guide every detailed aspect of a mission. However, as missions become more complex, there are many scenarios (particularly in the marine and ground domains) in which operators are unable to communicate with these uncrewed platforms in real time (due to adverse environmental conditions, regulatory restrictions on communications in ecologically sensitive areas, active interference by adversaries, or the desire to remain covert), making it challenging to know the health and status of the platform, and to recalibrate and update mission and control parameters on the fly. This becomes challenging especially when the uncrewed platforms are deployed for long-endurance and long-range missions. Fortunately, significant technical advances in onboard computing power and enhanced sensors offer a pathway to a level of autonomy that can overcome such communications limitations. Predictive maintenance algorithms and digital twins of health and status are becoming essential in preventing unexpected failure and extending the lifespan of uncrewed platforms.

In this paper, we present insights into the importance of predictive maintenance and provide examples of implementation of predictive maintenance in uncrewed platforms such as robotic combat vehicles (RCVs). Moreover, we will discuss how hybrid artificial intelligence (Hybrid-AI) techniques rooted in probabilistic models serve as a foundation for predicting health and status of uncrewed platforms.

INTRODUCTION

Uncrewed platforms provide a bevy of advantages for operators, Warfighters, and decision-makers across the Department of Defense (DoD) and industry abroad. For one, their ability to engage adversaries and navigate hazardous terrain (both underseas and on the ground) far beyond the forward line of troops offers protection to Warfighters operating remotely. Second, their payloads support the deployment and usage of a wide range of command, control, communications, computers, intelligence, surveillance, and reconnaissance (C4ISR) capabilities, providing intel on environments, adversaries, and other information far beyond the natural reach of humans. Finally, combining their combat and reconnaissance capabilities, as well as their overall mobility in rugged terrain, these uncrewed platforms can dramatically reduce the overall labor requirements on humans. However, as uncrewed platforms are increasingly used in hostile, complex, and dynamic environments, maintaining these vehicles to ensure that they are mission capable becomes increasingly difficult. Due to long-endurance and long-range missions in harsh environments, hardware system (sensors and actuators) performance of these uncrewed platforms can deteriorate and due to unreliable communication, the operators will not be aware of these performance changes in the platform affecting the mission objectives. Therefore, it is essential to have health monitoring/predictive maintenance

---

[*] Scientist, Charles River Analytics, 625 Mount Auburn St. Cambridge, MA 02138.
[†] Senior Scientist, Charles River Analytics, 625 Mount Auburn St. Cambridge, MA 02138.
[‡] Senior Scientist, Charles River Analytics, 625 Mount Auburn St. Cambridge, MA 02138.

https://doi.org/10.52202/075106-0009

and adaptive behavior capabilities on board uncrewed platforms to successfully complete missions in such environments.

Fortunately, thanks to advances in the modern sensing and computing capabilities on board these uncrewed platforms, these predictive analytics are attainable. However, there are several key technical challenges to address for such analytics to be feasible and useful for the end user. First, modern uncrewed systems like robotic combat vehicles (RCVs) will have modular mission payloads (MMPs) in which various capabilities can be swapped in and out. Furthermore, these systems will have other lowest replaceable units (LRUs) that can be changed based on the upgrade cycle/schedule of the system. As a result, an analytic designed for one set of MMPs and LRUs must be *adaptable* to various combinations of capabilities and components. Furthermore, the analytic must *learn* and *evolve* through use, meaning that it must incorporate any new information and data about the system without having to rely on a software engineer completely rewriting the analytic. Second, data sources on uncrewed systems will often be sparse and limited, meaning that the solution must be robust to partial or infrequent information. This requires that the solution be *robust* enough to reason well under uncertainty and avoid overfitting to spurious trends or red-herring correlations. Third, the solution must be *computationally compact* as edge-computing environments on uncrewed systems are limited in hardware and communication bandwidth. For instance, and RCV computing platform may have as little as 20 kilobits (20kb) of wireless bandwidth depending on signal strength, and much of that bandwidth is designated for remote control of the vehicle. Finally, the solution must build trust with the end user. This requires that the solution can *explain* its outputs to the end user in an intuitive way and is particularly important because an end user who does not understand or believe the output of an analytic will not use it, regardless of how accurate or performant its underlying computation.

CURRENT APPROACHES

In reliability engineering and condition-based maintenance (CBM+), expert models are useful for building digital twins that can provide highly accurate and faithful predictions of real-world systems and are useful for systems with known dynamics. We have seen examples of this when building physics-based models for assessing structural damages of generic structures to predicting the remaining life of lithium-ion batteries.[1,2] However, these approaches are often brittle and difficult to maintain as operating conditions and environmental dynamics keep changing. For instance, a physics model of an RCV with a combination of lidar and radar along with an aerial drone platform will differ wildly from a physics model of an RCV that has an almost identical suite of components but swapping out the drone platform for a 9 mm turret.

Data-driven algorithms based on techniques in artificial intelligence and machine learning (AI/ML), specifically in deep learning and reinforcement learning, are used for predictive maintenance. However, these methods require huge amounts of training data to learn the performance models of various components on board uncrewed platforms. Also, these data-driven models are often seen as "black boxes" and lack explainability, making it difficult for the end user to build trust or explain the outcome of these AI/ML models. A 2021 survey by Theissler et al. showed that most papers published in areas of predictive maintenance rely on deep learning approaches centered around a labeled dataset.[3] We have seen work from Li et al. that shows promise in detecting the remaining useful life for vehicle's power system and work from Manoharan et al. that yields similar results.[4,5] However, these methods do not work in the context of assessing vehicle control stability, as neural network models, while highly optimized, are difficult to generalize across domains. Ragone et al. use a physical model that simulates a vehicle powertrain to predict battery state of charge based on an electrochemical–thermal model of lithium-ion batteries.[6] While this method works well for a well-scoped component like the battery, this approach would become enormously difficult to sustain and generalize across multiple components for a vehicle.

Furthermore, we have seen in the past few years the emergence of powerful large language models (LLMs) that have had success in generative AI.[7] However, these models are enormous in size, requiring 500 pflops/s of computation to train and involving thousands of GPUs. Even after training, these models frequently take up to 200 GB of memory to load, which far exceeds the computational capacity of any RCV platform.

Work that combines expertise and data-driven approaches has attempted to extend physics-based models to inform deep-learning and has seen success in determining the natural degradation of machine components in a vacuum.[8] However, this approach still has limitations. For one, failure modes that do not follow the natural degradation (i.e., sudden perturbations like environmental impact or battle damage) will not be captured. This is because physics-of-failure models cannot capture these sudden dynamics easily, and training data for deep-learning models will usually lack these scenario-specific datapoints. Furthermore, this approach is largely shoehorning a physics approach into a data-driven approach and lacks the compactness, adaptability, and ease of implementation of alternative Hybrid-AI approaches.

HYBRID-AI APPROACH USING PROBABILISTIC PROGRAMMING

In this paper, we describe a more holistic and natural approach to Hybrid-AI for predictive maintenance rooted in probabilistic programming. Probabilistic programming is a computer programming paradigm that raises uncertainty to a first-class programming language construct and automates the processes of parameter learning and statistical inference, including evaluation of complex queries against the model. This enables the modeler to combine domain expertise, in the form of hard-coded prior probability distributions, with data, in which a model can learn summary statistics and parameters directly from a dataset. PP builds on probabilistic graphical models (PGMs), greatly generalizing their capabilities in three ways: (1) Instead of specifying model structure in a domain-specific language, PP models are specified in an ordinary programming language that is augmented with two simple, intuitive additional *sample* and *observe* statements. Writing x = sample(a, p) is equivalent to sampling $x \sim p(x| \ldots)$ and recording the sampled value, along with its probability, in a maplike data structure at the address a, while writing observe(a, p, v) is equivalent to evaluating $p(x = v| \ldots)$ and similarly recording the observed value and its probability at a. (2) Instead of enforcing a static model structure, probabilistic programs can encode arbitrarily complex model structures, including those with stochastic choice (in which future model state depends on the value of random choices) and open-universe structure (in which the model may include arbitrarily many random variables).

This interplay between incorporating domain knowledge with data is crucial because it provides probabilistic programs with (1) *adaptability*, as probabilistic programs are naturally modular and composable, meaning that with minimal sets of changes, they can be repurposed to fit new operating contexts of an uncrewed platform; (2) *robustness*, as probabilistic programs can rely on prior domain knowledge in the absence of data and also encode that knowledge as probability distributions with confidence bounds, enabling reasoning under uncertainty; (3) *compactness*, as these models do not require detailed understanding like a rich physics simulator, nor terabytes of data to train; and finally (4) *explainability*, in which probabilistic programs are easily traceable, and an end user can gain an understanding into the logical underpinnings of a probabilistic model.

https://doi.org/10.52202/075106-0009

CASE STUDY EXPERIMENTS AND RESULTS

Robotic Combat Vehicle

As part of our latest work with the Army Applications Lab, we conducted a series of experiments on a simulated dataset related to an RCV. This builds upon our previous work on this program, where we demonstrated diagnostics and health monitoring on a surrogate vehicle using probabilistic programming.[9] In this case study, we demonstrated that we could predict future health and status of a vehicle at least 100 minutes into the future with error bounds that are reasonably accurate so that the predictions are useful for the vehicle operator and decision-makers higher up the chain of command. In this experiment, we worked with the Army as well as domain experts on the vehicle to create a dataset representing the gradual, and then sudden, breakage, of a bogey-beam that supports the track and suspension system of an RCV, and we observe this gradual decay in the form of a changing angle of the bogey-beam relative to the ground.

For this experiment, we were able to quickly design and implement a probabilistic program representing a dynamic Bayesian network and we used a sequential Monte Carlo (SMC) inference algorithm to predicting with this model. In this model we encoded domain expertise representing the relationships between the observe bogey-beam angle and the health and status of the bogey-beam, along with a historical dataset of angles representing nominal operating conditions of the bogey-beam.

Our results showed that we were able to predict 100 minutes into the future on a non-noisy dataset with an average error of 6.1 minutes, and when predicting up to 30 minutes in advance, we had an average error of 2.7 minutes. Additionally, when we added noise to the data, equivalent to the hard-coded rate of change of the bogey-beam's angle, our error rate went up to 13.1 minutes when predicting up to 100 minutes, and 6.0 minutes when predicting up to 30 minutes. We visualize our prediction results in Figures 1 and 2.

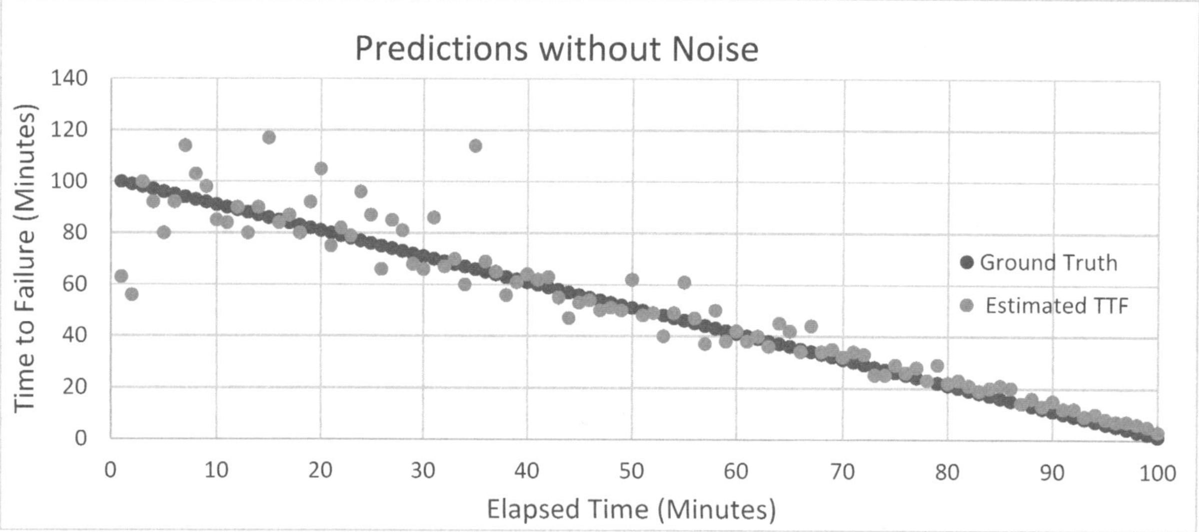

Figure 1. Results of Predictions without Noise. The blue line indicates the ground-truth remaining life of the bogey-beam, which degrades in perfect linear fashion as time elapses. Orange dots indicate the predicted remaining useful life of the bogey-beam at each time step.

https://doi.org/10.52202/075106-0009

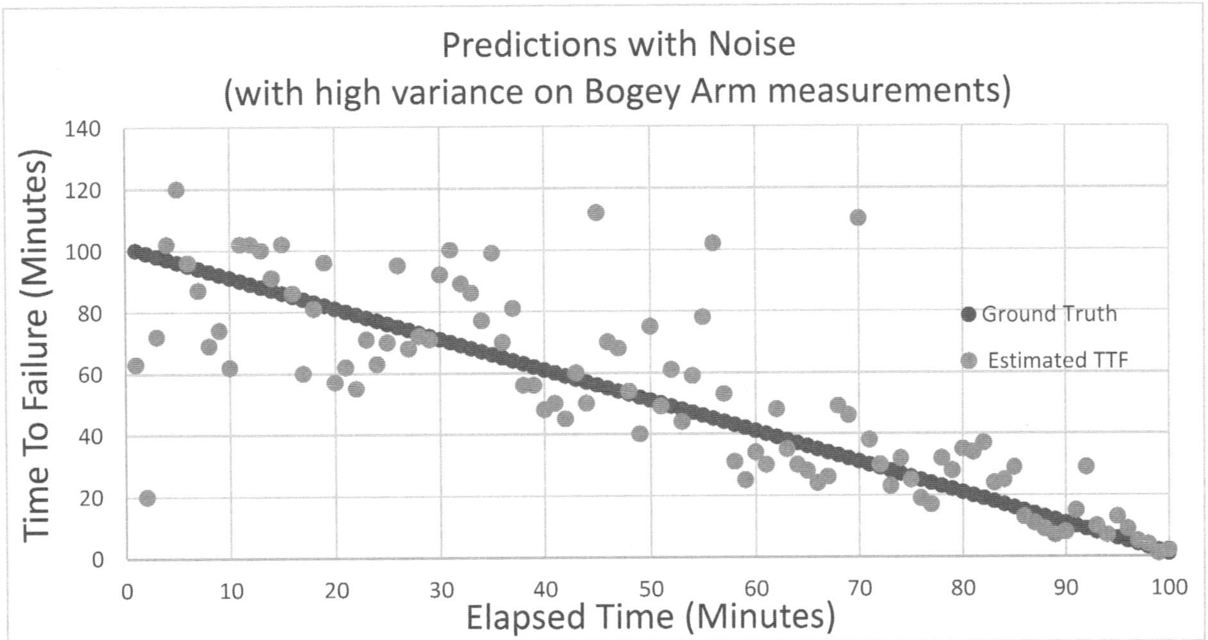

Figure 2. Results of Predictions with Noise. Compared with Figure 1, the results are more erratic initially, but within a 30-minute prediction window, we see that the error rate starts converging to be similar to that of the noiseless scenario. Also, the noise induced in this scenario was deemed to be much higher than that of a real-world sensor.

In terms of compactness and performance, this prognostics model takes up less than 1 MB of memory. Moreover, we have demonstrated previously that over 100 probabilistic models can run on less than 2 GB of memory on a Jetson Nano hardware system. Furthermore, this approach shows how a probabilistic model can compactify streams of sensor data into a single prediction of health and status, making its output (one number) suitable for transmission in a low-bandwidth signal environment.

CONCLUSION

In this work, we demonstrated the power of PP-based Hybrid-AI as a tool for predictive maintenance at the edge for uncrewed systems. We extended domain expertise to relate the observed angle of a bogey-beam for a track and suspension system to the likely health and status state of the vehicle. We also leveraged historical data to learn summary statistics representing the nominal operating conditions of the bogey-beam. This approach has been demonstrated on both a simulated dataset as well as real-world data, to show both health monitoring and predictive analytics, and can be easily extended to an RCV thereby providing Army Warfighters with a diagnostics tool that enables operational decision-making. Specifically, by notifying the RCV operator of a future failure in the bogey-beam, the operator can adjust their driving to compensate, by either reducing their speed in rough terrain, or by changing their steering to compensate for a steering angle bias. They can also inform a higher-level Commander on whether to bring the RCV back to the motor pool for repair. This in turn not only reduces the downtime for RCV repair, but also improves the ability to support longer range missions, as operators can drive RCV further beyond the forward line of troops, knowing the Hybrid-AI prediction capabilities can accurately gauge vehicle health and status. At the same time, commanders at higher level echelons can plan out tactical engagements and missions where RCVs are deployed across a wide geographical area, knowing that they will have awareness and insights into future fleet level availability of capabilities

https://doi.org/10.52202/075106-0009

REFERENCES

[1] M. Montaru, S. Fiette, J.L. Koné, and Y. Bultel, "Calendar ageing model of Li-ion battery combining physics-based and empirical approaches." *Journal of Energy Storage*. Vol. 51, 2022, 104544. https://doi.org/10.1016/j.est.2022.104544

[2] T.G. Ritto, and F.A. Rochinha, "Digital twin, physics-based model, and machine learning applied to damage detection in structures." *Mechanical Systems and Signal Processing*. Vol. 155, 2021, 107614. https://doi.org/10.1016/j.ymssp.2021.107614

[3] A. Theissler, J. Pérez-Velázquez, M. Kettelgerdes, and G. Elger, "Predictive maintenance enabled by machine learning: Use cases and challenges in the automotive industry." *Reliability Engineering & System Safety*. Vol. 215, 2021, 107864. https://doi.org/10.1016/j.ress.2021.107864

[4] P. Li, Z. Zhang, R. Grosu, et al. "An end-to-end neural network framework for state-of-health estimation and remaining useful life prediction of electric vehicle lithium batteries." *Renewable and Sustainable Energy Reviews*. Vol. 156, 2022, 111843. https://doi.org/10.1016/j.rser.2021.111843

[5] A. Manoharan, K.M. Begam, V.R. Aparow, and D. Sooriamoorthy, "Artificial neural networks, gradient boosting and support vector machines for electric vehicle battery state estimation: A review." *Journal of Energy Storage*. 55, 2022, 105384. https://doi.org/10.1016/j.est.2022.105384

[6] M. Ragone, V. Yurkiv, A. Ramasubramanian, B. Kashir, and F. Mashayek, "Data driven estimation of electric vehicle battery state-of-charge informed by automotive simulations and multi-physics modeling." *Journal of Power Sources*. Vol. 483, 2021, 229108. https://doi.org/10.1016/j.jpowsour.2020.229108

[7] D. Narayanan, M. Shoeybi, J. Casper, et al. "Efficient large-scale language model training on GPU clusters using Megatron-LM." In: *SC21: International Conference for High Performance Computing, Networking, Storage and Analysis*. 2021, pp. 1–14. https://ieeexplore.ieee.org/document/9910057

[8] M.A. Chao, C. Kulkarni, K. Goebel, and O. Fink, "Fusing physics-based and deep learning models for prognostics." *Reliability Engineering & System Safety*. Vol. 217, 2022, 107961. https://doi.org/10.1016/j.ress.2021.107961

[9] K. Lu, S. Cvijic, D. Dewhurst, J. Gorman, R. Hyland and J. Templin, "Democratizing AI for condition-based maintenance leveraging probabilistic programming for symbolic reasoning. In: *2023 Annual Reliability and Maintainability Symposium (RAMS)*. 2023, 1–6. dhttps://doi.org/10.1109/RAMS51473.2023.10088272

PAYLOAD DELIVERY TO OFFSHORE WIND TURBINES USING UNMANNED HELICOPTERS

Christoph Eck,* Carlo Zgraggen†, Markus Birrer‡, Anders la Cour-Harbo§

Numerous wind turbines have already been installed offshore for green and sustainable electricity production. Also new wind parks with steadily increasing dimensions are currently installed along many shores in many countries. However, the long-term and smooth operation of each wind turbine - under often harsh and challenging environmental conditions - requires frequent inspection and maintenance work. The technical crew, repair equipment, and spare parts are often provided by vessels traveling for several weeks within the wind parks. Nevertheless, shortages of replacement parts and time-consuming ship cruising and docking can create substantial downtimes. Here, unmanned industrial helicopters with high payload capacity can deliver various repair items to reduce down-times and minimize losses due to repair work. This presentation is based on the ADD2Wind research program where an unmanned, autonomously flying helicopter is being used for payload delivery to the nacelle of a wind turbine. Initial flights have been demonstrated and recorded in June 2023 showing the performance of the aircraft during approach, delivery, and homing flight. Beside video sequences from different perspectives real flight data is being analyzed. To achieve flight permission, the unmanned helicopter is equipped with an emergency flight termination system (FTS). The payload section is used with an electric winch to lower the payload on the nacelle without landing the helicopter on the nacelle.

INTRODUCTION

The maintenance of offshore wind turbines is challenging in multiple ways. Transportation of the repair crew and required spare parts is often time consuming and floating warehouses are both energy- and labor-intensive. Warehouses onshore are much less cost intensive and can be operated 24/7. But the transportation of spare parts to the offshore wind turbines on the other hand is also challenging. Within the research project "Airborne Drone Delivery to Offshore Wind Turbines" ADD2Wind the focus lies on an unmanned, industrial, autonomously flying helicopter with an integrated winch to transport and deliver spare parts to the nacelle of an offshore wind turbine while staying in hover flight during the delivery process and return to the shore afterwards. Initially flights were tested on a simulated platform described in section 2 where various flight data is shown. Various sensors onboard the helicopter have been used to monitor the payload delivery. This

* Prof. Dr. El.-Ing., Lucerne University of Applied Sciences and Arts, IME Competence Center Mechanical Systems, Technikumstrasse 21, 6048 Horw, Switzerland.
† BSc. El.-Ing., Lucerne University of Applied Sciences and Arts, IET Competence Center Autonomous Systems and Robotics, Technikumstrasse 21, 6048 Horw, Switzerland.
‡ BSc., El.-Ing., Aeroscout GmbH, Hengstrain 14, 6280 Hochdorf, Switzerland.
§ MSc. Math, Ph.D, Associate Professor, Aalborg University, Automation and Control at Department of Electronic Systems, Fredrik Bajers Vej 7C (C2-214), 9220 Aalborg, Denmark

https://doi.org/10.52202/075106-0010

includes inertial sensors, forward and downward looking cameras, and laser scanner to receive accurate distance information. These initial flights were followed by BVLOS flights described in section 3 and flights to a real wind turbine with delivery of a payload sample on its nacelle described in section 4. Section 5 summarizes the research program and shortly describes the process to achieve flight permission both in Switzerland (CH) and Denmark (DK).

Figure 1. Offshore wind park in Denmark.

Figure 1 shows part of an offshore wind park at the cost of Denmark. The ADD2Wind research project is based on an industrial, unmanned helicopter equipped with a flight control system (FCS) allowing autonomous lift-off, path following, and landing.

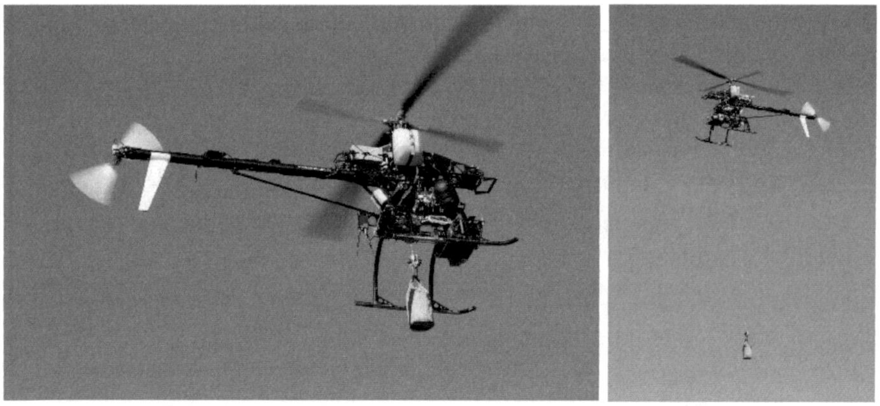

Figure 2. Aeroscout Scout B1-100 UAV helicopter equipped with payload and payload winch.

Data communication is based on a long-range WLAN network during the flight allowing video feedback to the ground control station (GCS) where the operators handle both aircraft and payload. The unmanned helicopter Scout B1-100 UAV produced by Aeroscout GmbH as shown in Figure 2 with its classical main and tail rotor configuration provides multiple benefits in this operation compared to manned helicopters as well as muti-rotor UAVs/drones. Besides the vertical lift-off and landing capability and the benefit to hover above the nacelle during payload delivery the helicopter as a substantial wind gust robustness due to its rapid helicopter swash-plate configuration which decouples in windy conditions the main rotor plane from its body movements, allowing stable and precise hovering. Multi-rotor UAVs have to adjust the complete airframe to stabilize in windy conditions. In addition, during cruise flight, the fuel consumption is considerably reduced compared to multi-rotors UAVs and an optimal traveling speed can be selected. Some of the main specifications for the Scout B1-100 UAV helicopter are given in Table 1. Further flights are planned with the

https://doi.org/10.52202/075106-0010

Scout B-330 UAV which provides a flight endurance up to 3 hours and a payload capacity up to 30kg.

Once the wind park and its individual wind turbines have been mapped in its spatial spreading, repeated flights can be pre-programmed and flights can be repeated very accurately for each individual wind turbine even under low visibility conditions. This is also a major advantage compared to manned helicopter flights. Furthermore, flights at night are possible. With a future flight endurance up to 3 hours, even offshore parks far away from the coast can easily be reached.

Table 1. Aeroscout Scout B1-100 UAV main specifications

Main rotor diameter	3.2m
Main rotor speed	860 rpm
Maximum Lift-off weight	85kg
Gasoline engine	100ccm (2 stroke)
Fuel tank	10l
Payload capacity	18kg
Flight endurance	1.5h

The Scout B1-100 UAV helicopter has been equipped with a payload winch allowing to lower the payload on the nacelle of the wind turbine and hoist the release mechanism back after delivery. In emergency situations, the payload connection can be cut off from the ground control station command. Multiple video and data links allow instant monitoring of the flight status, the payload delivery process, and situational awareness.

2. TESTING EINVIRONMENT WITH A SIMULATED NACELLE

In order to prepare and test the various components of the UAV installation, including data and video links, payload winch, laser scanner, forward and downward looking cameras, transponder, flight termination system, manual backup and manual override, hovering precision, delivery time required, as well as team coordination, the nacelle of a wind turbine has been built as a simulation model shown in Figure 3.

Figure 3. Aeroscout Scout B1-100 UAV helicopter hovering above a simulated nacelle.

https://doi.org/10.52202/075106-0010

During the flights, videos as described above are used to document the payload delivery. The center of the nacelle has been marked with a visual pattern which allows to measure the precision of payload delivery. The fence of the simulated nacelle also should reflect the real situation of the wind turbine which is also surrounded by the protection railing. In general, the simulated nacelle was built in smaller scale making the payload delivery even more challenging.

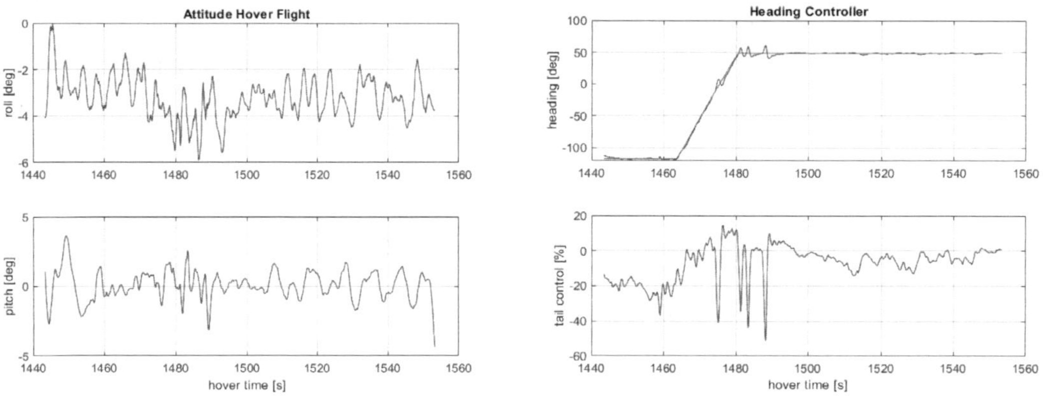

Figure 4. Attitude control during payload delivery

From the flights above the simulated nacelle the data sequence during the payload delivery has been extracted. The payload delivery took about 2min to lower the payload bag and to hoist the mechanism afterwards. Figure 4 shows the attitude data of the helicopter during the delivery process. Depending on wind gusts, minor changes in pitch and roll angle reach 2..5°. For the heading angle it is relevant to point the helicopter to wind direction. Also wind gusts have a significant influence on the tail control signal up to 40…50%.

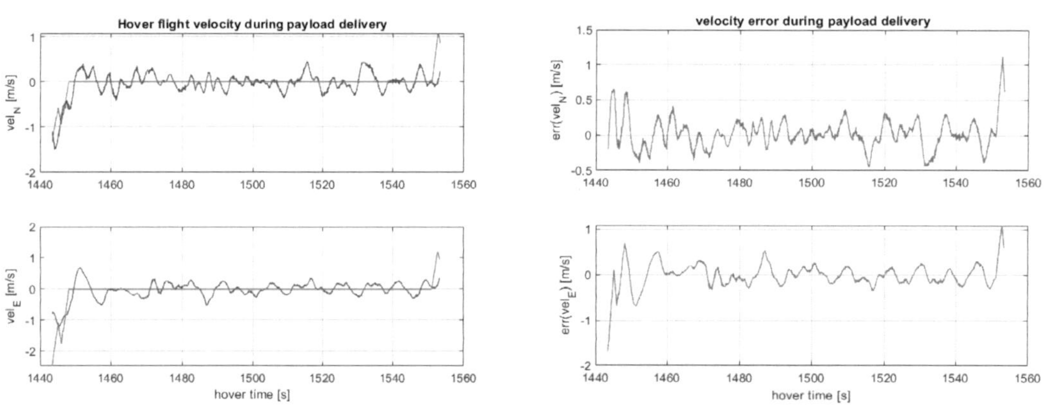

Figure 5. Velocity control during payload delivery

In Figure 5 the absolute UAV velocity, the reference velocity, and the velocity error in north and east direction are shown. Even under windy conditions, the velocity error remains below 0.2…0.3m/s which allows a highly precise payload delivery with minor position changes.

Figure 6 shows the absolute flight position during payload delivery. Once the target position is reached the reference position remains constant. At this instant also velocity control changes to position control of the aircraft. The cartesian standard deviation achieved is approximately $\sigma_{north} \approx 0.51$m and $\sigma_{east} \approx 0.86$m. This rough position accuracy gives an initial estimate of payload delivery under real conditions. However, it is noted that the final payload delivery accuracy also depends from the payload movements. Once the payload is let down, the payload can start to swing below the aircraft. In order to avoid these oscillations, the payload winch works rather fast to place the payload once the target position has been reached. All this testing could well be done with the simulated nacelle on ground.

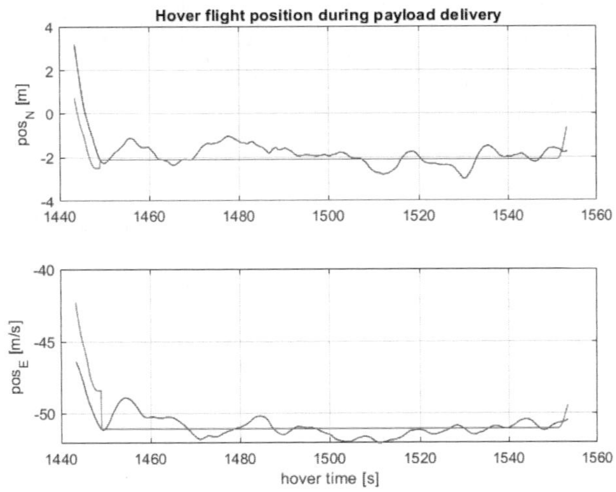

Figure 6. Position control during payload delivery

3. TEST THE AIRCRAFT FOR FLIGHTS BEYOND LINE-OF-SIGHT (BVLOS)

The intention of the ADD2Wind project is to finally demonstrate the payload delivery to off-shore wind turbines. This will require to perform flight beyond line-of-sight (BVLOS) and to achieve the related flight permission. This section summarizes the BVLOS flights performed in Denmark in June 2023.

Figure 7. BVLOS flights performed in Denmark

The BVLOS flights took place from Hans Christian Andersen Airport in north direction with distance of approx. 5km as shown in Figure 7. During the departure flight several flight circles have been added in order to test and validate the video and data link performance under all around conditions from the aircraft to the ground control station. During the flight more than 86 signals have been recorded. The flight took place at 100m altitude above ground. The number of GPS satellites in view were about $\#_{GPSsat} \approx 20$. The recorded standard deviation in latitude and longitude position error was achieved at $\sigma_{lat/lon} \approx 1.2 \dots 1.3\text{m}$. The cruise flight took place at $v_{cruise} = 10\text{m/s}$.

 https://doi.org/10.52202/075106-0010

Figure 8. BVLOS flights including flight circles and return flight

The flight circles as shown in Figure 8 have been performed with a radius of $r_{cruise} \approx 18$m and can clearly be seen in the heading data. The same radius was used when starting the return flight. The flight trajectories for departure and return flight are within a distance of five meters which was well covered. The return flight was preprogrammed as a direct line within the designated airspace.

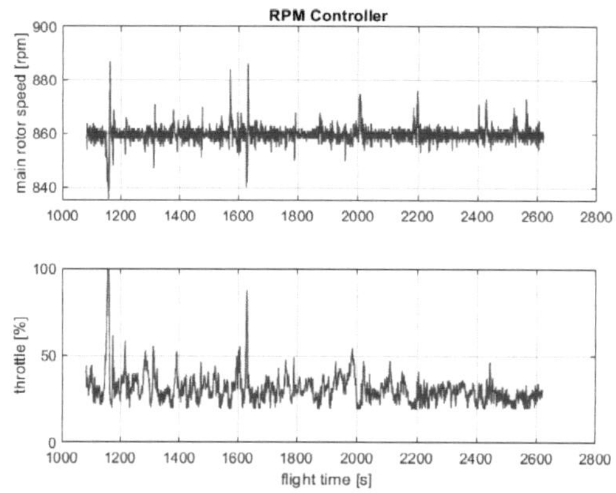

Figure 9. RPM control during BVLOS flight

Figure 9 shows both the main rotor rpm signal and the gas input. After the initial hovering period when all instrumentation is checked before departure, the maximum gas setting is reached to initiate the climbing process. While in cruise flight the power consumption drops significantly to 35…40% which is a major advantage of the classical helicopter configuration.

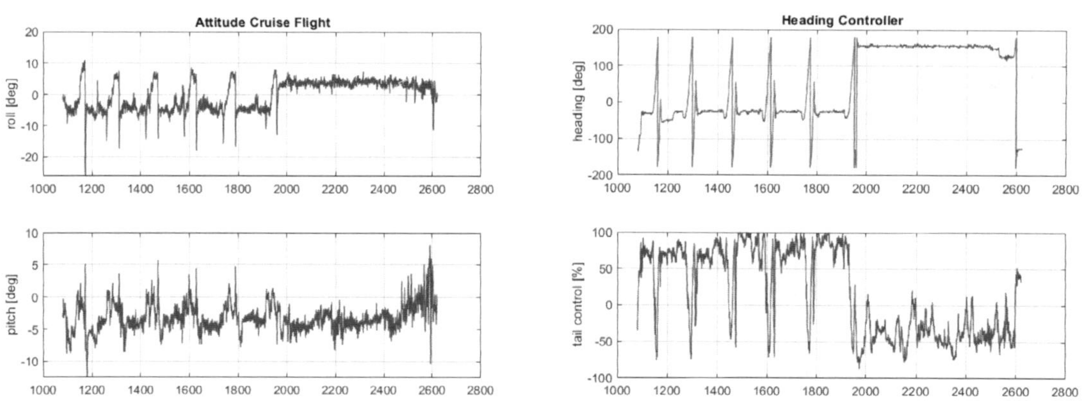

Figure 10. Attitude control during BVLOS flight

Figure 10 shows the attitude angles during the flight. While the pitch angle is more or less contant, the roll angle shows the slope of the aircraft during the flight circles. Also the influence from the side wind can be seen in the roll angle which changes from -5deg in departure flight to +5deg in return flight. The side wind also affects the tail rotor control which can reach 100% control authority in departure flight and -60% in return flight.

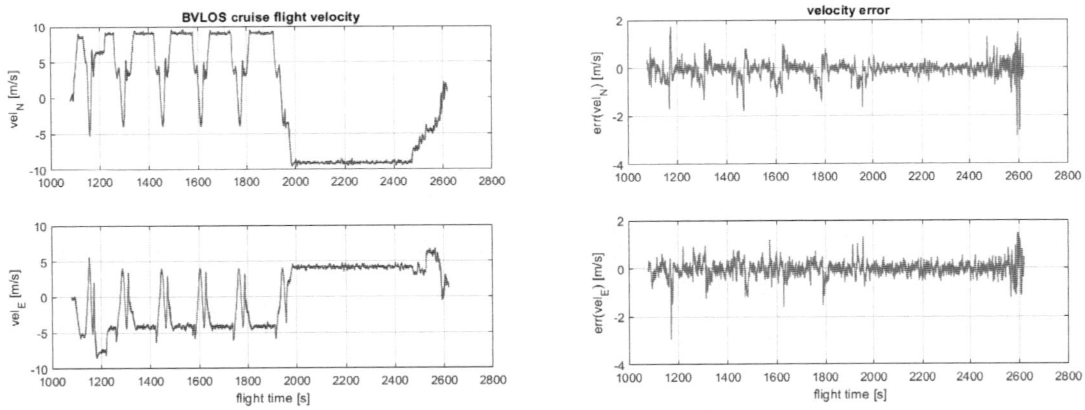

Figure 11. Velocity control during BVLOS flight

The velocity data is shown in Figure 11. Again, the flight circles in departure flight are visible while the return flight is at constant north and east velocity. The standard deviation of the velocity error is only $\sigma_{vel,north} \approx 0.35$m/s and $\sigma_{vel,east} \approx 0.31$m/s but will also depend on the wind conditions in general.

4. DEMONSTRATION OF THE PAYLOAD DELIVERY AT WIND TURBINE

End of June 2023 the payload delivery to a wind turbine has been demonstrated and recorded from various camera perspectives.

Figure 12. Demonstration video of payload to a wind turbine

The video as shown in Figure 12 has been processed by Anders la Cour-Harbo, associate professor and director of drone research lab at Aalborg University, Denmark. Anders is also initiator and head of the ADD2Wind research project.

https://doi.org/10.52202/075106-0010

The delivery of the payload was shown on a real wind turbine with its nacelle at 100m above ground. The unmanned helicopter did approach the wind turbine from behind. The downward looking camera was used to position the helicopter above the nacelle and to monitor the process of payload delivery. For safety reasons during initial flight tests the blades of the turbine have been fixed. This would have allowed the aircraft to also increase forward speed in case of sudden engine power loss or other irregularities. In addition, a camera on the nacelle could monitor the approach of the aircraft towards the nacelle. Finally a small electrical drone could record the flight following the main UAV.

Figure 13. Approach and payload delivery with the Scout B1-100 UAV helicopter (courtesy Aalborg University, Drone Research Lab)

The flight termination FTS82-H Mark II, system has been provided by Loxar and was installed to immediately stop the ignition of the gasoline engine in case of an uncontrollable situation. In Figure 13 the approach of the nacelle as well as the process of payload delivery has been captured from the accompanying camera drone.

Figure 14. Retracting the release mechanism after payload delivery (courtesy Aalborg University, Drone Research Lab)

After delivery the payload on the nacelle the small release mechanism has been retracted before starting the return flight as shown in Figure 14 (small item below the helicopter in left-side image).

5. SUMMARY AND ACKNOWLEDGEMENT

The previous sections describe the step-by-step procedure to successfully demonstrate the payload delivery to a wind turbine using an unmanned, autonomous helicopter. The work was provided within the ADD2Wind research project, led by Anders la Cour-Harbo and compliments go the complete research team both in Denmark and Switzerland.

A considerably amount of work was done to finally achieve flight permission. The submission to the CAA includes:

- SORA (Specific Operation Risk Assessment),
- OM (Operations Manual),
- ERP (Emergency Response Plan),
- ConOps (Concept of Operation),
- checklists, and
- AFM (Aircraft Flight Manual).

It is expected that unmanned robots can play a significant role for offshore wind parks for inspection, maintenance, and repair in near future. A combination of different sizes of UAVs can even increase the flexibility of required services.

https://doi.org/10.52202/075106-0010

MANNED-UNMANNED TEAMING (MUM-T): HOW FAR CAN I TRUST MY TEAM MEMBER AND ON WHAT IS THIS TRUST BASED?

Laura Samsó Pericón*

We are living a thriving moment in history, not only because everything is going exponential in terms of human development and technology, but because it could and will change how cross-domain/sector systems inter operate, communicate, how missions are ran and how decisions are taken and executed to achieve tactical superiority in the defense domain or, better performance, in commercial applications.

Autonomy propels interconnected systems, but the AI frontier introduces also a symbiotic blend of human and machine capabilities. This manifests as AI-driven Manned-Unmanned Teaming (MUM-T), leveraging Human-Machine Teaming and neuro interfaces. The core is the collaboration of human operators and autonomous drones, forging tactical supremacy. Fusing human intuition with AI precision enhances target identification and engagement. The core is the collaboration of human operators and autonomous drones, forging tactical supremacy. Fusing human intuition with AI precision enhances target identification and engagement.

In this evolving landscape, the integration of MUM-T systems has emerged as a pivotal paradigm. The paper delves into the dynamic realm of AI and Human-Machine Teaming enabled MUM-T uncrewed systems, encompassing innovative tactics, cutting-edge stealth technologies, wearable innovations, novel materials, advancement in sensor fusion, swarming intelligence and the transformative influence of exponential technologies such as Brain-Computer Interfaces (BCI) – neuro-interfaces, augmented reality and quantum technologies to name a few.

MUM-T systems have ushered a new era of commercial and defense capabilities, but they also bring forth a host of challenges and vulnerabilities that demand careful consideration. Those range from interoperability complexity, cybersecurity and vulnerability, Electronic Warfare (EW) threats, quantum technologies, ethical and legal concerns, operational concepts, Human-Machine Interfaces (HMI), training and skills, situational complexity, public perception and acceptance, reliability and redundancy.

This paper will show the state of art of MUM-T and it will also elucidate, through a use case, the pivotal role of a holistic approach in orchestrating the coordination between these elements, the challenges and vulnerabilities of integrating an HMI and AI enabled MUM-T capabilities in joint multi domain operation.

* R&D Cyberdrones, laurasamso.com

https://doi.org/10.52202/075106-0011

INTRODUCTION

"Can you really trust Artificial Intelligence (AI)?", this is one of the topics the experts discussed about during one of the sessions in the Annual World Economic Forum (WEF) meeting back in January 2024. If we go further into the details: How far can I trust my Team member and on what is this trust based?

A new geopolitical balance is undergoing across the world, testing the strengths and weaknesses of our societies and, at the same time, it offers a battleground for new technologies.

We are in the path of a major transformation in many ways, leveraged by the rise of new developments and its acceptance by the society, from workforce to logistic chains, economic challenges to how wars are fought.

The recent conflicts around the globe have seen an increasing military expenditure by countries worldwide, with rising needs for intelligence, surveillance, and reconnaissance (ISR) operations, analysis and autonomy; integrating manned and unmanned systems into all aspects of warfare and using AI as the glue for all those operations. But not only this, we are in the brink of a major leap, the symbiosis of human-machine in the realm of warfare.

Back to the initial question, when it comes to trust. On what is this trust based when human and machine team to enhance battlefield performance? How to ensure officers trust AI decisions and how to ensure the decision taken by the Human-Machine Teaming (HMT) is not the result of a hacking event? Both questions as slightly different but come to a convergence in the sense that both questions require analyzing human-human teaming into the realms of sociology and psychology (cognitive realm).

To really understand what an autonomous system, HMI in this case, is doing, it is not enough just to observe what the platforms are doing, but to really understand and have a view of the different transparent stages of the decision-making process. It is also key to know which are the components of a MUM-T ecosystem, so to be aware of the weak points that could be used to gain advantage of whole system. With that, an "AI behavior issue" could be isolated and contained, at best.

Uncertainty in contested environments adds another layer of complexity that only transparency between autonomous system and humans can ensure an effective way to exchange between them.

When it comes to MUM-T systems, there are few definitions, such as those real-world applications that involve one or more human and machine elements in which each is capable of some degree of autonomous actions, but their overall goal is shared and thus, some degree of cooperation is also assumed (Reference 1).

But how is it possible for this type of systems to work?

Autonomy and AI propel Human-Machine Teaming (HMI) and MUM-T operations, together with other secondary components or technologies such stealth, wearables, communications to name a few.

The paper uses the definition provided by NATO where it defines it as a system that decides and acts to accomplish desired goals, within defined parameters, based on acquired knowledge and evolving situational awareness, following an optimal but potentially unpredictable course of action.

The system's operator defines its parameters and objectives, but the system directs itself via artificial intelligence (AI). In the case of uncrewed aerial systems, an autonomous drone can

https://doi.org/10.52202/075106-0011

communicate, reroute, and make decisions based on the data it gathers about its environment, eliminating the need for continual human operation. Applying autonomy to a network of drones also enables drone swarming, allowing for the drone network to execute missions as a single, focused entity.

As it is mentioned in the Blueprint for Autonomy publication, the autonomous future is composed by five foundational building blocks: motivation, technology, airworthiness, operations, and integration.

In these type of MUM-T systems and scenarios, where the blend with human and machine offer so much performance operations enhancement, but also weak points, could your team member betray you? And would you have the means to spot it and up to what point?

The following sections will give the reader an overview about MUM-T ecosystem, its advantages and threats and how trust between MUM-T arena could be build.

MUM-T: THE NEW BATTELFIELD ARENA

MUM-T can be described as a cross dimensional arena, expanding current capabilities, providing a new combat strategy, planning and a collaborative environment. The United States Army Aviation Centre (USAACE) defines MUM-T as "*The synchronized employment of soldier, manned and unmanned air and ground vehicles robotics, and sensors to achieve situational understanding, greater lethality, and improved survivability*" (Reference 2).

Other bodies such the IEEE defined MUM-T as the "*coordinated use of manned and unmanned systems to achieve a common goal*". (Reference 3).

A high level MUM-T architecture is composed by different dimensions such as air, ground, sea and space in the future.

Given several facts such as the increase of different types of unmanned systems being used in different domains and that every nation design and procures UAS (Unmanned Aircraft Systems) themselves, interoperability needs to be ensured between the different NATO community platforms and systems.

This may enhance decision-making and mission effectiveness, offering new different tactical chances which could make a difference in the development of future strategies.

Table 1. Levels of Interoperability (NATO's STANAG 4586).

Level 1 Indirect receipt of UA-related data
Level 2 Direct receipt of ISR/other data where 'direct' covers reception of the UA data by the UCS when it has direct communication with the UA
Level 3 Control and monitoring of the UA payload in addition to direct receipt of ISR/other data
Level 4 Control and Monitoring of the UA, less launch and recovery.
Level 5 Control and Monitoring of the UA (Level 4), plus launch and recovery functions.

https://doi.org/10.52202/075106-0011

MUM-T configurations will offer different scenarios with advantages and capabilities and will also have an impact in the level of trust assumed. In this basic scenarios different usages could be envisaged such a small platform can be utilized to send sensor data to the manned aircraft or to some other node of the ecosystem, to deliver weaponry, when in coupling mode with a manned aircraft it could be used as a decoy. At the same time, the different nodes can launch attacks through different means overwhelming or confusing the defense opponent systems.

Apart of all the technological constraints, human factors and risks posed by MUM-T applications will become prominent when designing and developing the future aircrafts, manned or unmanned. The pilots/crew workload will be under stress in those type of scenarios and it is important to design not to exceed the human limits but also thinking in how AI can leverage that and which information in an HMI operations will be shared between human-machine and how decisions will be taken according to that.

OPERATION ENABLERS: TECHNOLOGY ADVANCEMENTS

We are currently at the brink of another exponential technology leap, where several advances are already converging into a major step, where reality and augmentation barriers become blurry.

AI and autonomy will revolutionize how wars are fought, symbiosis between humans and machines will become common, new sensors, novel materials and developments in stealth technologies will improve the survivability and operability rates; and quantum communications and processing will achieve the next leap.(Reference 10), (Reference 11).

But as advances are positive to grow as a society they normally come with drawbacks. In that case, all these enablers will be and are prone to be influenced, posing a risk/threat not only to the sole MUM-T operation but also to the civilians.

In addition, pilot (human) decisions whether to trust the AI companion or not, could also become influenced by parameters such the type of operation, the possible damage that would cause. But then, as a second derivative, are they able to detect and assess a misbehavior from the AI if it would be hacked/influenced?

Cognitive neuroscience of human decision making takes special relevance and AI will ensure speed in decision making as it analyzes all the information going through, but then the trust dilemma will appear.

Find below different of those operation enablers.

Stealth Technology

The beginning of stealth technology can be traced back to the Cold War era, when the US military recognized the need of this capability in order to penetrate enemy defenses undetected.

Petr Ufimtsev, a Soviet physicist, is considered the father behind the stealth technology. In 1962 he published the paper Method of Edge Waves in the Physical Theory of Diffraction, where he described the mathematical rationale for the development of those technologies. In the 1970s, his work was translated and reused in the US to produce the first operations stealth combat aircrafts, the Lockheed F-117 fighter and the Northrop B-2 Spirit strategic bomber.

There key principles that govern the design and implementation of stealth technology are focused in reducing the detectability of the platforms by radar, infrared, visual and other sensors.

Shape and Geometry

https://doi.org/10.52202/075106-0011

The Radar Cross Section (RCS) is a measure of how detectable an object is by radar. Stealth technology aims to minimize the RCS by using special materials and shaping techniques that scatter or absorb radar waves.

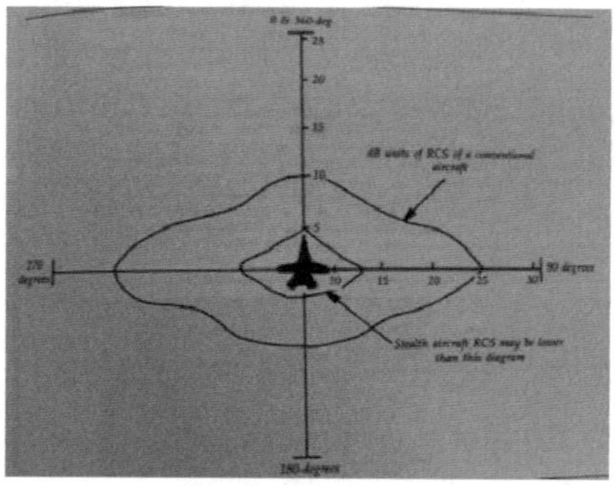

Figure 1. Airplanes Radar Cross Section (RCS).

Noise reduction: Stealth design aimed to minimize noise emissions making them harder to be detected by acoustic sensors.

Signature reduction: Design that uses special coating, absorbing paints, and materials together with techniques to reduce reflections and scattering of electromagnetic waves.

Sensor Fusion and Electronic countermeasures: Stealth vehicles employ various electronic countermeasures to confuse or jam opponent radar and communication systems.

Stealth technology could be applied to the different domains: airborne platforms, naval vessels and ground vehicles.

Wearables in the Battlespace

Another layer is the soldier wearables and portable military technologies that enhance physical capabilities, help in communications with the command centers, keep situational awareness, reducing risks in the battlefield and also monitoring their health.

As new technologies emerge and NATO countries increase their investments, the global soldier wearable technology market size is expected to reach US$ 3.77 billion in 2033. The market is projected to reach US$ 3.77 billion by 2033, with a growth rate (CAGR) of 1.8% during 2024-2033.

Those technologies range from exoskeletons, smart textiles, night vision technologies, biometric monitoring devices, communications equipment, navigation and GPS devices, and tech-fused armor.

https://doi.org/10.52202/075106-0011

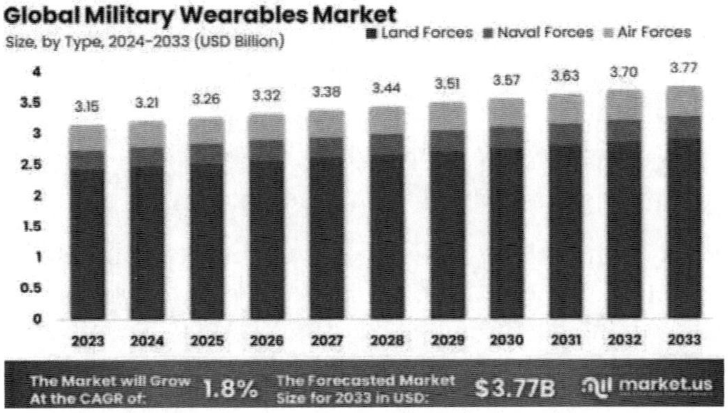

Figure 2. Military Wearables Market.

These are also used to identify soldiers and track their movements, analyze data from different sources, and share information about their surroundings, including maps, enemy positions, and mission objectives in real time with the different team members from the different dimensions, not only on ground.

Other Technology Enablers

Recent geopolitical events together with the low cost of COTS are democratizing more and more the use of uncrewed aerial systems during those conflicts in the form of swarm of drones to attack the enemy. Those drones could be used to overwhelm the defense capacities (as a decoy) and also to attack.

Those platforms are equipped with different sensors together with AI for autonomous navigation and obstacle avoidance. Sensors can be such E/O cameras, radar, LiDAR, IR. AI algorithms process this data in real-time, allowing drones to detect and avoid obstacles, navigate complex environments, and plan optimal flight paths.

Computer Vision and Object Recognition: Autonomous drones rely on computer vision techniques and object recognition algorithms to interpret visual data captured by their cameras.

Machine Learning and Deep Learning: This empowers drones to make precise, real-time decisions based on the data they collect.

Swarm Intelligence: Autonomous drones can operate collectively in swarms.

Sensor Fusion: Autonomous drones integrate data from a myriad of sensors, using sensor fusion techniques.

Automated Mission Planning and Execution: possess the capability to autonomously devise and execute missions based on predefined objectives.

Energy Efficiency and Autonomy: AI algorithms aid in optimizing energy consumption, enabling drones to undertake tasks for longer durations without frequent recharging.

Safety and Collision Avoidance: Ensuring safety is paramount in autonomous drone operations. AI-driven collision avoidance systems to identify and evade potential collisions with other drones, aircraft, or obstacles.

https://doi.org/10.52202/075106-0011

HUMAN-MACHINE TEAMING

Teaming is defined as a collective action targeted to reach supposedly joint and shared objectives. It is normally defined as a "distinguishable set of two or more people who interact dynamically, interdependently, and adaptively towards a common and valued goal/objective/mission, who have each been assigned specific roles of functions to perform, and who have a limited life-span membership" (Reference 2).

Human brain has 86 billion neurons and according to some studies (Reference 14) we do not use it to the full potential. A telepathic conversation with your manned-unmanned team members will be soon a reality in the war arena. As Former Deputy Secretary of Defense Robert Work, who led DoD's 3rd offset, summarized trends with military technology as follows:

The coin of the realm during the Cold War was armored brigades, mechanized infantry brigades, multiple launch rocket system battalions, self-propelled artillery battalions, tactical fighter squadrons, among others. Now, the coin of the realm is going to be learning machines and human-machine collaborations, which allows machines to allow humans to make better decisions; assisted human operations, which means bringing the power of the network to the individual; human-machine combat teaming; and the autonomous network. (Reference 4).

The first developments took place approximately 40 years ago: this technology is capable of reading brain activity and decipher thoughts to carry out man-machine interaction tasks. Tasks that ranged from controlling devices such as a wheelchair up to medical interventions for cognitive rehabilitation. In the last 10 years, some of this research has become available to the public through different companies that have developed wearable, comfortable and accessible brain sensing devices that start to be part of our daily lives.(Reference 13).

Many predict the singularity is coming closer and some even have already a date in mind: between 2025 and 2030. It is expected a convergence between biotechnology and other technologies such holographic, visual devices, AI.

Human-Machine Interface (HMI) techniques use bioelectrical signals to gain real-time synchronized communication between the human body and machine functioning. Research has proved that this technology permits the simultaneous control of multiple functions, UAS sensors and platforms in that case, improve the speed of communication and enhance common situational awareness.

Since at least 2007, the military have been researching what it is called "synthetic telepathy" (Reference 5) type of non-invasive Brain Computer Interfaces (BCI) to be used by the warfighters.

In 2012, DARPA issued a $4 million grant to build a non-invasive "synthetic telepathy" interface by placing sensors close to the brain's motor centers to pick up electrical signals — non-invasively, over the skin.

Three years later, a paralyzed woman was able to steer virtually an F-35 with an implantable small microchip; and later another person was able to also receive signals from the simulated aircraft. Brain signals can be used to command and control various types of aircrafts.

Elon Musk's Neuralink, recently became a reality and the first person to receive is currently under study. Those Brain-Computer Interfaces (BCI) are considered the initial step towards the "singularity", melding human and machine. As BCIs transitions from basic research to a more operational and commercial applications it will be important to lay the basis for regulations and policies (Reference 6).

Ongoing R&D efforts include many dimensions of technology, in particular, there is an increasing focus on human-machine collaboration for improved decision making, including human-computer interaction (HCI) and cognitive teaming, assisted-human operations, swarm control and operation, and advanced manned and unmanned combat teaming.

All these new developments will come with associated vulnerabilities and risks, and ethical and legal responsibilities. In particular, cyber threats will show new points of failure, adversary access to information and new areas of exposure, manipulate, deceive or harm the rest of the team members. HMI technology will be transformative during combat and reconnaissance operations but capabilities will depend largely on its fidelity and reliability during those operations. How far you can trust your team members and what is this trust based on will be a key for future successful operations.

Future R&D on this topic must work towards integrating AI & human cognitive models, advance human-agent feedback loops, optimize trust/transparency, and advance sensor/data decision model.

TRUST AND TRANSPARENCY OPTIMIZATION AMONG MUM-T TEAM MEMBERS

As humans and machines will collaborate cognitively and think together, new challenges from merging AI, autonomy and HMI technologies will arise and form a new set of vulnerabilities. Trust or lack of it is one of those and it will be key for the success of the missions (Reference 7).

When humans build trust while working for common goals or in social interaction, uncertainty is reduced, stress levels go down and the likelihood to be successful in whatever endeavor they work it, is increased. At the same time, humans spend less cognitive resources to deal with the topic.

Trust in AI and its effects could also be compared to the infancy steps of unmanned aircrafts, where an accident could have damaged the whole efforts in order to launch the market of services and applications. A small AI error with catastrophic results could jeopardize public acceptance. With that in mind, a stepped approach towards trusting AI and its interaction with humans must be put in place.

In the context of AI, trust refers to the decision to use a technology based on the perception that it will reliable perform as expected in pursuit of shared goals (Reference 3).

Cyber-attacks will put not only technology but trust in general on the spot. Several types of trust issues could arise: deficit of trust in HMI technology, deficit of trust on the decisions taken by the different team members once human-machine system works together and the general threat of cyber-attacks.

Trust issues between the different team members such as potential erosion of unit cohesion, unit leadership among other critical interpersonal and inter platforms relationships will arise with unknown consequences.

How could we approach those trust issues?

One topic is the technology itself that the operator has to trust in. If not, any outcome of this technology will be doubted.

Another one, how to trust the final decision taken by the symbiosis of human-machine interaction? Layering the decisions taken through cognitive methods could be one way to segregate and go through a process of intention analysis.

https://doi.org/10.52202/075106-0011

An overall prerequisite would be to have effective cyber measures in place to safe guard the involved technological systems and the final result of the operation.

When humans take decisions they do it based in different thoughts such: How critical it is? Do we have time to decide? Is it life endangering? Will there be collateral damage? Who else in the chain will be impacted? How important is to have success here? Do we need to follow the rules? How public will perceive that? Could this endanger our relations with other nations?

Based on that above, humans may trust less or more the AI to take over in each scenario or situation. So it can be envisaged that in a mixed operation, trust in AI can vary ranging from pilot in the loop to fully autonomous.

In addition, once the pilot decides to trust the AI; relying in one AI only may not be enough to ensure operations will be risk free. As many security organizations have such the policy, it may be needed also to have a main AI umbrella that would make sure the secondary AI is not misbehaving due to an overtake by a nefarious attack.

ENTANGLED CYBERWAR AND AI TRUSTWORTHINESS

Soon manned and unmanned fighters will jointly cooperate in real scenarios leveraged by AI and HMIs teaming capabilities. The recent conflict with Israel and Iran is a perfect example on how a swarm of drones were used to overwhelm the defense capabilities of Israel, to distract them. Once the system collapsed, long range missiles were launched by Iran over Israel targets.

Not so far in the future, we can envisage an expanded scenario containing all the elements commented in the sections above.

We will analyze the following use case that is elaborated in more detail in my presentation:

Imagine a joint multi domain mission where different nations work together, having all of them different human-machine teams and connected not only to their concrete AI, but to a global AI that would make sure the mission is successful. It will be also important to define what attributes make a mission "successful", so the AI will make sure it is keeping the operatives safe.

A unit is sending a team with one or several loyal wingmen unmanned aircraft to assess the enemy forces and send advance information to the rest of the unit, requesting additional resources if required and attack.

The use case describes a scenario in a highly contested environments characterizing current military missions, where uncertainty is high and new threats appear, making it difficult to assess a conflict situation. In order to generate a response, the commander is required to make assumptions and inferences. At some point there are so many and they come so quick that a human cannot properly process and handle them.

There is also uncertainty whether a situation will develop into a wider conflict or not, so decisions need to be taken into the same moment it happens.

Humans find it difficult to accept uncertainty, acting on pretended knowledge which may lead to overconfidence, accepting risks that they would not if they would really understand the situation.

Military have developed dedicated methods for conducting their decision making processes (ex. NATO Allied Doctrine for Operational Planning). Every commander at every level goes through a decision making process and depending on the level of the commander, and the complexity of the mission, the process is becomes more difficult. This process can be very time consuming leading to the inability to handle the situations or conflicts properly.

https://doi.org/10.52202/075106-0011

Intelligence systems can help to overcome the vulnerabilities of human decision making, the diversity and complexity of conflict situations, the information and technology means employed in warfare and the amount of information needed to be processed in real time.

The previous section brings up some questions that have to be answered for this concrete use case:

How critical it is? Do we have time to decide? It is life endangering? Will there be collateral damage? Who else in the chain will be impacted? How important is to have success here? Do we need to follow the rules? How public will perceive that? Could this endanger our relations with other nations?

There are several ways to take over manned and unmanned aircrafts through their sensors, connections and communications by a cyber-attack. HMI systems are as exciting as delicate, a hacker could gain access and hack the rest of the ecosystem including taking over the AI system, which is the core of the operation and mislead the human pilot. Furthermore, in joint multidomain missions, the hackers could get to other nations ecosystem. The difficulty with a human in the loop is to be able to detect an anomaly in system behavior at all.

So how to ensure there is a process in place to follow and to evaluate the level of trust or uncertainty of AI decisions? At the level of human-machine, there should be a secured redundancy AI system, that is not connected to the same network, that would evaluate the behavior of the MUM-T AI. This segregated AI would take over in case it detects abnormal behavior or some type of attack to any of the MUM-T ecosystem components.

In addition to that, a human-AI decision level process should be in place. Some research follow a stepped out approach to assess the trust and transparency of the AI interaction (Reference 6), (Reference 8), (Reference 9).

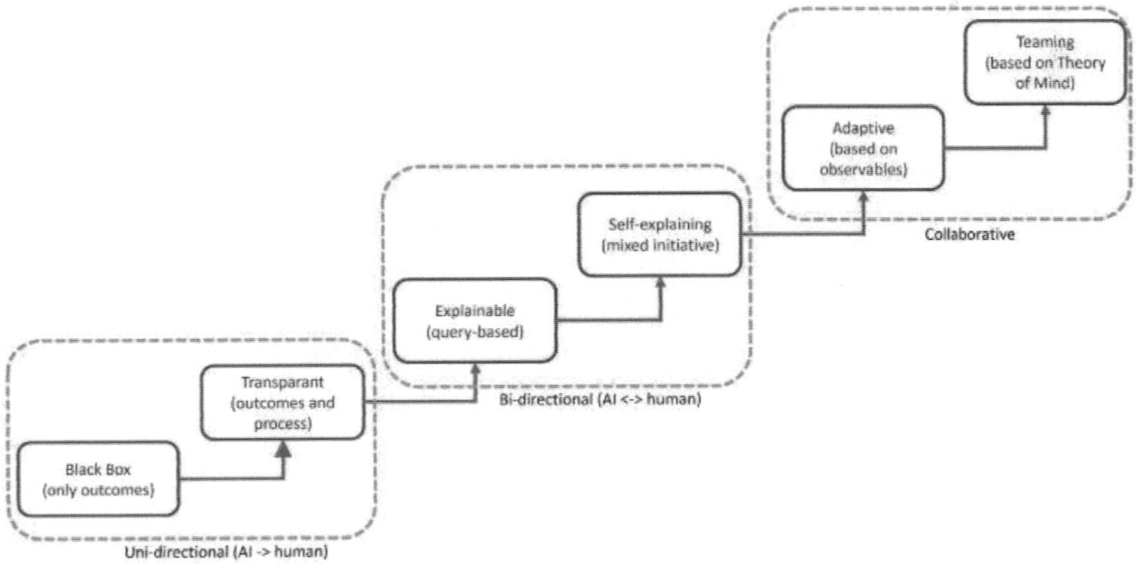

Figure 2. Levels of human-AI collaborative decision making.

https://doi.org/10.52202/075106-0011

This human-AI collaborative decision making approach will also help to enhance the transparency of those systems. Figure 2 shows a stepped approach towards fully AI autonomy going from the basic step, where AI is considered as a black box up to human-AI teaming, achieving the highest step.

In the use case presented above, the mission may develop towards a more complex conflict scenario, requiring lots of data processing capability and high speed decision making. The loyal wingman (Reference 12) sent to inspect the field in advance may encounter different challenges while performing their intended mission of acquiring pre battle field data. The risk of collateral damage may be a factor as the opponents base is close to a residential area but it is a target of high priority. Additionally, international regulations have to be adhered to and public perception should be kept under control.

It could occur that meanwhile this mission is progressing a cyber attack is launched, to defeat the MUM-T AI and/or its components, in order to gain control of the whole ecosystem: loyal wingman sends wrong positioning of the opponents, to send its own loyal wingman to attack friendly forces,… It will be then the moment when a supervisory AI would be needed to inspect all the MUM-T decision making steps applying a human-AI collaborative approach.

CONCLUSIONS

MUM-T architectures will be the future of how wars will be fought, posing advantages over the adversaries but also challenges.

There is still more research required towards integrating AI & human cognitive models, refine the stepped approach human-machine feedback loops, optimize trust vs transparency, how to have a common ground, and develop new advances in sensor/data acquisition decision models.

In parallel, it will be key to be aware if decisions taken by HMT are the result of hacking event, as a MUM-T ecosystem is filled by different nodes prone to be taken over.

ACKNOWLEDGMENTS

Thank you to all that made it possible my mentor Jere, Jim Atkinson and Enrico Mollenhauer for their support and inspiration.

REFERENCES

[1] VVAA, "Manned-Unmanned Teaming: Research and Applications Panel", HFES DARPA 2021.

[2] Col. Livio Rossetti, "Manned-Unmanned Teaming", NATO JAPCC Magazine. 2021

[3] Gp Capt (Dr) D. Kumar, "MUMT – AD: Exploring to dominate in a contested environment", 2023.

[4] VVAA, , "Brain Computer Interfaces: US Military Applications", RANDT Corp 2020.

[5] J.M Rickli, "Neurotechnologies and Future Warfare", RSiS 2020.

[6] Lit. Col R. Ichaso, "Neurotechnology. Artificial Intelligence – Human Symbiosis in Fighter Aircraft End of the Fighter Jet Era or a New Evolution?", JAPCC NATO, 2022.

[7] VVAA, "Human-AI Cooperation to Benefit Military Decision", S&T NATO

[8] VVAA, "Humans and Autonomy: Implications of Shared Decision Making for Military Operations", US Army Research Laboratory Aberdeen Proving Ground US, 2017.

[9] G. Kuczynsk, "US Military Decision Process: Organizing and Conducting Planning", 2023.

[10] VVAA, "Unmanned Aircraft Systems: Roles, Missions, and Future Concepts", GAO 2022.

[11] Dr. R. O'Toole, "FY Annual Report 2023", Operational Test and Evaluation, 2024.

[12] Clay J. Humphreys, Major, USAF, "Optimal Control of an Uninhabited Loyal Wingman", Airforce Institute of Technology, WPAF Ohio, 2016.

[13] VVAA, "Above and Beyond: SoA of UAS Combat Aerial Systems and Future Perspectives", Instituto Affari Internazionali, 2023.

[14] C.Caruso, "New Field of Neuroscience Aims to Map Connections in the Brain", Harvard Medical School, 2023.

https://doi.org/10.52202/075106-0011

ASTM AC377 AUTONOMY DESIGN AND OPERATIONS IN AVIATION

WHITE PAPER

ROLES AND RESPONSIBILITIES FOR OPERATIONAL CONTROL IN THE AGE OF INCREASINGLY AUTONOMOUS FLIGHT

IN THIS WHITE PAPER

CONTRIBUTORS

Mark Blanks, Wing

Stephen Cook, Chair, ASTM AC377

Jonathan Daniels, Praxis Aerospace Concepts International

Anna Dietrich, AMD Consulting

Tom Gunnarson, Wisk

Phil Kenul, TriVector Services

Andy Lacher, NASA

Adam Morrison, Streamline Designs

As the aviation landscape evolves with increased automation and autonomy, the traditional roles and responsibilities of operators, pilots, and other stakeholders need to be re-examined. In light of emerging technologies and systems, it is crucial to reassess accountability, authority, and responsibility of all parties involved in aircraft operations. In this context, this paper explores the current responsibilities of Operators (i.e., the entity responsible for initiating, conducting, and terminating a flight) and the impact of increasingly autonomous systems on the aviation industry. This paper addresses the changing nature of accountability and the potential for shared responsibility and accountability between operators, OEMs, and other stakeholders in this dynamic environment.

CURRENT OPERATIONAL FRAMEWORK

In an aviation context, an Original Equipment Manufacturer (OEM) builds products—such as hardware and software—and an Operator uses these products to conduct flight operations. "Operator" in this sense has a very specific meaning and is not a euphemism for a "pilot"; instead it is the organizational entity that exercises the authority for initiating, conducting, and terminating flight—an authority known as "operational control" as defined in 14 CFR § 1.1[2]. Operational control is accomplished through establishment of procedures, training, qualification of personnel, quality control of aircraft, and is typically exercised by pilots, dispatchers, and other personnel that are associated with the entity.

Organizational Operators take many forms ranging from airlines to air ambulance services to single pilot/aircraft operations. Regardless of the configuration, an Operator is charged with maintaining operational

[1] For simplicity, this paper's language is framed around United States operations, however individuals can also be operators and may hold several roles relative to the concepts under discussion.

[2] Per 14 CFR § 1.1, Operational control, with respect to a flight, means the exercise of authority over initiating, conducting or terminating a flight.

https://doi.org/10.52202/075106-0012

control over their flight activities. In essence, this means the Operator must establish a detailed framework for how, and by whom, operations are conducted such that safety and regulatory compliance are maintained throughout all phases of flight. In an air carrier organization (e.g. 14 CFR § 135 or 14 CFR § 121) the operational control framework is typically exercised by a tiered series of controls from strategic leadership—such as the Director of Operations and/or Chief Pilot—approving procedures and methods of operation for tactical implementation by certificated dispatchers and pilots who exercise direct control over individual flights.

In practice, an OEM will design, build, and test a product that subsequently gains design and production approval by aviation regulatory authorities with any applicable operational limitations. These products may come with a manual which describes the general use of the product. However, it is the Operator who must determine how the product fits within the organization's specific operations and then develop procedures and training for its safe and effective use within their environment. The Operator has discretion to determine when, where, and how flights may be conducted as long as they fall within the performance and regulatory limitations of the product and operation.

SHIFTING LANDSCAPE

Over the past decade, the increasing automation of aircraft systems has begun to shift the landscape of how aviation systems are operated in the national airspace. Standing assumptions based on direct pilot command and control for individual aircraft and direct oversight by operating entities are now routinely being challenged with systems that are able to shift elements of operational decision making from humans to automated system capabilities. This includes elements that range from flight planning to contingency management and everything in between. This increasing automation[3] now allows the potential for a single human being to safely oversee the simultaneous operation of numerous aircraft and handle only things that automation cannot process independently. Likewise, the health and safety of the Operator's aviation system/operation may now be monitored through automated tools and approaches that were previously impractical with non-automated operations.

Additionally, distributed or federated services, systems of systems, and the rapid introduction of automation to aviation including advanced technologies like artificial intelligence and machine learning are bringing innovations to an industry that prioritizes safety through deliberateness of change over speed of innovation. This manifests as a clash of cultures between technology developers, regulators, and legacy aviation stakeholders who perceive risks and responsibilities from dramatically different viewpoints. Another significant factor in this new risk-reward conversation is the overall opinion of the general public – from those who receive benefits from increasing autonomous aviation (e.g., drone delivery customers) and from those who may be averse to it (e.g., those who don't want drones flying over their houses).

Increasing automation of operational decisions, disparate approaches to risk management, and entirely new technologies raise questions and issues around existing assumptions regarding the role of an Operator (i.e., the entity operating an aviation system). The next section explores some of these

[3] It is beyond the scope of this paper to address the technical maturity of the technology associated with this shifting landscape.

https://doi.org/10.52202/075106-0012

challenges to highlight issues that need to be addressed to move this next phase of the aviation industry ahead; capturing the increased safety, efficiency, and use cases that autonomy can facilitate.

ISSUES

Moving toward a future where automation can make decisions and take actions that were previously performed solely by humans raises several issues to be considered. Typically, humans making operational decisions were affiliated with the Operator to include pilots, dispatchers, maintainers, etc. With the shift to increasingly autonomous flight, automation developed by OEMs will effectively be making operational decisions and executing actions that have traditionally been the responsibility of the organizational entity responsible for operating the flight. The following issues need to be considered:

1. Authority, Responsibility, and Accountability
2. Command Authority and Responsibility of the Pilot
3. Design Assurance vs. Operational Safety Assurance
4. Training, Qualifications, and Licensing

Issue 1: Authority, Responsibility, and Accountability

As automation takes on more safety-critical functions during flight, questions arise regarding which organizational entity is accountable for these actions. The shift to increased automation is likely to impact the authority, responsibility, and accountability of the Operator. Consider the following definitions from ASTM's 2022 technical report Regulatory Barriers to Autonomy in Aviation [4,5]

ACCOUNTABILITY	AUTHORITY	RESPONSIBILITY
The obligation to answer for an action taken by a responsible entity	The power to give orders and/or make decisions.	The obligation or duty to carry forward an assigned task to its successful conclusion; this is closely coupled to the authority.

In today's operations, automation is typically given limited authority and responsibility for very specific functions (e.g., fly-by-wire, auto-pilot, auto-trim). In many cases (e.g., auto-pilot, auto-trim) this authority can be revoked during flight by the pilot flying[6]. Throughout, the pilot in command remains accountable for all of the operational actions taken during the flight[7]. In general, today's automation is given the authority and responsibility for relatively straightforward tasks that don't require significant perception and judgment to execute. In the future, the complexity of tasks associated with automated decision-making is expected to increase significantly (Figure 1). In the future, automation may have significantly greater responsibility and authority to make decisions and act, delegated by humans.

[4] AC 377, Technical Report #3, "Regulatory Barriers to Autonomy in Aviation", 2022 ASTM.

[5] Ibid. As identified in AC 377 Technical Report #3, the term "liability" was determined to be out of scope and is therefore omitted from this paper.

[6] The pilot flying is the pilot operating the flight controls of the aircraft. FAA, Roles and Responsibilities for Pilot Flying (PF) and Pilot Monitoring (PM), SAFO 15011, 11/17/2015, https://www.faa.gov/sites/faa.gov/files/other_visit/aviation_industry/airline_operators/airline_safety/SAFO15011.pdf.

[7] 14 CFR 1.1.

This raises the question as to who is accountable; the entity that developed the automation or the entity that configured and is operating the automation?

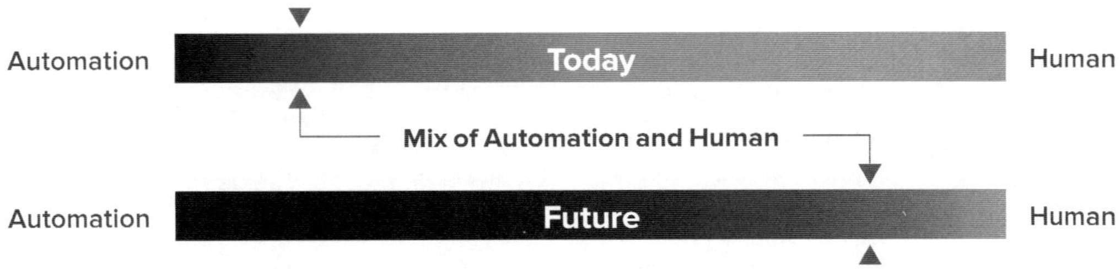

Figure 1. Notional Description of the Scale of Accountability for the Performance of a Function

In general, it seems logical that accountability can only reside with a human or human-based organizational entity. Through procedures documented in the Operator's operations specification, which is approved by the regulator, automation can be delegated the authority and responsibility to act. By today's regulations, the Operator is the entity that must maintain operational control, thus it would seem logical that they will remain accountable for those actions since they are responsible for procuring, configuring, maintaining, developing procedures for, and operating the automation systems.

For example, an Operator can delegate aircraft maintenance to a third-party organization. Today, Operators (e.g., airlines) often rely on third-party aircraft maintenance providers to perform a variety of maintenance tasks. The work that is outsourced can vary widely in scale and in scope, such as: servicing a particular component, overhauling an engine, or performing a D-check for an entire fleet of aircraft. Even when a task is performed by a third-party maintenance facility, the Operator is still responsible for the repair shop's compliance with the Operator's approved policies, procedures, and requirements. Thus, the Operator is still accountable for ensuring that the appropriate maintenance was performed.

If the Operator is accountable for operational decisions made by automation, they will need a mechanism by which they can ensure oversight and be able to provide guidance to the automation. In addition, some degree of transparency in how the automation makes decisions and how configuration changes impact its behavior will be required. The automation will have to be developed such that the Operator can appropriately configure it to adhere to their procedures as documented in their operations specification and be able to respond appropriately given the authority and responsibility which has been granted. Accordingly, the automation must be capable of meeting an Operator's procedures. The capabilities of a traditionally type certified aircraft are well known and documented, however the capabilities of an automated system may not be quite as clear depending on the method of design approval. There is some burden here on the OEM to define the limits of performance and capability of the automation they have developed.

Standards bodies like ASTM could lead the way in developing appropriate performance-based requirements and standard practices to guide manufacturers of automation systems that ensure that Operators are able to provide the necessary oversight and guidance. Standardizing this boundary between the creation efforts of the manufacturer and the operational obligation of the Operator is critical for clearly defining the authority, responsibility, and accountability roles and ensuring that the

https://doi.org/10.52202/075106-0012

Operator has appropriate transparency and can provide the necessary guidance to continue to fulfill their operational control obligations.

Some questions to consider:

- Who is accountable for operational decisions made by automation, the entity that created the automation or the entity that configured the automation?
- How should the interface between the manufacturer and the Operator best be defined for an automated system that makes operational decisions?
- What are the requirements for transparency and configurability of automation to ensure that the Operator can maintain operational control matching their accountability?
- Can automation be accountable or can only persons (consistent with the 14 CFR 1.1 definition)[8] be accountable?

Issue 2: Command Authority and Responsibility of the Pilot

During an inflight emergency requiring immediate action, only the pilot in command can make a decision to deviate from the rules and established procedures if they determine it is required to meet that emergency.[9] They are also accountable for explaining that decision.[10] How might this command authority endure as we increasingly automate flight operations?

Today, the Operator exercises operational control through the actions of humans who are hired, qualified, and trained to follow procedures established by the Operator which are consistent with regulations and are approved by the regulator. This includes the pilot in command and the pilots who are at the controls of the aircraft (i.e., the pilot flying). As routine aircraft control moves towards increasingly autonomous systems, the human role in the performance of many flight functions is increasingly reduced. In many cases, the human is over-the-loop, monitoring the automation system performing the function. The human can provide guidance and oversight that will impact the system behavior while automated processes have the authority to act without human actions or additional authority. Even with technology today, the pilot in command will likely remain accountable for the actions taken on the flight deck even if they are not the pilot flying. As automation matures, it can assume many—and potentially in the future, all—of the duties of the pilot flying, but not necessarily the duties, authorities, responsibilities, and accountabilities of the pilot in command, especially those related to deviating from established rules and procedures. It's important to note that in current regulations the authority to deviate from established regulations and procedures during flight is only given to the pilot in command, not to the Operator organization as a whole.

Predefined contingency actions could potentially be developed that deviate from rules and procedures. Perhaps predefined contingency actions can really be thought of as procedures to be included in an operational specification approved by the regulator? This would create the potential of Operator established contingency actions, where the Operator specifies the action in advance and thus potentially exerts authority over those actions, but not in a tactical sense.

[8] Per 14 CFR 1.1, Person "means an individual, firm, partnership, corporation, company, association, joint-stock association, or governmental entity. It includes a trustee, receiver, assignee, or similar representative of any of them."
[9] 14 CFR § 91.3b.
[10] 14 CFR § 91.3c.

As automation becomes more sophisticated, a clearer articulation of the line between control authority/responsibility/accountability and command authority/responsibility/accountability, could provide a path forward for regulatory approval while also potentially helping inform design assurance and operational assurance decisions. This is consistent with the recent European Union Aviation Safety Agency Opinion which draws a clear distinction between control and command authority and makes it clear "a human is always in command".[11]

IMAGE COURTESY OF NASA

Some questions to consider:

- What situation awareness information must the pilot in command maintain as automation increases?
- How does the pilot in command remain in command if automation is effectively the pilot flying?
- How does the pilot in command exert command authority during flight for increasingly autonomous aircraft?
- How does command authority change if the pilot in command is responsible for more than one flight at a time?
- Does the pilot in command's authority have to be "real time"? To what degree can configuring automation behavior ahead of time be considered exerting command authority?
- Can the deviation from rules decision be made by the pilot in command a priori (i.e., by configuring the desired response before the event occurs)?
- To what degree can a pilot in command still perform their duties if they are not able to ever be in direct control of the aircraft?
- Can increasingly autonomous aircraft systems reach a point where a pilot in command is no longer required and instead the Operator exerts command authority through system configuration or other means? If so, can the "pilot" requirements in the operational regulations be effectively exempted in their entirety?
- To what degree can automation ever be given the authority to choose to ignore regulations (i.e., exercise 14 CFR 91.3b) or is that an inherently human role to execute command authority? Is technology mature enough to be programmed to appropriately perceive the situation and make the judgment to deviate from rules and procedures?
- As automation becomes more sophisticated, do we need to do a better job clearly articulating the line between control authority and command authority?
- How should operational requirements levied on the pilot be addressed?

Issue 3: Design Assurance and Operational Safety Assurance

As more operational decisions are being made by automation with less direct human involvement, there are more operational safety assurance processes that are now performed by systems. Traditional engineering system safety processes that are used for aircraft certification (e.g., ASTM F3230, SAE ARP 4761, SAE ARP 4754A[12]) and software assurance standards (e.g., RTCA DO-178C[13], ASTM F3201) may not

[11] Available from EASA, https://www.easa.europa.eu/en/document-library/opinions/opinion-no-032023

[12] Available from SAE International, https://www.sae.org.

[13] Available from RTCA, https://www.rtca.org.

https://doi.org/10.52202/075106-0012

conveniently translate to the realm of operational safety where operational experience and contingency management (e.g., adapting to traffic, weather) are key contributors to safety. In the aviation context, design assurance is the mechanism by which safety is achieved through the design, construction, and production of various systems (e.g., aircraft). Operational Safety Assurance is the mechanism by which safety is achieved through the design of processes and procedures for operational decision-making and the training and execution of those procedures. As operational decision execution shifts from a human-centric process to automation, some aspects of safety assurance that are currently part of Operational Safety Assurance will need to blend with aspects of Design Assurance.

Additionally, regulators typically rely on safety oversight systems such as implementing requirements for Safety Management Systems (SMS) among Operators and Original Equipment Manufacturers. Does the current SMS paradigm hold for increasingly autonomous aircraft systems, where operational decisions may be allocated to software? If operational procedures change (e.g., in response to a safety concern), how do these requirements flow to the air system and how does the OEM deploy the software update in an expeditious manner?

Without a clear path forward, there is a risk that progress towards increased safety and the greater utility of aviation will be stymied over concerns of "who is responsible?" if something were to go wrong related to tasks traditionally performed by the Operator or pilot in command which are now programmed into the aircraft and/or ground automation software.

IMAGE COURTESY OF NASA

Some questions to consider:

- Are existing software and design practices sufficient for highly automated systems that make operational decisions?
- Are operational procedure changes necessary for autonomous aircraft? How can we adapt current operations for new aircraft functionality?
- How can industry standards promote design assurance?
- Under what circumstances should automation be allowed to deviate from regulations? Who has authority in emergency situations when the pilot is replaced by automation?
- Would keeping the line clear between design assurance and operational assurance provide a path forward for approval?
- How can automation deviate from regulations without a pilot in command?

Issue 4: Training, Qualifications, and Licensing

Training, qualification, and licensing are other important areas that face challenges related to the increasing use of automation. For most aircraft operations today the pilot in command requires some form of certification that validates they are competent and qualified to safely conduct a flight; whether available certifications are appropriately scoped for increasingly autonomous aircraft is a question that needs to be collaboratively answered by aviation stakeholders.

Pilot certification is generally a tiered system where more complex operations require higher levels of pilot qualifications, experience, and training. Similarly, flight operations that involve the carriage of people or property for compensation or hire also generally require higher levels of pilot qualifications and medical fitness.

https://doi.org/10.52202/075106-0012

Similar to pilots, Air Carrier Operators conducting flights for compensation or hire are subjected to organizational certification requirements. Generally speaking, these are organizations with more than a single employee conducting scheduled service operations, cargo transport, external load operations, agricultural or other various types of commercial flights. Air Carrier Certification (e.g., under 14 CFR Part 135 and Part 121) requires mature processes and training programs and specifies clear roles and responsibilities for the people that comprise the Air Carrier organization.

For both pilots and Operators, there are rigorous certification requirements that span from knowledge and practical testing to organizational procedures, record-keeping, and ongoing training. Many of these requirements are rooted and designed with an assumption of not just human involvement in the regulator-accepted processes but of an individual human serving as the ultimately responsible party.

As automation is introduced and assumes some of the role of the human pilot or organizational Operator for completion of certain functions that are otherwise certificated, traditional training, qualification, and licensing processes are no longer appropriate.

Some questions to consider:

- As an example, if an airline transport pilot must demonstrate a minimum visual acuity within the prior six-month period to conduct a commercial flight operation, how does this translate to an automated system that does not rely on human sight?
- If automation systems solve a challenge traditionally addressed by pilot qualification or medical fitness through different approaches, are the relevant pilot qualifications or medical fitness no longer relevant? What requirements must these systems meet to be similarly certificated?
- If Operator certification requires specific "stick and rudder" skills for human crewmembers, how does this translate to automation systems for which human control input skills may have no safety benefit?
- What is the need for functional breakdown and splitting out of human and aircraft system knowledge, skill, and decision-making requirements that can be tested and approved?
- How will the diversity and potential rapid evolution of automation, along with the general opacity of complex systems, affect pilot training and qualification?

RECOMMENDATIONS

Increasingly autonomous aircraft systems and flight operations, enabled by emerging technologies, have led to a shifting landscape in which traditional assumptions and practices are being challenged. In this paper, we summarized the shifting landscape and identified four pertinent issues. Based on this work, the recommendations are:

A. Reassess Roles and Responsibilities: Given the evolving aviation landscape with increased automation, it is crucial to re-examine the roles and responsibilities of OEMs, operators, pilots, and other stakeholders. This re-assessment should consider the changing nature of accountability, authority, and responsibility in the context of emerging technologies and autonomous systems.

B. Clearly Define Accountability for Actions Taken by Automation: As automation takes on more safety-critical functions, it is necessary to clarify the accountability for operational decisions made by

https://doi.org/10.52202/075106-0012

automation. While automation may have increased responsibility and authority, accountability must ultimately reside with a human or human-based organizational entity. Procedures should be established to delegate authority and responsibility to automation while system capabilities must be able to ensure oversight, transparency, and configurability to align with the operator's procedures and fulfill operational control obligations.

C. Establish Standards and Guidelines: Standards bodies like ASTM can play a crucial role in developing guidelines that ensure the necessary transparency and oversight by Operators so that they can fulfill their operational control responsibilities. Future and incumbent operators should be part of the standards-making efforts and can assist the standards bodies as they seek to develop guidelines in three recommended areas:

- Standards (performance-based requirements and standard practices) to guide automation system manufacturers so that they ensure transparency and configurability of automation and facilitate the necessary oversight and guidance by Operators.
- Standardizing the boundary between manufacturers and Operators to clarify authority, responsibility, and accountability roles, enabling Operators to maintain operational control and fulfill their obligations.
- Performance-based standards to identify methods for ensuring automation can perform the intended operational functions to acceptable levels of safety.

D. Address Pilot Command Authority: With increasingly autonomous aircraft systems, there is a need to clarify the command authority and responsibility of pilots. In this context, we should carefully consider and accordingly assign where the final decision authority resides, as high levels of automation may impact the pilot's ability to exert command authority. Any changes to pilot final decision authority will likely have an impact on operational control.

E. Assess the Existing Certification Processes: As operational decision execution shifts from a human-centric process to automation, some aspects of safety assurance that are currently part of Operational Safety Assurance will need to blend with aspects of Design Assurance. This would require a careful assessment of current certification, airworthiness, design assurance, and operational approval processes for systems, Operators, and personnel. This could include establishing consensus standards that 1) define additional performance-based standards for intended functions that were previously performed entirely by humans; 2) provide guidance/requirements for continued operational safety and maintenance for increasingly autonomous systems; and 3) redefine qualifications and training, if needed, for operational personnel.

F. Develop a Net Risk Assessment Framework: As the boundary between Operational Safety Assurance and Design Assurance blends together, there is a need for a framework in which total net risk impact and mitigation can be appropriately quantified. Strategic, tactical, and technical risk mitigations need to be accounted for, as do any risks that are being removed or introduced through the use of increased automation.

Moving forward, the aviation community can realize the promise of safe, increasingly autonomous aircraft operations through careful attention to the standards that will provide technical rigor and accountability for operational control.

https://doi.org/10.52202/075106-0012

ASTM INTERNATIONAL HEADQUARTERS

100 BARR HARBOR DRIVE
P.O. BOX C700
WEST CONSHOHOCKEN, PA 19428-2959
USA

TEL +1.610.832.9500
FAX +1.610.832.9555
WEB GO.ASTM.ORG

 https://doi.org/10.52202/075106-0012

SYSTEM SAFETY PROCESS FOR ARTIFICIAL INTELLIGENCE AND MACHINE LEARNING IN UNCREWED SYSTEMS

Wayne E. Devoid,[*] and J. Timmothy Middendorf[†]

Systems and functions augmented with Artificial Intelligence/Machine Learning (AI/ML) promise to offer significant improvements over human operated functions. Assuring the successful and safe integration of AI/ML into everyday life is a challenge. Uncrewed systems being developed with AI/ML components have unique safety concerns for the system developer, product user, general public, and society at large. While robust system and software safety processes exist to identify and mitigate general risk, AI/ML components must be developed and integrated using a specialized and focused safety process. This specialized system safety process is required to characterize, analyze, and mitigate the unique aspects of AI/ML elements. This paper presents an overview of an AI/ML system safety process, how AI/ML system safety augments a robust software safety program, and will highlight unique safety characteristics of a system guided by AI/ML.

INTRODUCTION

Automobiles controlled by an Advanced Driving System (ADS) using unreliable AI/ML object classification models to avoid obstacles could easily result in fatalities or injuries to occupants of the ADS vehicle, occupants of other vehicles, and pedestrians, or property damage. The same can be said for uncrewed aerial systems using on-board AI/ML models to initiate a payload release. While system and software safety processes identify and mitigate general risk, AI/ML systems/components must be developed and integrated using specialized and focused safety processes. These specialized system safety processes are required to characterize, analyze, and mitigate the unique aspects of AI/ML elements.

The System Safety Process for AI/ML that is presented in this paper was developed by leveraging years of extensive background in system safety and software safety to identify specific areas where the specialized AI/ML domain challenges the traditional system and software safety paradigm. Subject matter experts developed specific analyses to identify, assess, and mitigate hazard sources directly attributed to AI/ML components. These analyses are elements of an overall AI/ML System Safety Process that effectively targets safety deficiencies in a system or system of systems hosting AI/ML components.

This paper presents an overview of an AI/ML System Safety Process that focuses on detailed consideration of the design and operational domains defined for the AI/ML system/component and the data used to train and test the AI/ML model.

[*] Senior Aerospace/Research Engineer, A-P-T Research, 4950 Research Drive, Huntsville, AL 35805.
[†] Senior Aerospace Engineer, A-P-T Research, 4950 Research Drive, Huntsville, AL 35805.

https://doi.org/10.52202/075106-0013

BACKGROUND

System Safety is a specialized discipline within System Engineering that is used to achieve an acceptable level of safety risk during all phases of the system's life cycle. Both government regulators and the system's program management set risk requirements and limits that must be met to assure public safety. System Safety professionals supporting the program will lead an integrated system safety program to assure the system meets specified safety risk levels.

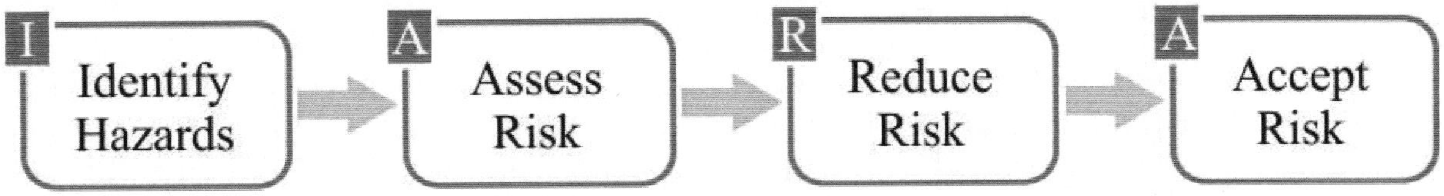

Figure 1. IARA: Iterative system safety process for mitigating risk

Figure 1 presents a general System Safety model that includes the four key System Safety elements: Identify Hazards, Assess Risk, Reduce Risk, and Accept Risk, or IARA. APT developed and successfully applies the IARA model in government and commercial programs to achieve acceptable levels of safety.

If the system under study contains software that controls safety critical functions, it is common for a program to define software safety efforts in a Software System Safety Program Plan. Software safety is, therefore, a critical subset of a robust system safety program. Hazard sources and mitigations identified employing the IARA model standard as part of a software safety process are then fed back to the system safety program for disposition.

Although AI/ML components interface to the overall system and are embedded in a system's software, it is a mis-held belief that AI/ML components can simply be addressed by the Software Safety Program. There are specific nuances AI/ML components present that software safety is not capable of capturing and must be addressed with a specialized system safety process. In addition, software engineers who are assigned to perform software safety may not be well versed in examining the validity of data used to train AI/ML models. Statisticians and data scientists must be experienced with analyzing key characteristics such as data domain and interfaces to the overall system. This necessitates a closer examination by safety engineers trained in data analytics of AI/ML components and assures any nondeterministic* outputs do not contribute to nor generate a hazard.

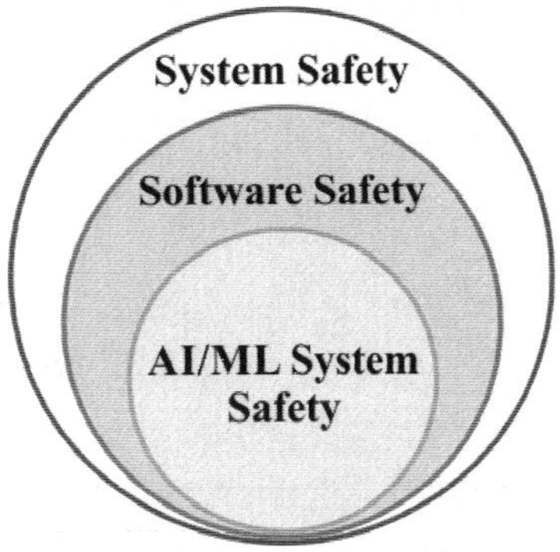

Figure 2. AI/ML system safety is a critical subset of a robust software safety program.

* A nondeterministic algorithm is an algorithm that, even for the same input, can return different results on different runs. This makes validating acceptable results difficult since there is no single consistent answer.

https://doi.org/10.52202/075106-0013

The Venn diagram shown in Figure 2 illustrates how AI/ML system safety fits into the overall system safety program architecture. It is important to note that while the last function of the IARA system safety model is to Accept Risk, the final action of the software safety and AI/ML system safety process is to verify and validate that mitigation measures reduce risk. Any residual risk after all mitigation techniques have been properly applied must flow up to the system safety process to be reassessed or accepted.

UNIQUE SAFETY CHARACTERISTICS WITHIN AI/ML

AI is a software element or model that, for a fixed set of inputs, can produce nondeterministic outputs, approximations, or multiple valid solutions. These outputs, produced by the AI component, are modeled predictions that the system will use to initiate an action or function. ML is an application of AI that entails the process of training the AI model with a known dataset of inputs and outputs to "learn," solve problems, and produce an output to narrowly defined problems with a limited scope. Therefore, when a fielded system is used in a real-world application, the trained AI model should make valid predictions within the operating domain using previously unseen inputs. Non-AI/ML software routines have definitive input and deterministic output with clear boundaries defined to certify the system within. The uncertainty in AI/ML component output makes current software safety processes ineffective in capturing hazard sources.

AI/ML components cannot use traditional safety risk assessment procedures that consider the deterministic severity and probability of a mishap to produce a risk rating. This is because the probability of a misclassification is challenging to determine with an AI/ML model and a misclassification by itself is a hazard source that may contribute to the severity of yet another hazard. New processes are required to quantify risk as a critical component of AI/ML system safety.

A unique characteristic of AI/ML that warrants specialized analysis is the dataset used to train and test the model. The data must be accurate, representative of the operating domain, and in sufficient quantity to adequately teach the AI/ML model. The dataset must have high fidelity, accuracy, and robustness within the operating domain. Datasets must also be examined for outliers that manifest themselves as data anomalies. The AI/ML component must also be examined for its ability to safely process these data anomalies. The quality of the dataset used directly corresponds to how well the AI/ML component performs.

In addition to examining the datasets for accuracy, the scope of the operating domain must be clearly established. It is imperative that the data used to train the AI/ML model covers the defined scope of the operating domain and must represent the operating domain established for the product/system functions. Any diversion between the operating domain and the dataset can cause significant anomalies in the AI/ML model results and therefore introduce safety risk.

GENERAL AI/ML SYSTEM SAFETY PROCESS

The APT AI/ML System Safety Process is designed to provide specialized analyses necessary to identify, assess, and mitigate unique characteristics and risks associated with AI/ML implementation. Figure 3 presents the APT AI/ML System Safety Process.

At the program level, it is imperative to determine the operating domain of the system, secure a robust operating domain data pipeline for training and test data, and ensure the proposed AI/ML model is sufficiently expressive. Focusing on identifying the operating domain and data source at the outset of the program scopes the AI/ML implementation and helps focus AI/ML model developers within the System Safety Program.

AI/ML components are typically developed based on either a ***Data First Framework*** or a ***Requirements First Framework***. In a Data First Framework, actionable decisions are made based on findings in a large set of pre-collected data. Data First AI/ML models are developed in the intelligence, business, logistics, and maintenance domains. In a Requirements First Framework, a requirement is identified before data are collected to train and test an AI/ML component. Requirements First AI/ML models are well suited for military weapons systems. While the AI/ML system safety process is not employed differently for either framework, it is instructive for safety personnel to know whether the genesis of an AI/ML component is data or requirements driven to help focus the subsequent analyses and risk management.

Starting with the baseline AI/ML component framework (Data First or Requirements First), a Level of Rigor (LoR) analysis is then performed to scope the AI/ML system safety process. LoR describes the depth and breadth of analyses and verification activities necessary to provide sufficient confidence that the AI/ML component and its safety-related functions will perform as required. Following an LoR determination, the risk management cycle can commence with an understanding of how detailed analyses must be to properly manage risk.

https://doi.org/10.52202/075106-0013

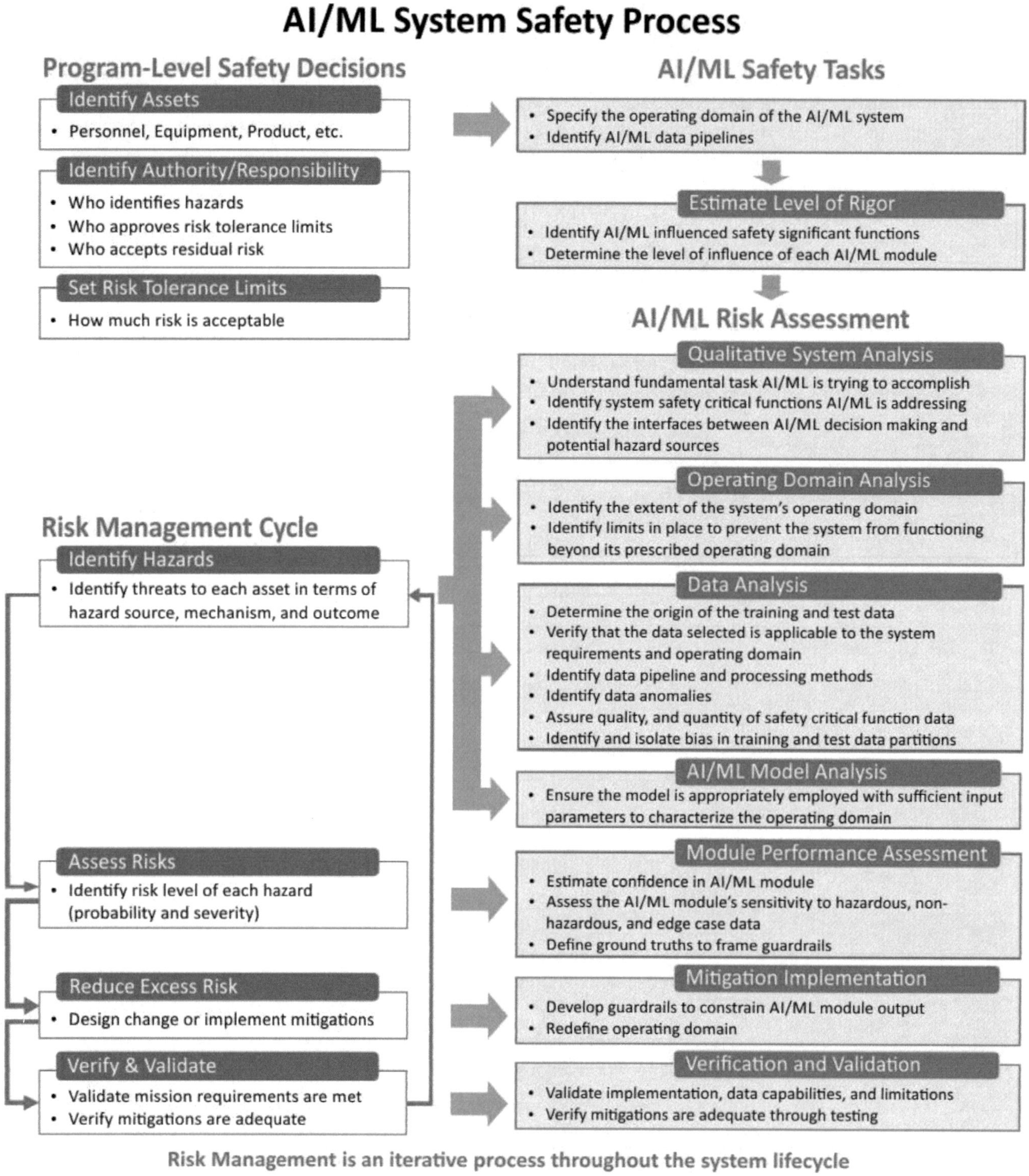

Figure 3. AI/ML specific tasking performed as part of a risk management process.

https://doi.org/10.52202/075106-0013

Identify Hazards

The first step in the risk management cycle is to identify hazard sources posed by the AI/ML component. These hazard sources, if not mitigated at the AI/ML component level, will be missed at the software safety level and may induce additional hazard mechanisms and outcomes. To help identify AI/ML component hazard sources, APT employs four main analyses: Data Analysis, Qualitative System Analysis, AI/ML Model Analysis, and Operating Domain Analysis. These analyses target the four critical components of AI/ML component development. Any deficiencies in an AI/ML model, dataset used, or error in defining an operating domain will introduce hazard sources.

Qualitative System Analysis – This analysis examines the overall integration of an AI/ML component into its parent system. The analysis identifies what the AI/ML component is designed to accomplish, how results from the AI/ML component influences system safety critical functions and identifies interfaces between AI/ML component decision making and potential hazard sources.

Operating Domain Analysis – This critical analysis identifies the extent of the system's operating domain. Significant hazard sources may be introduced by an AI/ML component that is called to operate outside the operating domain for which it has been trained. This analysis provides verification that the training and test data selected are applicable to the system requirements and operating domain. Ultimately, it is the goal of this analysis to verify that guardrail limits are in place to prevent the system from functioning beyond its prescribed operating domain.

Data Analysis – This analysis determines the genesis, quality, and quantity of the training and test data and verifies that the data, datasets, and pipelines selected are applicable to system requirements. Examining the process used to smooth the data will identify issues with dimensionality, causation, and data mining. Finally, there is potential to identify data anomalies that may have introduced errors in AI/ML model training that result in hazards.

AI/ML Model Analysis - This examination of the AI/ML model will provide insight as to the robustness of the results. A properly designed model will neither have too few layers and nodes (underfitting) nor have too many layers and nodes (overfitting). The AI/ML model will also be probed to see how it handles both hazardous and non-hazardous input.

Hazard sources identified and collected as part of these four analyses are documented and tracked as part of the software safety or program level system safety efforts. The benefit is that the entire system safety program is now capable of identifying and mitigating the hazard source that leads to a safety critical mechanism and outcome from all sources within the system. The depth and breadth of each analysis is determined by the LoR analysis and will contribute to the cost and manpower requirements assessment necessary to complete the AI/ML system safety process.

Assess Risk

It is well understood in software safety that software code error, by itself, is not considered a hazard. Rather, the software will return a result or command that instantiates a hazard mechanism and ultimately a hazardous outcome. To complicate matters, it is difficult to quantitatively assess risk due to software errors since the probability of occurrence is hard to quantify.

AI/ML components are even more troublesome in that their results are nondeterministic. The key is to understand the range of output and determine if all output within that range is properly bounded. If results from an AI/ML component are not properly examined downstream, there is potential for considerable risk.

Module Performance Assessment – This analysis identifies interface and dataset cases where erroneous inputs or outputs to and from an AI/ML component could introduce entire system-level or subsystem-level errors. The analysis will identify those interface errors that could cause a

https://doi.org/10.52202/075106-0013

system/sub-system mishap. A Module Performance Assessment will also define ground truths to frame guardrail target functions for future risk reduction options.

In some instances, the program will designate a desired success rate of the AI/ML component. Results of this testing can sometimes be factored into the risk assessment as a probability of occurrence; however, it is important to note that even a slight deviation in operating domain can cause significant deviations in the success rate of an AI/ML component.

It is imperative that software safety engineers collaborate with AI/ML safety engineers when performing a Module Performance Assessment. AI/ML safety engineers can provide useful data on what the potential mis-characterization rate of an AI/ML component may be, and software safety engineers can determine if the results of an AI/ML component will contribute to a safety critical hazard. Hazard sources generated by the AI/ML component that contribute to significant risk are identified at this stage. Those hazards are forwarded to the risk reduction phase of the risk management cycle.

Reduce Excess Risk

Risk reduction solutions for AI/ML components will generally follow two approaches, dataset updates and guardrails. Data updates will correct issues identified with training datasets that may not adequately represent the operating domain or include all expected hazards. Guardrails are usually software controls that prevent the AI/ML component from providing an output that is outside specified parameters or deviates from an established ground truth. A risk reduction solution may follow several approaches to either eliminate or mitigate an AI/ML safety risk.

Verify and Validate Implementation

The last phase of the risk management cycle is to verify and validate that the risk reduction techniques and implementation are applied properly. AI/ML system safety engineers will validate mitigation implementation, data capabilities, and limitations of the AI/ML component and then verify that the mitigations applied are adequate through testing.

Once verification and validation have been completed, the AI/ML system safety engineers will circle back to the hazard identification phase of the risk management cycle and ensure the risk posed by the AI/ML component is acceptable.

CONCLUSION

This AI/ML System Safety Process provides a robust, systematic approach to manage system safety risks for an AI/ML based product that dovetails with current industry accepted System Safety Program standards. The AI/ML system safety process can be tailored for product-specific applications and can be implemented to manage AI/ML safety risks at the beginning of product development, later in product development effort, and after product deployment to encompass the entire system lifecycle. This AI/ML System Safety Process bridges the gap between AI/ML safety and the system safety and software system safety domains, ensuring a robust yet cost-effective AI/ML system safety process to identify, assess, reduce, and accept system risk.

An Introduction to The Economics of Flying Beyond Visual Line of Sight (BVLOS)

Darryl C. Jenkins

Drone Economics, LLC.

Author Note

I have no conflicts of interest to disclose.

Correspondence concerning this article should be addressed to: Darryl Jenkins at airjenkins@aol.com

https://doi.org/10.52202/075106-0014

Abstract

This paper explores the necessary conditions for Beyond Visual Line of Sight (BVLOS) drone operations and the concept of a single pilot controlling multiple drones simultaneously. These two factors are crucial for many segments of the commercial drone industry to achieve profitability and long-term sustainability. While the paper introduces these factors, it also acknowledges that the sufficient conditions are entirely dependent on regulatory frameworks. However, the rapid advancement of technology in this industry makes the realization of the discussed concepts highly probable.

For BVLOS operations to become profitable, several factors must align. The paper delves into these factors, highlighting the evolving role of pilots over the next decade. Although there will still be a pilot in command, their responsibilities will undergo a drastic transformation. The economics of BVLOS revolve around cost reduction and revenue maximization. Rather than training more pilots, the key to success lies in leveraging sophisticated software that minimizes the need for human intervention.

Drones have already begun to revolutionize industries by enabling closer, more cost-effective, and automated aerial inspection and data collection. However, most current commercial and public safety drone applications are limited to flying within the visual line of sight (VLOS) of the pilot or ground observers. This constraint significantly hinders the full economic potential of drone services. The paper presents and defends the fundamental axiom that the economics of commercial drones begins with BVLOS operations. Furthermore, it introduces and discusses nine essential laws governing the economics of commercial drone BVLOS operations.

Key words: BVLOS, economies of scale and scope, One to N flying, fundamental economic laws

Introduction

Expanding drones into routine BVLOS operations promises immense economic upside across major industries. Technological innovations now allow specially equipped drones paired with advanced software to fly more autonomously over far longer ranges. Missions can span hundreds of miles collecting imagery, sensor data, and carrying cargo across restricted, hazardous, or previously inaccessible terrain. The economics of delivering these solutions scales favorably.

As regulators continue opening airspace access and infrastructure for managing BVLOS improves, entirely new drone use cases become financially viable. Vast linear infrastructure like pipelines, railroads, power lines and bridges snake thousands of miles but have been enormously expensive to inspect thoroughly until now. Current VLOS operations within these market segments are not sufficient to scale and significantly reduce costs. Remote area delivery services, enhanced border security and police operations, detailed land management mapping efforts, and sophisticated commercial data analytics stacks will finally take off economically speaking through advanced BVLOS activities.

https://doi.org/10.52202/075106-0014

Realizing this full potential depends on the drone industry maturing to meet escalating BVLOS operational demands at affordable pricing. Pilots need upgraded training and credentials for spatial orientation. Command and control systems must evolve for managing extensive fleets efficiently across broad stretches of sky. Sensors, battery technology, weatherized airframes and night operation capabilities require enhancement as well.

The economics clearly suggest the upside for increasingly autonomous drones traversing hundreds of miles in unfettered flight is monumental. As regulations and maturing technology lower barriers, expect BVLOS drone adoption to achieve escape velocity growth. The following section explores the financial factors and use cases driving this expansion in detail. While the FAA is doing more with waivers, the industry is still a long way from profitability, and more is needed. As wide scale BVLOS operations are currently unavailable, this paper is future looking.

Disruptive Technology

Disruptive technologies are innovations that significantly alter the way industries or markets operate, often by displacing established market leaders and transforming the way people live and work. These technologies typically start off as inferior to existing solutions in terms of performance or features, but they offer unique benefits such as simplicity, convenience, or lower costs. As disruptive technologies mature and gain market share, they can have a profound impact on the cost structures of industries and the prices paid by consumers.

One of the key characteristics of disruptive technologies is their ability to drive down costs over time. Initially, these technologies may be more expensive than established solutions due to the research and development costs associated with bringing them to market. However, as adoption increases and economies of scale come into play, the costs of production and implementation tend to decrease significantly. This cost reduction can be attributed to factors such as improved manufacturing processes, increased competition, and the spread of knowledge and best practices.

As costs fall, disruptive technologies become more accessible to a wider range of users, further accelerating their adoption and impact. This phenomenon can create a virtuous cycle, where increased demand leads to even greater cost reductions, making the technology even more attractive to potential adopters. Over time, the cost advantages of disruptive technologies can become so significant that they completely reshape the economic landscape of industries, forcing established players to adapt or risk being left behind.

https://doi.org/10.52202/075106-0014

Economies of Scale and Scope

This section discusses the economic concepts of scale and scope and are vital to understanding all the concepts in the chapter. In assessing the financial viability underpinning specialized commercial drone business models, two pivotal intertwined economic concepts stand paramount influencing potential sustainability – economies of scale and economies of scope. Together these correlated forces describe how per-unit costs behave as production volume and operational breadth expand, with immense implications determining ultimate profitability prospects separating successful ventures from unsuccessful participants across this intricate emerging industry.

Economies of scale refer specifically to the economic advantages where per-unit production costs for a firm decline as total output grows larger thanks to that enlarged scale. This reflects efficiencies around fixed initial research, platform development and infrastructure investments getting spread across greater numbers of saleable units - driving down breakeven pricing. Basically, aggregated volume production allows amortizing upfront fixed expenses over more items – meaning firms that can scale up market share save money through expansion dividends impossible for smaller firms lacking equivalent output volumes over which to distribute incurred costs (see Figure 1).

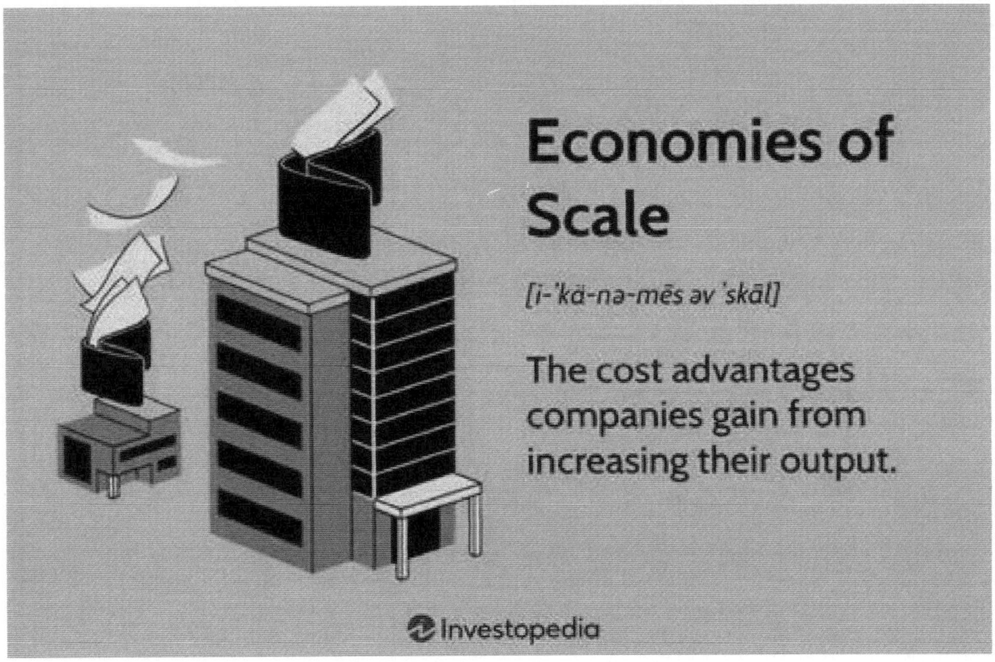

https://doi.org/10.52202/075106-0014

Figure 1: Economies of Scale [1]

A drone manufacturer able to sell 1,000 units of a specialized camera payload for example can distribute associated non-recurring engineering, certification, factory calibration, and production start-up expenses across those 1,000 payloads - enabling competitive market pricing and profits unworkable selling just 10 payloads which lack volume scale sufficient to recoup those fixed costs. Service companies enjoy parallel benefits leveraging fleets, analytics software, and insurance over large numbers of mission hours. Scale drives lower nominal costs.

While closely related, economies of scope distinctly describe economic advantages where a company produces/offers a wider range of related products or services thereby lowering individual item costs through operational synergies. This means cost efficiencies materialize from "scope" diversification itself as increased breadth allows leveraging common components, expertise, supply chains and platforms across an array of applications – essentially cross-subsidizing through resource fungibility where focused specialty firms lack flexibility spreading investments broadly (see Figure 2).

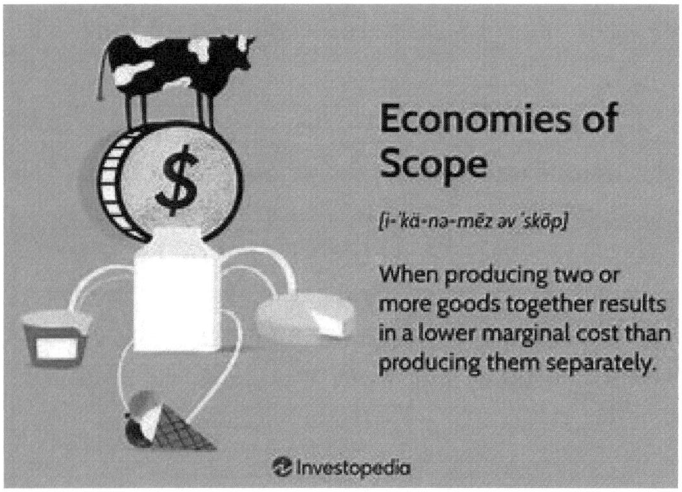

Figure 2:Economies of Scope [2]

For drone service operators this proves equally crucial as specialized aircraft, sensors, data links and other hardware can get redeployed on diverse customer missions from agriculture to energy to public safety once acquired - defraying individual application costs impossible for niche suppliers concentrating on targeted singular industries alone. Scope synergies multiply portfolio value.

In both scale and scope contexts, commercial drone economics prove uniquely sensitive to expansion factors driving down or raising per-item/per-hour costs because specialized vertical equipment, inspection expertise, and advanced flight logistics prohibit casual commoditization. Only operators leveraging scale production volumes or scope application synergies sustain

---

[1] Economies of Scale Types - Search Images (bing.com)

[2] Economies of Scope Example - Search Images (bing.com)

against competition absent those flexibility advantages innate to fragmentation. In financial projections analysts must consider implications from potential fleet scaling and transferrable reusability determining ultimate competitiveness. Gearing up output volume or operational breadth offers sustainability - staying small invites revenue erosion absent serious differentiation.

Fundamental Axiom and the Laws of Drone BVLOS Economics

The fundamental axiom of drone economics states drone economics on a large scale, begins with BVLOS operations. We believe this axiom applies in general, but not specific terms.

The Fundamental Axion of BVLOS Drone Economics:

Drone Economics begins with BVLOS.

From an economist's perspective, the drone industry begins with the advent of Beyond Visual Line of Sight (BVLOS) operations. While Visual Line of Sight (VLOS) drones have proven useful for a variety of applications, such as aerial photography and short-range inspections, they are inherently limited in terms of their range, autonomy, and scalability. In contrast, BVLOS drones, which can operate independently over long distances without direct human supervision, offer the potential for significant economies of scale and scope that can transform the economics of the drone industry and unlock a wide range of new use cases and business models.

Several factors contribute to the economies of scale in BVLOS drone operations. First, the ability to operate drones over longer distances and for extended periods allows operators to serve a larger geographic area and customer base with a single drone, reducing the need for multiple drones and operators. Second, the increased autonomy of BVLOS drones enables them to execute missions with minimal human intervention, reducing labor costs and increasing operational efficiency. Finally, the standardization and automation of BVLOS drone operations allow for more consistent and predictable mission outcomes, reducing the risk of errors and increasing the overall reliability of drone services.

The economies of scale achievable through BVLOS operations have the potential to fundamentally reshape the drone industry and its role in the broader economy. By reducing the cost of drone services and expanding their reach and capabilities, BVLOS operations can open new markets and applications that were previously uneconomical or impractical. This could include large-scale

https://doi.org/10.52202/075106-0014

infrastructure inspections, long-range cargo delivery, precision agriculture, and remote sensing for environmental monitoring and resource management.

Moreover, the economic benefits of BVLOS drone operations extend beyond the direct cost savings and revenue opportunities for drone operators. The widespread adoption of BVLOS drones has the potential to drive significant productivity gains and cost efficiencies across a wide range of industries, from transportation and logistics to construction and energy. By providing faster, cheaper, and more flexible solutions for data collection, asset monitoring, and service delivery, BVLOS drones can help businesses optimize their operations, reduce their costs, and improve their competitiveness.

From an economist's perspective, the drone industry begins with the advent of BVLOS operations, which offer the potential for significant economies of scale that can transform the economics of the drone industry and unlock a wide range of new use cases and business models. By reducing costs, expanding capabilities, and driving productivity gains across multiple sectors, BVLOS drones have the potential to become a major catalyst for economic growth and innovation in the years ahead.

There are at least nine fundamental economic drivers underpinning the viability and growth of BVLOS drone operations. These include:

1. Drones designed for Beyond Visual Line of Sight (BVLOS) operations will be significantly more expensive than traditional Visual Line of Sight (VLOS) quadcopters.
2. The role of AI software will take on increasing importance.
3. The core value of BVLOS drones derives from enhanced data capture (imagery, sensor data) and logistics (transport, delivery) efficiency.
4. For BVLOS drone operations focused on non-logistics use cases, the greatest value lies not directly in the aerial data capture itself, but rather the sophisticated analysis and insights extracted from the torrents of imagery, topological and other sensor data accumulated across expansive flights.
5. The key value proposition is the post-processing algorithms and analytical techniques applied to rich drone-gathered datasets that ultimately enable superior decision-making, risk detection and predictive modeling across industries from agriculture to insurance.
6. Certain market opportunities are only financially viable given the expanded operational radii enabled by BVLOS. Delivery, surveying, security, and agricultural use cases rely on beyond line-of-sight ranges making otherwise unprofitable jobs feasible.
7. Shifting one pilot from operating a single drone to coordinating many drones reduces per-unit labor costs while enabling each pilot to generate higher total revenue.
8. Increased altitude expands visible range significantly, enabling wider area data capture and reduced need for imaging overlap. This drives higher resolution insights paired against operational costs and hardware capabilities.
9. As flight scales change so do business models and the end customer. BVLOS markets dynamics changes substantially.

BVLOS Economic Law #1: Drones designed for Beyond Visual Line of Sight (BVLOS) operations will be significantly more expensive than traditional Visual Line of Sight (VLOS) quadcopters.

We start out with this as it is so important. To justify their higher cost, BVLOS drones must be built with more robust and durable components, enabling them to withstand the rigors of extended missions and harsher environmental conditions. Furthermore, to achieve economic viability, these drones should be designed with a longer lifespan in mind, allowing their initial costs to be amortized over a period of seven to ten years, which is considerably longer than the typical lifespan of consumer-grade VLOS quadcopters.

One of the issues with VLOS, is there are no operational economies of scale. No matter how often a task is performed, it must be short lived due to battery constraints. Because of this, costs remain constant over time and destroys the disruptive nature of the service. While there are many VLOS markets, and these will have their own life, many will also be lost to BVLOS due to economies of scale.

A hardened drone is one with significant redundancy and a platform made to withstand various weather conditions. BVLOS drones represent a significant advancement in drone technology, enabling longer-range missions and more autonomous operations. However, these capabilities come at a higher cost compared to traditional Visual Line of Sight (VLOS) drones, such as consumer-grade quadcopters. The increased costs associated with BVLOS drones can be attributed to several factors, including more advanced sensors and onboard processing systems, ruggedized designs to withstand harsh environmental conditions, and the need for reliable long-range communication links.

Given the substantial upfront costs of BVLOS drones, it is essential to consider their economic viability over an extended period. To justify the investment, these drones must be designed and built to last, with a target lifespan of seven to ten years. This longevity allows the initial acquisition costs to be amortized over a more extended period, making the investment more financially feasible for organizations and businesses seeking to leverage the benefits of BVLOS operations.

To achieve this longer lifespan, BVLOS drones must be engineered with durability and maintainability in mind. This involves using high-quality, robust components that can withstand the wear and tear of regular use, as well as designing the drones in a manner that facilitates easy maintenance and repairs. Additionally, the software and firmware of these drones should be updatable and adaptable to keep pace with evolving technologies and changing mission requirements over their extended service life.

https://doi.org/10.52202/075106-0014

In addition to the drones themselves, a robust command and control center is crucial for the successful operation of BVLOS drones over their extended lifespan. The command-and-control center serves as the central hub for monitoring, managing, and directing the drones during their missions. It must be equipped with reliable, high-bandwidth communication systems to maintain constant contact with the drones, as well as advanced software for flight planning, real-time data processing, and decision support. The command-and-control center should also be designed with redundancy and failsafe measures in mind to ensure continuity of operations even in the face of technical issues or external disruptions. Investing in a well-designed, durable command and control center is essential for maximizing the long-term economic benefits of BVLOS drone operations.

By building BVLOS drones with a target lifespan of at least seven to ten years, manufacturers and operators can help ensure that the upfront costs are effectively amortized over time. This approach not only makes the adoption of BVLOS drones more financially attractive but also encourages the development of more sustainable and long-lasting drone solutions. As the use of BVLOS drones continues to expand across various industries, from agriculture and infrastructure inspection to public safety and logistics, the importance of designing these drones for long-term economic viability will only continue to grow (see Figure 3).

Initial Cost	**Depreciation in Years**	**Annual Cost**
$250,000	3	$83.333
$250,000	4	$62,500
$250,000	5	$50,000
$250,000	6	$41,667
$250,000	7	$35,714
$250,000	8	$31,250
$250,000	9	$27,778
$250,000	10	$25,000
$250,000	11	$22,727
$250,000	12	$20,833

Figure 3: Annual Cost of Capital Expenditures

One of the key strategies for reducing the unit costs of Beyond Visual Line of Sight (BVLOS) drone operations is to maximize the utilization rate of the drone fleet. A high utilization rate refers to the proportion of time that a drone is actively engaged in revenue-generating missions, as opposed to being idle or undergoing maintenance. By ensuring that BVLOS drones are consistently deployed on a high volume of missions, operators can spread the fixed costs of drone ownership, such as capital expenditure and overhead, across a larger number of flight hours, thereby reducing the cost per mission.

There are several ways in which BVLOS drone operators can achieve a high utilization rate. One approach is to focus on serving a diverse range of customers and applications, rather than relying on a single use case or industry vertical. By diversifying their customer base, operators can minimize the impact of seasonal or cyclical fluctuations in demand and maintain a more consistent stream of missions throughout the year. This can involve actively seeking out new market opportunities, developing customized solutions for different industries, and building a flexible and adaptable drone fleet that can accommodate a wide range of mission requirements.

Another strategy for increasing utilization is to optimize the scheduling and routing of BVLOS drone missions. By carefully planning and coordinating the deployment of drones, operators can minimize downtime and ensure that each drone is efficiently utilized. This may involve the use of advanced scheduling software and algorithms that can dynamically assign missions to available drones based on factors such as location, battery life, and maintenance status. By streamlining the mission planning and execution process, operators can reduce the time and cost associated with each flight and maximize the number of missions completed by each drone.

In addition to optimizing the utilization of individual drones, BVLOS operators can also achieve cost efficiencies by leveraging economies of scale. As the size of the drone fleet grows, operators can spread fixed costs such as infrastructure, software development, and training across a larger number of units. This can lead to lower unit costs and improved profitability, particularly for operators that can achieve a high level of standardization and automation across their fleet.

However, it is important to note that achieving a high utilization rate for BVLOS drones also requires significant investments in infrastructure, maintenance, and support services. Drones must be regularly inspected, serviced, and updated to ensure optimal performance and reliability, and operators must have access to a robust network of charging stations, maintenance facilities, and spare parts. The costs associated with these support services can be substantial, and operators must carefully balance the benefits of increased utilization against the ongoing expenses of fleet management.

Achieving a high utilization rate is a key strategy for reducing the unit costs of BVLOS drone operations. By maximizing the proportion of time that drones are actively engaged in revenue-generating missions, operators can spread fixed costs across a larger number of flight hours and improve their overall profitability. This requires a focus on diversifying customer bases, optimizing mission scheduling and routing, and leveraging economies of scale. However, operators must also carefully balance the pursuit of high utilization with the need for robust maintenance, support services, and compliance with safety and regulatory requirements. By striking the right balance and continuously optimizing their operations, BVLOS drone operators can unlock significant cost efficiencies and drive the widespread adoption of this transformative technology (see Figure 4).

Annual Cost	Annual Hours Flown	Hourly Cost
$25,000	350	$71.43
$25,000	450	$55.56
$25,000	550	$45.45
$25,000	650	$38.46
$25,000	750	$33.33
$25,000	850	$29.41
$25,000	960	$26.32
$25,000	1050	$23.81
$25,000	1150	$21.74

https://doi.org/10.52202/075106-0014

$25,000	1250	$20.00

Figure 4: Hourly Unit Costs

BVLOS Economic Law #2: The role of AI software will take on increasing importance and will ultimately be more important than the platform.

The platform is merely a delivery mechanism. The platform is only partially a function of where the value is created. The role of artificial intelligence (AI) software in (BVLOS) drone operations is becoming increasingly critical, and its significance is expected to surpass that of the drone platform itself soon. As BVLOS missions become more complex and autonomous, the AI software responsible for guiding the drones, processing data, and making decisions will play a central role in determining the success and efficiency of these operations.

One of the key reasons for the growing importance of AI software in BVLOS drones is the need for real-time decision-making in dynamic environments. During BVLOS missions, drones must be able to navigate autonomously, adapt to changing weather conditions, avoid obstacles, and respond to unexpected situations without human intervention. Advanced AI algorithms, such as machine learning and computer vision, enable drones to process vast amounts of sensor data, identify patterns, and make split-second decisions to ensure safe and effective operation.

Moreover, AI software plays a crucial role in enabling BVLOS drones to perform complex tasks and gather valuable insights. For example, in precision agriculture, AI algorithms can analyze high-resolution imagery captured by drones to identify crop health issues, optimize irrigation and fertilization strategies, and predict yields. In infrastructure inspection, AI software can automatically detect and classify defects, such as cracks or corrosion, in bridges, pipelines, or wind turbines, saving time and reducing the risk of human error.

As the capabilities of AI software continue to advance, it is expected to become the primary differentiator in BVLOS drone solutions. While the physical drone platform remains important for ensuring reliable and durable operation, the AI software will be the key driver of mission performance, data quality, and overall value generation. This shift towards software-centric solutions will likely lead to a greater emphasis on developing and refining AI algorithms specifically tailored for BVLOS drone applications.

The growing importance of AI software in BVLOS drones also presents new challenges and considerations. Ensuring the reliability, security, and transparency of AI algorithms will be critical to building trust and acceptance among regulators, operators, and the public. Developers will need to prioritize robust testing and validation processes to ensure that the AI software can handle a wide range of scenarios and edge cases, while also being resilient to potential adversarial attacks or data manipulation attempts.

Furthermore, the increasing reliance on AI software in BVLOS drones will require a new set of skills and expertise within the drone industry. Operators and maintenance personnel will need to be trained not only in the physical aspects of drone systems but also in understanding and troubleshooting AI software components. This will likely drive demand for interdisciplinary talent, combining knowledge of aeronautics, robotics, computer science, and data analysis.

The role of AI software in BVLOS drone operations is poised to become more important than the drone platform itself. As BVLOS missions become more autonomous and data-driven, the performance and capabilities of the AI software will largely determine the overall effectiveness and value of these solutions. While this shift presents new challenges and requirements, it also opens up exciting opportunities for innovation and growth in the drone industry. By prioritizing the development and integration of cutting-edge AI software, BVLOS drone manufacturers and operators can position themselves at the forefront of this transformative technology trend.

BVLOS Economic Law #3: Value Creation

The drone industry is entering a transformative era propelled by advancements across two frontiers - enhanced aerial imagery through better pixel capabilities and the ability to rapidly deliver products to consumers' doorsteps. Drones today are equipped with high-resolution cameras and sensors that can capture strikingly detailed visual data from the skies, analyzing everything from crop health to the progression of construction projects through sophisticated pixels. However, the practical use cases of drones extend beyond just data gathering. Advanced logistics functionalities like longer flight times, heavier payload capacities, and integrated mapping/GPS systems now allow drones to facilitate timely last-mile delivery operations. Whether it's a retail giant like Walmart transporting a new pair of shoes or a medical warehouse delivering urgently needed pharmaceuticals, drones integrated with inventory and logistics networks promise delivery speeds and efficiencies hitherto unheard of without dependence on roads. While pixels have dramatically expanded the reconnaissance and analytical abilities of drones, rapid logistics promises to unleash their disruptive potential in sectors relying on the movement of goods. With both facets maturing in tandem, drones are transitioning from visual tools to becoming integral components of automated, aerial delivery infrastructure.

BVLOS drones has unleashed promising new value creation opportunities across various industries. Underpinning this potential is the ability of BVLOS technology to enormously expand data capture and logistics capabilities compared to existing methods. From agriculture to insurance to commerce, drones promise leaps in informational and operational efficiency even as technological and regulatory vehicles for their capabilities are refined. The key to realize is that while drones themselves represent a fascinating new technology, the true value comes from their role as enabling platforms. Specifically, these remote piloted and autonomous aircraft vastly expand the timely accessibility of hard-to-reach locations for data gathering or delivery activities. By offering a means to access places people or ground vehicles cannot reach, they provide a window into minimizing uncertainty and expanding operational reach in fields ranging from crop health analysis to search and rescue to transport. The value is not necessarily in the drones themselves, but in the otherwise unavailable vantages and connections they facilitate.

On the information side, this means massively expanding both the volume and geographical spread of image and sensor datasets firms can leverage to minimize uncertainty. For example,

https://doi.org/10.52202/075106-0014

insurers and underwriters are exploring how small but expansive BVLOS drone fleets might enable faster, higher resolution assessments of disaster-stricken properties remotely. Not only could this speed claims processing for clients by weeks, it also cuts claims processor labor costs. What one analyst could slowly examine on foot now gains multipliers in speed and scope thanks to airborne eyes, with only the cost and safety assurance of a drone operator at the helm. This better mitigates financial uncertainties for insurers while improving disaster responsiveness for policyholders (see Figure 5).

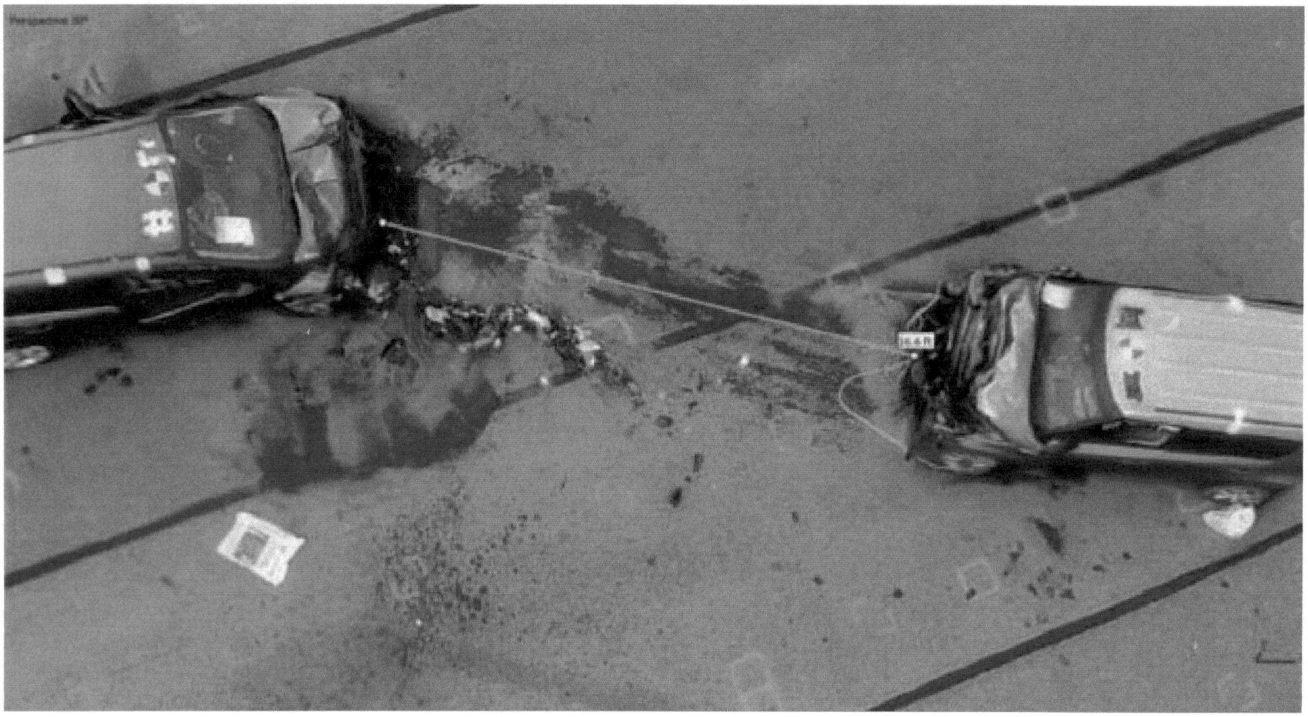

Figure 5: Drone inspecting a car accident[3]

Similarly, industries like mining, construction and agriculture lean on drone observation for functions from site engineering planning to crop pest identification. As drone cameras cover more ground at higher resolutions than inspectors traversing sites manually, both topographical and nanoscopic insights of sites are enhanced. This adds value through planning efficiencies, but also improved yields as threats to output are caught and contained earlier based on automated drone sweeps not possible previously. In vast field crop applications, where complete manual inspection is implausible, this enables entirely new preventative measures enhancing harvest reliability and abundance. Expanding pilots' visual range and frequency of these operations amplifies advantages.

On the logistical side, BVLOS introduces entirely new transportation connectivity options benefitting both commerce and society. Delivering everything from life-saving medical supplies to consumer packages in hard-to-reach locales enables markets otherwise lacking accessibility. Just as cell phones introduced communications where landline infrastructure could never connect profitably, drones now fill operational gaps. Last mile delivery needs from urban areas to rural

[3] drones inspecting a car accident - Search Images (bing.com)

communities can gain flexibility from air routes ranging beyond roadways mired in traffic. This expands firms' addressable market reach while offering faster fulfilment demanded by time-conscious shoppers. Even public entities benefit, with medical drone delivery promises for enhanced rural access to defibrillators, vaccines, or prescription refills at lifesaving speed.

While public concerns around factors like privacy, safety and nuisance shall persist around proliferating high-flying drone highways, the expansive latent demand for their services speaks to their value creative potential. Realizing this potential relies on steady advancement within still nascent drone technologies themselves and the air traffic coordination frameworks needed to enable enterprise-scale BVLOS adoption. Battery life constraints, durability limitations, software reliability and appropriate automation enabling reduction or elimination of constant pilot oversight during extended beyond line-of-sight journeys all pose surmountable but nontrivial scaling challenges in coming years. Yet as solutions emerge, so shall confidence in embracing drones among both private and public sector leaders needed to unleash their dormant economic contributions waiting overhead.

BVLOS Economic Law #4: Data capture at unprecedented scales

BVLOS drone operations are unlocking new possibilities across many industries by enabling aerial data capture at unprecedented scales. However, for non-logistics use cases, the true value lies less in the raw data collection itself and more in the sophisticated analysis and insights that can be extracted from the rich sensory inputs accumulated on these expansive flights.

In sectors like agriculture, infrastructure inspection, mining, and environmental monitoring, BVLOS drones equipped with high-resolution cameras, thermal sensors, hyperspectral imagers, LiDAR scanners, and other payload instruments can scan very large areas - hundreds or thousands of acres - in a single flight. This allows the rapid aggregation of immense datasets consisting of high-density photographic images, intricate topological maps, multispectral readings, heat signatures, and more. But without analytical techniques to process and translate this torrent of information into actionable intelligence, much of the utility is squandered (See Figure 6).

https://doi.org/10.52202/075106-0014

Figure 6: Hyperspectral image of a field[4]

For example, in precision agriculture, drones equipped with multispectral and hyperspectral sensors and AI-powered analytics engines can survey huindreds of farms in a single flight, analyzing vegetation health across vast acreages far faster than scouting on foot. However, for growers to respond to insights from the broad-scale aerial crop analysis and make data-driven decisions, the torrent of spectral readings and photographic imagery must first be turned into key agricultural metrics. Automated machine learning pipelines can extract plant health visualizations, yield prediction maps, variability zones, and targeted prescriptions for fertilizer, irrigation, and pesticides tailored to each individual farm. This enables rapid translation of the rich aerial inputs into actionable insights customized across the dozens of growers in the flight path.

Similarly, in solar farm inspection, BVLOS drones equipped with thermal and hyperspectral payloads can scan rows of solar panels over hundreds of acres in a single flight - but the true value emerges after analytics engines process the data to detect subtle heat signatures indicative of faulty panels and panels with abnormal efficiency issues difficult to spot through visual inspection alone. By overlaying these insights onto the GPS-tagged topological maps of the facilities, operators can rapidly pinpoint and fix problems at scale. And when monitoring critical infrastructure like pipelines, drones conduct extensive multi-sensor scans above expansive areas far more rapidly than ground patrols, augmenting static CCTV networks. Here, the key is leveraging algorithms to analyze long video segments from drone flights to automatically detect subtle anomalies - things like construction activity encroaching on rights of way, inconsistent

---

[4] hyperspectral iamge of a corn field - Search Images (bing.com)

vegetation growth indicating a leak, suspicious vehicular activity, and other threats that might otherwise slip through the cracks. This enables operators to selectively review only the most concerning events versus sifting through days of tedious footage (See Figure 7).

Figure 7: Drones Inspecting Solar Panels[5]

Unlocking such use cases requires an analytics system robust enough to handle rapid streams of heterogenous sensory inputs from large-scale BVLOS missions and isolate the true signals from an immense sea of noise. Combining computer vision, edge processing, modern cloud infrastructure, and purpose-built AI/ML algorithms tailored to each industry's challenges, these insights engines automate the conversion of torrential sensor data into intelligent observations that drive real-world decisions and workflows. And over time, as rich pools of analytical metadata accumulate from hundreds of flights over farms, solar facilities, pipelines, and other sites, new opportunities emerge to find deeper connections through historical comparative analyses, predictive modeling based on identified trends, and continuously tuned machine learning algorithms. Farmers can better predict optimal harvest dates based on crop development patterns mapped over seasons. Pipeline operators can reduce false alarms by distinguishing usual vegetation factors from suspicious encroachment. Solar managers can make capital investment decisions optimized to actual - not presumed - degradation rates across different equipment.

Unencumbered by the practical flight range and endurance limits inherent to small VLOS drone operations, BVLOS unlocks invaluable new aerial perspectives. But it is the maturation of the remote sensing capabilities and analytics engines purpose-built to interpret those rich vantage

---

[5] drone inspecting solar cells - Search Images (bing.com)

https://doi.org/10.52202/075106-0014

points that make the intelligence actionable at scale across the industries pioneering this new frontier. It is no longer enough to just reach new heights and capture more data. To fully benefit from BVLOS drone capabilities and ROI, the information torrent must be channeled into insights through sophisticated analysis - converting raw inputs into calibrated outputs that reshape workflows.

BVLOS Economic Law #5: The Key Value Proposition

The key value proposition unlocked by large-scale BVLOS drone infrastructure is not merely expanding data gathering capacity, but rather the sophisticated post-processing algorithms and analytical techniques applied to rich drone-captured datasets that ultimately enable superior decision-making, risk detection, and predictive modeling across industries from agriculture to insurance. Simply gathering vastly more aerial imaging and sensor data through BVLOS networks absent advanced systems to parse that torrential firehose of information into actionable intelligence would serve little intrinsic purpose. While expanding raw data gathering capacity promises invaluable material to feed analysis, successfully translating that potential into profitability and sustainability for drone operators relies upon extensive interdisciplinary cross-pollination with artificial intelligence, cloud computing, and data analytical sciences.

To extract differentiation and defensibility amidst an increasingly crowded industry of longitude-latitude-timestamp data providers, BVLOS pioneers must emphasize proprietary systems integrating customized machine learning pipelines, analytics visualizations, simulation modeling capabilities, change detection protocols, and other specialized techniques to optimize post-processing software, streamline customers' decision workflows, and yield insight-revealing interpretations tailored to each use case and vertical. For example, in precision agriculture, while multispectral/hyperspectral sensors gathering aerial crop data at vastly larger scales via BVLOS provide invaluable foundations, translating those rich inputs into accelerated adoption relies upon downstream agronomic analytics engines generating smart prescription maps that farming customers can integrate into real-world fertilizer, irrigation, harvest planning operations. To produce such actionable intelligence algorithms must filter noise, isolate signal, reveal variability, render predictions - culminating in decisions not just observations.

Similarly, vast pipelines for regional solar energy producers resist adopting BVLOS services gathering thermal data on hundreds of thousands of panels absent customized machine learning to pinpoint damage, estimate production impact, and prescribe maintenance actions location-by-location to optimize yield. For industrial mega-scale cattle feedlots, raw aerial monitoring serves limited standalone utility without AI rapidly flagging behaviors indicative of illness outbreaks and modeling transmission risks, so mitigation responses protect investments at margins of pennies per animal. Across security-conscious enterprises spanning mines to mixed use developments encompassing billions in hard assets, beyond simply accumulating surveillance footage intelligence systems must alert BVLOS customers to anomalous patterns indicative of theft, leaks, encroachments through analytics intentionally designed to parse signal from noise amidst expansive camera fields.

Unlocking such vertical use cases and delivering defensible value atop mere longitude-latitude-timestamp datasets demands investment not just in drone fleets, but interdisciplinary artificial intelligence teams designing customized neural networks, learning pipelines, reinforcement learning protocols, computer vision models, explainability frameworks and control mechanisms purpose-built for translating torrents of unstructured sensor data into structured insights that reshape decision workflows. Further value accrues by accumulating industry-specific corpuses of benchmark training data over successive missions and shared learnings across customers to continuously refine computer vision, model reliability, explainability and accuracy over time - seeking not just reactive insights but predictive intelligence and prescriptive decision recommendations.

Architecting specialized analytics engines around tailored machine learning capabilities purpose-built for individual verticals and use cases requires extensive investment - but powers differentiation. This unlocks the fuller value propositions by not just showing customers more observational data but telling them what it means and advising data-optimized decisions amidst complexity at scopes impractical for unaided human analysis. The business sustainability and dependability for BVLOS service providers ultimately relies upon pursuing this interdisciplinarity - not just aviation capability.

BVLOS Economic Law # 6: New accessible markets and use cases enabled by BVLOS drone operations

By expanding operational radii beyond constrained VLOS limitations, BVLOS drones unlock entirely new market opportunities. Applications from agricultural insights to emergency response rely on accessing locales previously beyond economic reach. The vastly expanded ranges introduce new economies of scale and scope, fueling services unviable under restricted VLOS caps. As you go through this exercise, consider how the market size for package delivery increases as the scale of BVLOS increases. When you consider 90% of the US population is within 10 miles of a Walmart store, this example takes on added meaning.[6] Here are the calculations showing how the market size for drone operations expands drastically as the operational radius increases from 1 mile to 10 miles, using the area of a circle formula (See Figure 8):

[6] 25+ Essential Walmart Statistics for 2024 | Fortunly

https://doi.org/10.52202/075106-0014

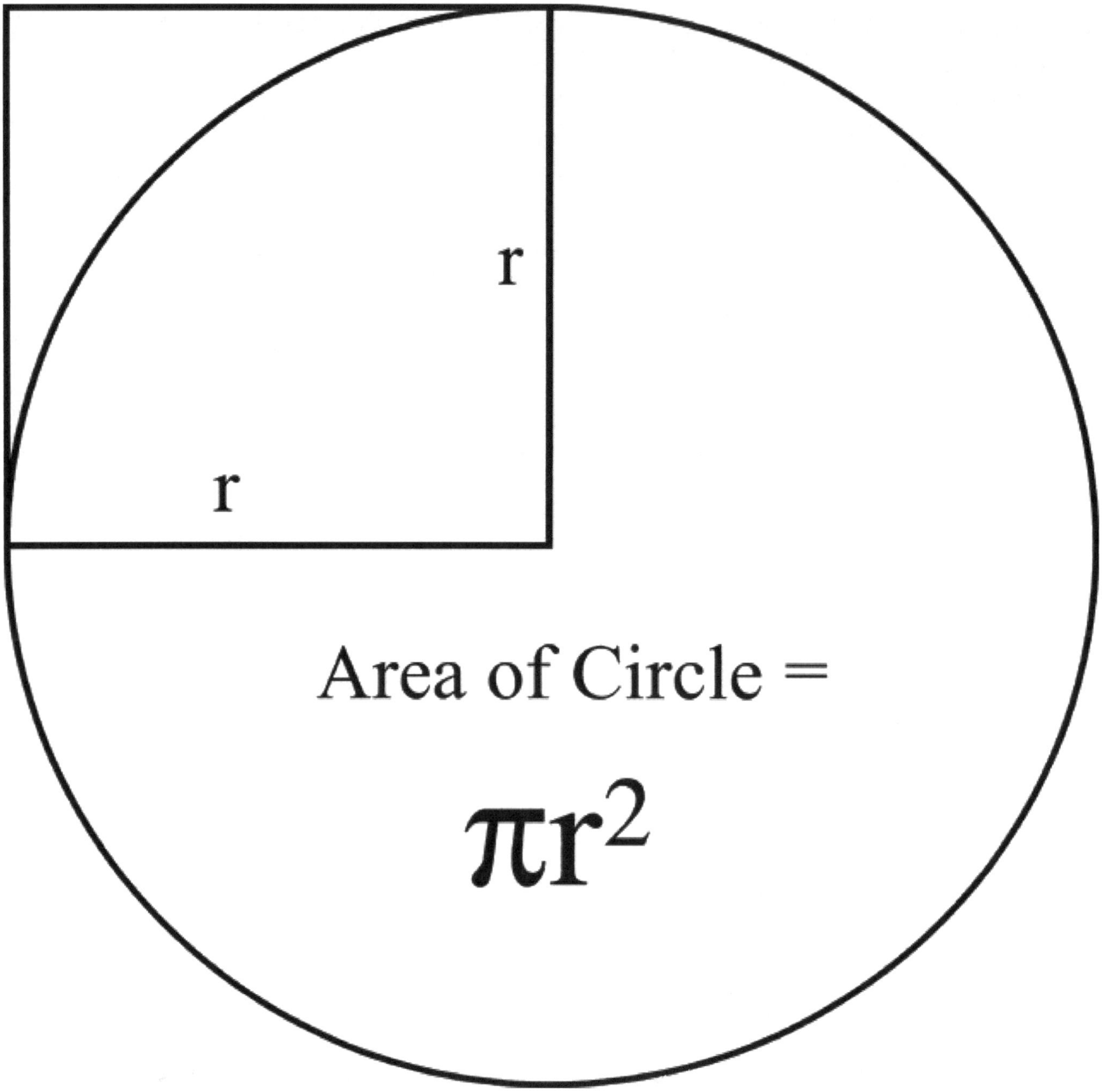

Figure 8: Area of a Circle[7]

1. Area of Circle Formula: A = πr2

Where:

A is the area

π is the constant approximately equal to 3.14

r is the radius

[7] showing how the area of a circle increase geometrically - Search Images (bing.com)

2. 1 Mile Radius

Area = π x (1 mile)2

Area = 3.14 x (1 mile x 1 mile)

Area = 3.14 square miles

So with drones only capable of a 1 mile radius the accessible market size is 3.14 square miles.

3. 10 Mile Radius

Area = π x (10 miles)2

Area = 3.14 x (10 miles x 10 miles)

Area = 314 square miles

When drones can operate across a 10-mile radius the addressable market area for drone operations rises to 314 square miles. Increasing the effective drone operational radius 10X from 1 to 10 miles expands the addressable market area by 100X from just 3.14 to 314 square miles thanks to the quadratic area of a circle geometric relationship.

This demonstrates the immense market potential unleashed when longer distance beyond visual line of sight drone operations become approved through advancing regulatory frameworks and operational capabilities over this decade. Even relatively modest linear range improvements geometrically multiply the aerial coverage and far wider market adoption possibilities from superior area access. The incremental area of coverage market size expansions going from a 1-mile drone operational radius up through a 10-mile radius is as follows (See Figure 9):

Mileage	Square Miles
1	3.14
2	12.57
3	28.27
4	50.27
5	78.54
6	113.10
7	153.94
8	201.96
9	254.47
10	314.56

Figure 9: Market size Increases as the Radii Increases

https://doi.org/10.52202/075106-0014

As shown, each additional mile of incremental linear drone range, enabled by advancing BVLOS functionality, allows access to exponentially larger surface areas for coverage across infrastructure inspection, delivery logistics, security patrol, agricultural survey, and other drone industry use cases. For example, vast expanses of farmland, forests and industrial facilities have seen manual surveying and assessment costs prohibitively high, locking knowledge value left unclaimed. Drones now offer promise in cost-effectively gathering imagery, topological and sensor data at regular intervals across properties spanning thousands of acres which foot and ground vehicles cannot regularly cover. This opens avenues for monitoring crop health patterns, detecting invasive species spread, and assessing equipment integrity across territories previously unobservable at sufficient frequency. Where sampling may have occurred annually at best, weekly unmanned sweeps extract multiples more analyzable data to guide land use decisions and risk detection.

Natural disaster response similarly relies on vastly expanding quick observational range to accelerate aid. As extreme weather events from hurricanes to wildfires afflict larger swathes of territory, being able to rapidly assess damage reports and send supplies promises dramatic human impact. BVLOS drone infrastructure offers route planning and assessment intelligence unbounded by impassable debris that could trap terrestrial responders. Further, payload delivery of essential medical supplies when hospitals are cut off by storms or separated fire fronts grows more urgent than ever with climate change. Only BVLOS capacities can make reaching these sites reliably to save lives and property feasible on mass scales.

From crop health to disaster response to border security and delivery services, BVLOS drones introduce entirely new operational scopes at cost structures finally viable enough to tap latent demand. Just as past innovations from rail to refrigeration similarly shrunk operational friction enabling economic activities for the first time, so shall drones multiply market opportunities. Realization depends on advances in battery life, aircraft durability and weight thresholds as well as air traffic coordination and even public acceptance to fulfill this proliferation potential. But the promising economics underpinned by expanded accessibility to new data and operational connection opportunities beckon compellingly as enablers now accelerate.

BVLOS Economic Law #7: The Economics of Transitioning from 1:1 to 1: N pilot to Drone Operations Ratios

The conventional VLOS restriction of one pilot to one drone in flight creates an intrinsic bottleneck on revenue. A single pilot operating a lone quadcopter or rotary drone within eyesight can only collect imagery or sensor data from that tight geographic circle in real-time before needing to land and recharge batteries. Even with near 100% utilization, total flight time caps out

https://doi.org/10.52202/075106-0014

around three or four hours per day optimistically due to the short flight times and the need to recharge batteries and drive to different locations. This will drastically change with BVLOS.

Conversely, trained BVLOS pilots licensed to manage multiple fixed-wing drones simultaneously can scale data collection and services provided exponentially higher. Instead of binding one pilot to just one set of eyes, advanced ground stations synthesize real-time drone telemetry, air traffic feeds, terrain maps and sensor footage from each asset simultaneously. When drones can operate for hours without direct eyes-on monitoring, one pilot overseeing a fleet of five to ten drones in staggered flight rotations essentially increases their productivity by an order of magnitude. New BVLOS economics emerge by spreading the sunk capital costs across these force multiplying factors. Consider the improvements in pilot cost per hour as we transition from one pilot per drone to one pilot per 10 drones.

Here are the calculations using a $100 per hour pilot wage:

1 Pilot: 1 Drone

- Pilot pay = $100 per hour

- Total drone flying hours per pilot hour = 1 hour

- Unit cost per drone hour = $100

1 Pilot : 10 Drones

- Pilot pay = $100 per hour

- Total drone flying hours per pilot hour = 10 hours (10 drones)

- Unit cost per drone hour = $10

  = $100 pilot cost per hour / 10 drone hours per hour

  = $100 / 10 drones

With a $100 per hour pilot wage, transitioning from 1:1 to 1:10 pilot-to-drone management ratio still reduces the unit economics by 10X - dropping from $100 down to $10 per drone flight hour. As the number of drones that can be flown safely increases, the unit costs of flying them trends towards zero as a limit function. This order of magnitude cost efficiency improvement remains highly compelling even at higher pilot salaries. It shows the immense financial upside of progressing to BVLOS systems where one pilot oversees larger drone fleets rather than just a single aircraft. The dramatically lower unit costs per flight hour promise vastly superior commercial viability potential.

https://doi.org/10.52202/075106-0014

Here are the incremental unit economics per drone flight hour with the pilot getting paid $100 per hour (See Figure 10):

Number of Drones Being Flown	Pilot Cost per Hour
1	$100.00
2	$50.00
3	$33.33
4	$25.00
5	$20.00
6	$16.67
7	$14.29
8	$12.50
9	$11.11
10	$10.00

Figure 10: BVLOS Pilot Costs

With a $100 per hour pilot wage, the cost savings going from 1:1 to 1:10 pilot to drone management ratios drops from $100 down to $10 per flight hour. Enabling coordinated oversight of larger fleets via improving BVLOS systems still produces monumental cost efficiency gains regardless of absolute pilot pay. The order of magnitude unit cost reduction promises vastly superior commercial scale potential helping drones proliferate mass adoption.

The revolution lies in transitioning pilots from visual scanners to meta-data managers through superior command and control systems. Skilled BVLOS pilots then optimize and choreograph drone fleets to execute larger combined missions that were simply impossible under constrained VLOS restrictions. While unit pilot costs are going down, the ability to earn incremental revenue goes up as the pilot to drone ratio goes up from flying one drone at a time to ten.

Here are the incremental revenue calculations as pilot productivity grows from flying 1 drone per hour to 10 drones per hour, with each drone contributing $300 per flight hour (See Figure 11).

Pilot Flying number of Drones	Potential hourly earnings per pilot
1	$300
2	$600
3	$900
4	$1,200
5	$1,500
6	$1,800
7	$2,100
8	$2,400
9	$2,700
10	$3,000

Figure 11: Revenue Potential per hour per pilot.

By enabling each pilot to scale their productivity from 1 to 10 revenue-generating drones per hour, total hourly revenue potential grows over 10X from $300 to $3,000 thanks to the increased fleet oversight capabilities that advancing BVLOS systems facilitate. This order of magnitude revenue growth multiplier promises tremendous commercial upside. There will be diminishing returns to scale at a certain point given the magnitude of the various submarkets.

A strict 1:1 ratio of one remote pilot per drone is only sufficient to cover flight operations themselves. Simply keeping a drone airborne and capturing raw aerial data serves little end purpose. To transform and analyze volumes of airborne sensor measurements into usable business intelligence requires an entire backend workflow.

Additional staff becomes essential across IT, data science, engineering, mapping, reporting and more to handle elements like:

1. Data Management/Storage - Processing troves of high-res imagery and lidar models from drones

2. GIS Specialists - Creating geospatial data layers and 3D terrain maps

3. Analysts - Extracting and formatting key infrastructure insights from the drone data

4. Application Developers - Building tools to annotate images and automate workflow

5. Client Services - Packaging actionable deliverables and driving adoption

When regulation allows a single pilot to control many drones in BVLOS scenarios, generating value from the drone hardware alone is insufficient. The real ROI stems from budgeting for the human analytics horsepower behind the scenes to elevate aerial data into knowledge products that influence better decisions. The labor, supervisory, cost recovery and fatigue issues make a dedicated 1:1 BVLOS drone structure completely impractical. BVLOS economics need regulatory changes to embrace command centers where teams of pilots coordinate fleets of drones using automation aids, simulations, and advanced data systems.

The amortization and financing dilemma for expensive fixed-wing BVLOS drones poses a surmountable economic obstacle. The more yearly flight time these assets log, the quicker their immense value gets realized through actionable aerial intelligence that outweighs the sticker shock. Early drone operations have largely depended on dedicated one-to-one ratios between remote pilots and aircraft. This caps revenue generation potential for pilots while driving up operating costs considerably. However, advancing automation and regulatory enablement of BVLOS functionality promises a transition to single pilots overseeing larger fleets of drones simultaneously.

 https://doi.org/10.52202/075106-0014

On the cost side, the ability for properly trained pilots to coordinate multiple drones rather than just one vastly improves per unit economics. Transitioning to an operating model of one pilot capable of coordinating dozens of drones executing automated routes or data gathering missions' slashes per drone operating costs substantially. This will achieve profit break-even points on far more drone applications.

Further, aggregating more drone operations under individual pilot direction allows those personnel to generate substantially higher revenue streams as force multipliers of their expertise. Much like master train conductors, BVLOS operations shall enable master drone operators to play more strategic oversight roles while monitoring automated navigation and data capture across vast terrain. The sole scenarios when dedicating one remote pilot to operate an individual drone remains financially and operationally viable are limited to two niche cases:

1. Low-Cost Drones for Short, Visual Line of Sight Missions

Commercial pilots utilizing very basic VLOS drones for short duration flights within eyesight, such as quick property inspections or simple aerial photography, can still recoup costs with a dedicated 1:1 staffing ratio. When drone hardware remains simple and flight time is measured in minutes rather than hours, the lack of scale does not yet cripple profitability potential. As soon as more expensive hardware, sensors, licensing, insurance, or specialty pilot skills enter the picture though, the model collapses.

2. Mission-Critical Specialist Operating Solo Drone

The only other viable 1:1 scenario is that of highly specialized drone operators who take on end-to-end mission responsibility themselves. For instance, expert geospatial drone photographers who pilot drones for high resolution imaging and handle all data analysis, client reporting and application development themselves can still generate substantial revenue off operating solo aircraft. Their niche skillset, efficiency and service value sustain a 1:1 operation. Similarly for software developers piloting drones with proprietary algorithms and analytical payloads, relying on no other staff allows protecting IP sensitivities. Outside these two exceptions though, single pilot oversight even during visual line of sight flights offer no economies of scale.

Any scenario involving sophisticated drone hardware, expensive sensor and analytical software payloads, liability and regulatory compliance, and any desire to fly longer BVLOS sorties renders the 1:1 pilot-drone ratio utterly impractical from cost and utilization perspectives. Only the two above categories see financial or operational logic favor the simplistic 1:1 approach given their niche economics. For all others, transitioning to team oversight of larger drone fleets promises vastly superior cost and revenue scalability.

https://doi.org/10.52202/075106-0014

Ultimately, transitioning BVLOS drone economics towards single pilot oversight of larger fleets promises exponential gains in both per unit cost reductions and revenue creating potential through force multiplication. Identifying these key economics drivers and reducing their inefficiency shall accelerate mainstream enterprise adoption as regulations enable expanded BVLOS functionality.

BVLOS Economic Law #8: Observational efficiency for drone imaging from higher altitude BVLOS operations

As drone technologies enable safer operations at expanding altitudes, their value proposition for aerial imaging gains compelling advantages. Specifically, increased flight elevation allows captured imagery to span wider geographic areas while also reducing the degree of image overlap needed to create contiguous mosaics during processing. By enhancing per-flight observational efficiency, higher vantage BVLOS drone infrastructure can exponentially escalate data productivity at even greater quality.

For example, if we increase the flight height being flown from 400 Feet Above ground level (AGL) to 3,000 feet AGL, the change in instantaneous field of view (IFOV) and coverage area per image is quite large. Instantaneous field of view (IFOV) is the angular cone of visibility from a camera at a given instant in time. It defines the ground area that is captured in an individual image frame. A wider IFOV means more ground area is covered in a single image (See Figure 12).

https://doi.org/10.52202/075106-0014

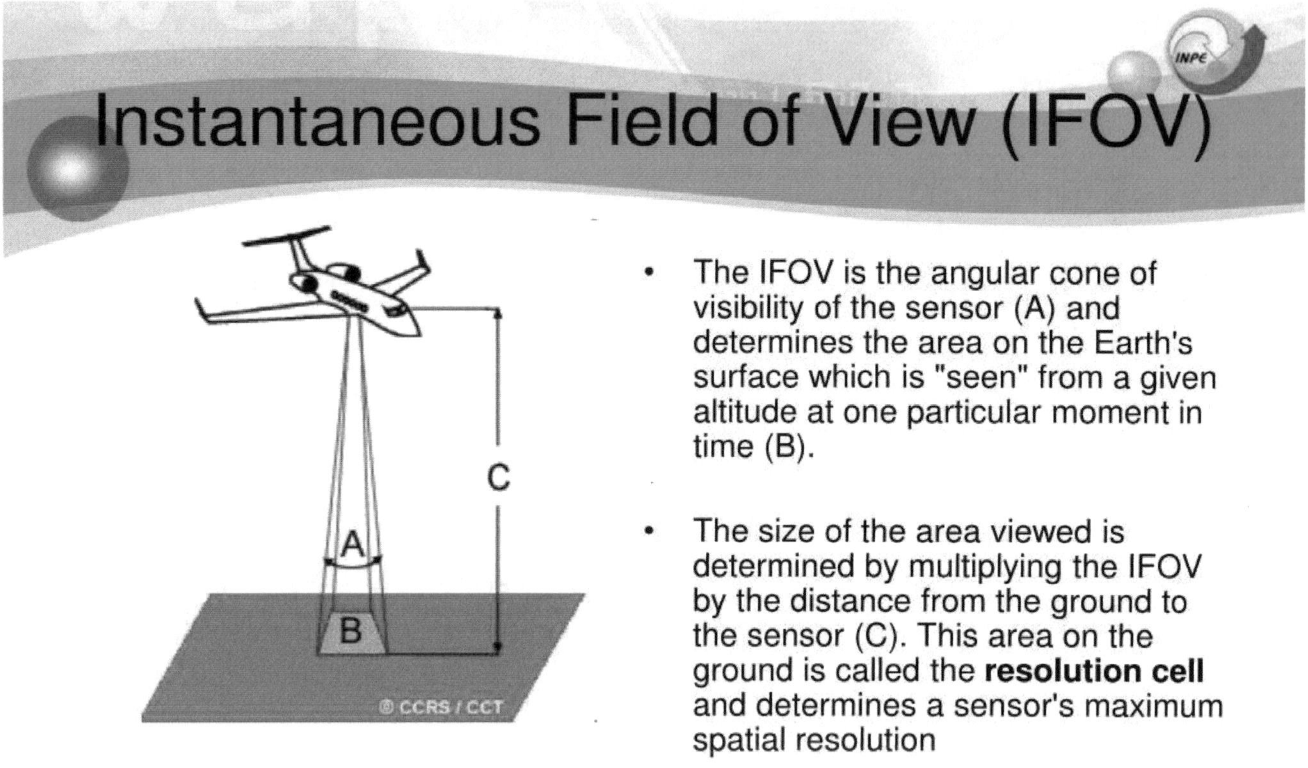

Figure 12: Instantaneous Field of View[8]

- At 400 ft AGL, the IFOV covers a relatively narrow area - let's still call it X sq ft
- At 3,000 ft AGL, the IFOV covers a much wider area. For a rough estimate:
  - As a rule of thumb, IFOV increases linearly with increase in height[9]
  - So at 3,000 ft (7.5 times higher than 400ft), the IFOV would be ~7.5 times wider
  - Thus each image covers ~7.5X sq ft rather than just X sq ft at 400 feet

The exact multiple depends on camera specifications, but the area covered by each image at 3,000 ft AGL is much larger than at 400 ft. For example, an area that takes 100 images to survey at 400 ft could be covered in roughly 13 images (100 divided by 7.5) from 3,000 ft. This allows

[8] drone Instantaneous field of view (IFOV) - Search Images (bing.com)

[9] Electro-Optical Imaging Systems (fas.org). The exact relationship is: IFOV is proportional to the tangent of the half-angle of the FOV 1. This means that as the height increases, the half-angle of the FOV decreases, and the IFOV increases proportionally.

much larger project areas to be mapped efficiently with fewer total images, but at a sacrifice of ground resolution due to being at a greater distance above the area. Altitude needs to be chosen judiciously based on coverage needed versus resolution required for the application. This allows faster sweeping of sites of interest, key for time sensitive applications like disaster damage assessment and precision agriculture field scanning. Further, expensive camera payloads can gather vastly more image data per mission as their observational range grows. This provides economies of scale on costly sensor investments by using them more actively through expanded purview. Further, added elevation enables construction of reliable imagery mosaics with significantly reduced side and front/back overlap between captured frames during missions. This overlap facilitates accurate stitching in post processing but requires additional flight time and therefore battery overhead to execute previously. The combined advantages amount to massive per unit area data generation improvements using the same piloting and imaging equipment simply elevated on appropriately certified drone airframes. Surveying sites of several square miles shifts from full day endeavors to executable in under an hour as altitude caps transition toward thousands of feet eventually. This proves compelling economics not just for commercial users like surveyors, but public sector applications as well. More regions can be assessed more quickly after disasters to coordinate response or for regular infrastructure integrity checks at heightened resolutions using the same equipment and manpower.

BVLOS Economic Law #9: As Flight Scales change so do Market Dynamics

In precision agriculture, drones equipped with multispectral sensors and AI-powered analytics engines can survey hundreds of farms in a single flight, analyzing vegetation health across vast acreages far faster than scouting on foot. However, for growers to respond to insights from the broad-scale aerial crop analysis and make data-driven decisions, the torrent of spectral readings and photographic imagery must first be turned into key agricultural metrics. Automated machine learning pipelines can extract plant health visualizations, yield prediction maps, variability zones, and targeted prescriptions for fertilizer, irrigation, and pesticides tailored to each individual farm. This enables rapid translation of the rich aerial inputs into actionable insights customized across the dozens of growers in the flight path. The sheer scale of coverage enabled by BVLOS not only unlocks new analytics use cases, but also enables business model shifts as services become viable across vastly larger geographic areas. Rather than selling piecemeal field-by-field or firm-by-firm as with most VLOS services, extensive BVLOS networks make it economical to sell annual county-wide or even statewide subscriptions to governments, cooperative extensions, industry associations, and major agribusiness conglomerates managing expansive land assets across entire regions. This same dynamic applies to infrastructure verticals, where owners of whole pipeline networks, railway systems, electrical grids and roadways become the anchor customers versus

https://doi.org/10.52202/075106-0014

individual site managers. The customers are fewer but bigger. To serve and support these large regional clients, the data aggregation, standardization, analytics, and machine learning algorithms do heavy lifting behind the scenes while delivering insights customized to each customer's assets. By tapping bigger tickets at national scales, the addressable market potentials grow substantially.

And over time, as rich pools of analytical metadata accumulate from hundreds of flights over dozens of farms, solar facilities, pipelines, and other sites across counties and even states, new opportunities emerge to find deeper connections through historical comparative analyses, predictive modeling based on identified trends, and continuously tuned machine learning algorithms. Regional farm collectives can better predict optimal harvest dates based on crop development patterns mapped over seasons across participating counties. Pipeline operators can reduce false alarms by distinguishing usual vegetation factors from suspicious encroachment across wider networks. Solar managers can make capital investment decisions optimized to actual - not presumed - degradation rates across different equipment based on insights gathered from hundreds of installations. Unencumbered by the practical flight range and endurance limits inherent to small visual line of sight drone operations, BVLOS unlocks invaluable new aerial perspectives at massively expanded scales. But it is the maturation of remote sensing capabilities and analytics engines purpose-built to interpret those rich vantage points that make the intelligence actionable across industries. It is no longer enough to just reach new heights and capture more data. To fully benefit from BVLOS drone capabilities and ROI, the torrential sensor data must be channeled into insights through sophisticated analysis - converting raw inputs into calibrated outputs that reshape workflows.

The keys to unlocking this potential are:

1) Advanced aerial capture technologies like high-resolution hyperspectral imaging which can reliably gather precise sensory inputs across diverse environments and conditions.[10]
2) Scalable cloud analytics leveraging modern GPU/TPU computing to rapidly process torrential streams of aerial data into useful metadata.[11]
3) Purpose-built machine learning algorithms continuously tuned on industry-specific benchmarks to transform sensor readings into decision-ready intelligence.[12]

These pillars work in conjunction to extract the full value from BVLOS drone missions as operations shift from mere data collection confined to individual sites toward networked services delivering regional insights that cannot be achieved with traditional methods. The superior speed, cost, quality, and analytic depth ultimately enable better decisions and outcomes.

---

[10] Hyperspectral imagery applications for precision agriculture - a systemic survey | Multimedia Tools and Applications (springer.com)

[11] GPTPU: Accelerating Applications using Edge Tensor Processing Units (arxiv.org)

[12] Machine Learning: Algorithms, Real-World Applications and Research Directions | SN Computer Science (springer.com)

The Promised Land of BVLOS Economics

The Promised Land of BVLOS Economics Rests on Taming Torrents of Sensor Data. BVLOS drone operations live or die based on the ability to cost-effectively manage immense volumes of aerial sensor data. Whether long distance drone missions focus on infrastructure monitoring, agricultural yield projections, environmental conservation efforts or public safety enhancement, the aircraft itself serves mostly as the ephemeral data collection vector. The true value getting extracted which justifies heavy investment in expensive fixed-wing or powered rotor drone platforms lies in downstream processing and analytics of their voluminous payload.

Generating Prescriptive Insights from Data

Pushing further beyond processing refinement brings us to the true holy grail and watershed economics promised by BVLOS innovation. Converting historically inaccessible or cost-prohibitive data flows directly into decision catalysts and automated action through advanced modeling techniques. This next level of analytical maturity unlocks substantial labor cost and risk reduction - while informing prescriptive business, operations and policy moves not otherwise apparent.

Consider our electrical grid transmission line example. Baseline analytics would accurately pinpoint component failures down to specific towers, arms or insulators driving enhanced repair crew efficiency and lowering outage impacts. But top-tier analytics solutions take this a full level deeper. Machine learning algorithms ingesting billions of data points across vast BVLOS monitoring territories uncover hidden year-over-year trends and correlations. Predictive failure signatures emerge at macro grid level. Planners enhance budgets for proactive component upgrades and vegetation clearing in grid segments exhibiting telltale high corrosion indicators. Engineers develop advanced composite materials for towers and connectors tailored to terrain prone to accelerated infrastructure decay based on environmental and operational factors. Real-time dashboards track ambient impacts of weather, load changes, voltage indicators on component stress and lifespan dynamically across the overall grid network. Renewed assets get purpose built for maximum longevity per their exact intended operating conditions. Outage levels fall while maintenance costs drop through proactive investments - all via BVLOS analytics allowing planners to monitor the massive grid territory comprehensively.

https://doi.org/10.52202/075106-0014

Figure 13: Power Line Inspection before drones[13]

This hints at the sheer economic upside lurking within BVLOS sensor data waiting to be harnessed through creativity. And how advanced algorithms and AI will transform many hands-off industries as complex modeling combines streaming BVLOS sensor inputs with traditional SCADA measurements and weather data to generate powerful predictive insights and mitigation paths.

Obstacles to Overcome

Realizing this full potential depends on the drone industry maturing quickly across a few pivotal domains:

i. Automated Data Curation - open-source labels, neural nets and quality control protocols allowing drones to interpret their own data accurately while embedding key metadata for context.[14]

ii. Enhanced Fleet Operations Tools - hardware, sensors and software supporting dynamic vehicle to vehicle / vehicle to ground communication

[13] power line inspection by drones - Search Images (bing.com)
[14] Cleanlab Raises $25M Series A to Automatically Increase the Value and Accuracy of the World's Enterprise Data Used by AI, ML, and Analytics Solutions | Business Wire

for smoother distributed missions encompassing both piloted and pilotless drone assets across vast connected skies.[15]

iii. Avionics and Control Stations- advancing reliability for severe operation scenarios spanning weather, magnetic interference, limited visibility, technical malfunctions, and adversary detection events. Reducing risk for both equipment and surroundings during abnormal scenarios.[16]

The goal is purpose-built drone technology moving beyond standardized hobbyist platforms to enable safe, advanced BVLOS use.

Conclusion - BVLOS Economics Depend Upon Mission Scale and Infrastructure Maturity

Truly realizing the immense latent value of wide area drone monitoring hinges upon maturing several pivotal supporting capabilities. Sophisticated aircraft and sensors merely provide the ephemeral means for gathering intelligence. All else being equal, the core economics boil down to infrastructure able to ingest, process, analyze, and act upon the resulting data flows across vast imagined geographies.

BVLOS drone operations promise immense capability advances once freed from the labor-intensive oversight models the industry relies upon presently. The future sees primarily pilotless aircraft managed by slim supervision teams aided by vehicle-to-vehicle collaboration, airspace automation services and resilient contingency logic. Onboard optics feed servers leveraging

[15] Ask an Expert: Capturing fleet impact from telematics | McKinsey
[16] On the Reliability of Interference Limited Unmanned Aerial Vehicles | Wireless Personal Communications (springer.com)

https://doi.org/10.52202/075106-0014

scalable cloud compute rather than just storage alone. Machine learning provides the true differentiator as software derives otherwise inaccessible insights from endless petabytes of aerial intelligence. Together, the collective economics pivot drastically once these forces align at scale. But developers caution the roadmap ahead remains strewn with obstacles. Hardware gets prohibitively expensive absent commitments for large fleet orders. Advanced negotiable autonomy still faces skepticism among conservative aviation authorities. Bandwidth costs continue strangling small operators lacking bulk access to vital satcom infrastructure. Insurance products now only nibble around the risk margins of complex BVLOS operational envelopes envisioned. And the skilled labor pool familiar with optimizing enterprise drone data outputs remains shallow.

Still pioneers press ahead undaunted. The sheer magnitude of aging infrastructure requiring perpetual monitoring at continental scale simply defies traditional methods absent around the clock aerial surveillance. And exponential tech improvements rapidly tackle barriers one by one. Expect determined teams to patiently coax regulators forward while building irrefutable proof cases demonstrating the immense latent upside. Advanced BVLOS drone operations constitute an inevitably technology once key infrastructure and trust foundations solidify. Those laying the groundwork today stand well-positioned to reap substantial rewards in the coming decade.

COLLABORATIVE AUTONOMY MEETS THE REAL WORLD

David Koelle*

The physical world is being increasingly served by autonomy, including self-driving taxi cabs, autonomous mining vehicles, and robots in warehouses and hospitals. While this is certainly a significant step in technical advancement, these robots and vehicles all operate independently. The ability for robots to work together is a powerful way for them to achieve more than anyone could achieve individually. Working together would allow for achieving high-level, multifaceted goals such as autonomous construction in remote locations, including in space; achieve military mission objectives; find and mitigate forest fires before they spread; and more. What considerations do we need to make for the interoperability of autonomous systems? In this presentation, we will discuss applied research and development work in multirobot autonomous systems that enable robots to know about each other's changing capabilities during the course of a job, manage expectations and exception conditions that can be caused by both intended and unintended consequences of the real world, and provide a way for robots to anticipate other robots' actions when communications are interrupted.

INTRODUCTION

The physical world is being increasingly served by autonomy, including self-driving taxi cabs, autonomous mining vehicles, and robots in warehouses and hospitals. These vehicles operate independently, or if they operate interdependently (such as warehouse robots or drone firework displays), they are part of a specific system that is designed to work in a choreographed manner.

One thing that we are not yet seeing is the ability for autonomous robots and vehicles from multiple manufacturers to have the ability to collaborate with other autonomous vehicles out of the box. However, given the growing presence of independent autonomous vehicles, the ability for such autonomy to have the ability to collaborate in ways defined at deployment time will satisfy several objectives that are currently out of reach but would introduce significant improvements to addressing real-world challenges.

For example, consider autonomy in agriculture. Autonomous farming machines perform regular tasks, saving the farmer time that they can spend on planning their operations and addressing immediate concerns. Autonomous tractors and other machines can collect data on soil moisture and other attributes, or perform weeding tasks. Uncrewed aerial vehicles (UAVs) can spray crops in specific areas. Precision farming enabled by these technologies increases crop production and reduces consumption of pesticides, water, and fuel. However, equipment is expensive, and the most likely use case for autonomous equipment on the farm is that investments into expensive equipment will be made over time. Farmers could purchase vehicles from multiple manufacturers with different capabilities over time, tailoring what they purchase to their specific agricultural operation; in

* Director of Engineering, Charles River Analytics, 625 Mount Auburn St., Cambridge MA 02138.

https://doi.org/10.52202/075106-0015

essence, they can build an ad hoc swarm of autonomous farming vehicles. There are currently no standards by which those vehicles will operate, and no planning capability for a farmer to design cooperative strategies for the vehicles to work together (e.g., "If the crop scanning tractor finds insect damage at a location, then prepare the UAV to deliver pesticide to this location.").

Or, consider autonomy that could assist in finding incipient forest fires, performing search and rescue, or assisting in disaster zones. In these cases, autonomy from multiple organizations, such as local, state, Federal, and nongovernmental organizations (NGOs), could bring their autonomy together, creating an ad hoc swarm of autonomous emergency response vehicles, but again there are no capabilities for such vehicles to interoperate or for experts to craft behaviors that let the autonomous vehicles build on each other.

CURRENT STATE OF THE ART

We are in the early days of autonomous multivehicle collaboration. Much of the work in this area resides in academic labs, where research into a variety of task allocation and planning techniques (e.g., centralized versus distributed task allocation, optimization approaches) is explored. These explorations are performed on well-known hardware with well-specified network connectivity in laboratory environments. Cooperative warehouse robots benefit from these optimization approaches, but warehouse robots consist of well-known vehicles working in a well-defined space. Drone firework displays are choreographed ahead of time using tools that are more similar to animation software than task planning tools and also use platforms that are known ahead of time.

FUTURE DIRECTIONS

What is absent from today's technology that we need to develop to support a future of interoperability for autonomous systems?

We are essentially seeking the development of an Internet of Things (IoT) but for autonomous, mobile systems that experience real-world challenges in a dynamically changing environment, such as a farm tractor that encounters unexpected mud on its way to a field, or a disaster response vehicle that loses communications with other vehicles.

This sounds like "IoT for swarms," and taking inspiration from existing IoT interoperability standards is a solid first step toward realizing such a capability. Today, IoT enables smart home ecosystems, industrial automation in areas such as predictive maintenance, and electrical grid efficiency and reliability. With IoT, one can establish an ecosystem of devices from multiple manufacturers (e.g., for the home, smart lights, thermostats, and security cameras) and specify how they connect and work together. These devices are often either passive data collectors, or simple actors that may open blinds or turn on lights, and the devices do not leave their home network. IoT for collections of autonomous vehicles will require more considerations. What happens if the vehicle has a problem connecting to its network? How does the collection of vehicles react when it encounters unexpected conditions in the physical world? How do we incorporate safety into autonomy? With vehicles that can move and be assigned to tasks in different locations around the world, how can the swarm's behavior optimize the physical presence (incorporating elements of time, fuel consumption, and safety) and task assignment of vehicles?

We believe that establishing the foundation for how autonomous vehicles, purchased over time and from different manufacturers to account for practical realities of cost and ownership, is an important step in defining the future of smart agriculture, search and rescue, environmental monitoring, space exploration, and military operations, and we welcome opportunities to work with industry and military leaders to continue to make dependable and trustworthy swarm autonomy a reality.

https://doi.org/10.52202/075106-0015

THE CHALLENGES OF BUILDING TRULY ROBUST ROBOTIC AUTONOMY - DEFINING THE LEVELS OF AERIAL AUTONOMY

Exyn Technologies*

Exyn Technologies presents a framework for Levels of Aerial Autonomy based upon the SAE Levels of Driving Automation in order to better define technical capabilities, maturity, and risks. Presented are definitions and a taxonomy for Aerial Autonomy which is used to map operator and aerial system capabilities to a discrete levels.

INTRODUCTION

Autonomy is the ability to think for yourself and be self reliant - the capacity of self-governance or self-determination. There are different arenas of autonomy, but the one we're most concerned with is personal autonomy -- the capacity to decide for oneself and pursue a course of action in one's life. That's ultimately what we're trying to achieve through our work in autonomous robotics: to create an artificial intelligence platform that can operate a robot in dangerous environments, complete multiple operations successfully, and take care of itself with minimal or no human interaction.

With Driverless Automation pushing into Level 4 and aiming toward Level 5, we were getting questions about our robot's level of autonomy and found we were lacking a clear set of guidelines on what classifies as Aerial Autonomy and defining the levels therein. Work has already been done around classifying autonomy for Unmanned Aerial Vehicles. But in its current state, aerial autonomy is placing the pilot further out of the loop, but still an essential element to operation, and we find the available level classifications inadequate. Additionally, it is becoming increasingly important for UAVs to fly beyond the operator's visual line of sight (BVLOS) for industrial applications, search & rescue missions, and Intelligence, Surveillance and Reconnaissance (ISR) for government operations. Level 4 & 5 autonomy is crucial for these applications to be successful.

PROBLEMS

Current Disjointed Level Definitions

Aspects of aerial autonomy have been siloed in individual organizations since their inception, creating their autonomy in a vacuum. As you'll see in the resources listed at the end of the submission, the problem of defining aerial autonomy has been discussed since ~2002, with each new iteration adding more straw to the pile. And while we acknowledge this is yet another level definition

* Vijay Kumar, Ph.D, University of Pennsylvania, GRASP Labs, Jason Derenick, Ph.D., Exyn Technologies, Denise Wong, Ph.D., Exyn Technologies, Justin Thomas, Ph.D., Exyn Technologies, Nick Lynch, Exyn Technologies, Alex Burka, Ph.D., Exyn Technologies, Justin Lehmann, Exyn Technologies, and Rachel Appel, Exyn Technologies.

 https://doi.org/10.52202/075106-0016

added to that pile, we aim to bring these together cohesively and have the larger UAV community collaborate and agree to them as an industry.

Defining Environmental Complexity

One of the biggest challenges we face when trying to classify levels of autonomy for aerial robots is nailing down what environment they will be able to fly in. Much like a self driving car, a robot can appear to be operating at quite a high level of autonomy if the environment is built to specific robot sensors or capabilities. This isn't so much an issue in the early stages of autonomy, but making steps through Level 4 and Level 5 autonomy will require clearer definitions to prevent false claims using a dubious setup.

Infrastructure

Driverless autonomy is built around road infrastructure that is ubiquitous around the world. Due to assumptions that come along with having such infrastructure (in this case, a largely known and mappable road network) the problem becomes much more tractable. The complexity of environments for aerial robots are generally less well defined, making it much more difficult to prepare an Autonomous Aerial Robot for every eventuality. Current aerial infrastructure includes air traffic control, airspace classifications, air corridors, altitude restrictions, FAA regulations, etc. However those mainly apply to high-altitude aircraft, whereas drones and UAVs might fly underground, throughout an urban environment, or only as high as a general electric tower for inspection (for example).

Rotor vs Fixed Wing

Autonomy, in the form of an auto-pilot, has been an aspect of fixed wing flight for quite some time. While the work we're doing deals mainly with multi-rotor flight -- a type of vertical take-off and landing aircraft -- we have to include fixed-wing in these levels for completeness.

AUTONOMY VS. AUTOMATION

Both terms are thrown around interchangeably, but in reality they are two completely different things. Automation has been around for quite some time, and corresponds to a machine performing a programmed action. For example, autopilot on a commercial airliner. It instructs the aircraft to stay on a chosen course and altitude but a pilot must be present to monitor a variety of other conditions to achieve a successful flight and landing.

Autonomy means that the system has choices to make free of outside influence. Current autopilot systems cannot detect approaching objects or storms and determine the safest and most efficient course to the destination. Future UAVs must be able to think and act on their own without human intervention to be cost effective at scale.

If we look at SAE's Levels of Driving Automation[1], we can see increasing levels of autonomy. At each level, the driver can remove themselves more and more from immediate actions and instead rely on the robot's best judgment.

https://doi.org/10.52202/075106-0016

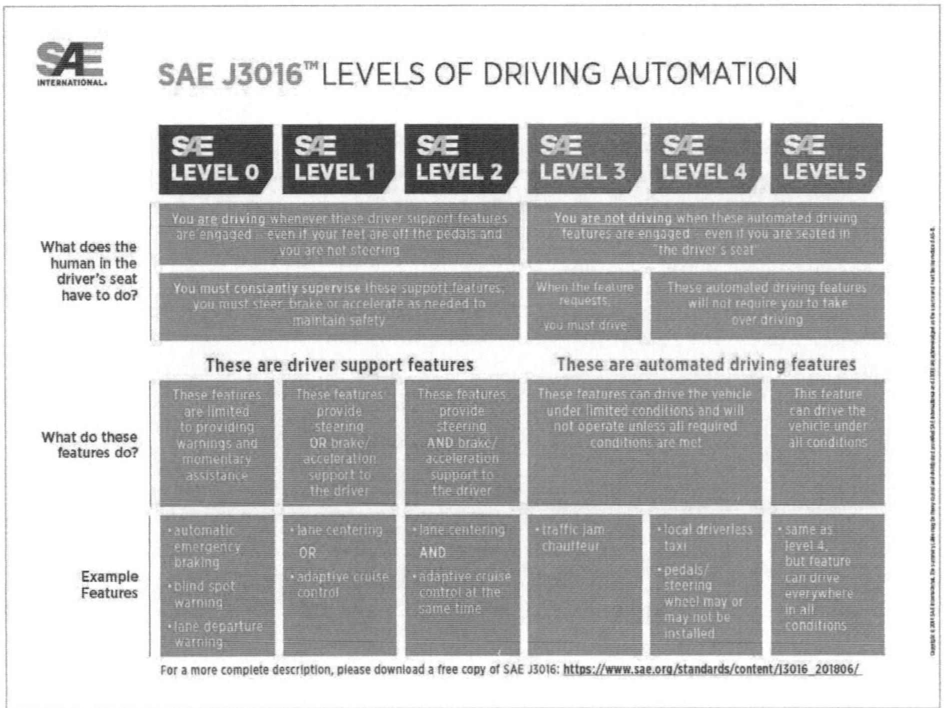

Figure 1. SAE J3016 Levels of Driving Automation

WHAT DOES TRUE AERIAL AUTONOMY LOOK LIKE?

- The robot must be aware of its own sensors and state (operational integrity)
- Decides to launch / complete a mission dependant on the health of the robot
- The robot must sense and react to its environment without human assistance
- The robot must perform its mission successfully without human assistance
- The robot must be able to replicate a successful mission without human assistance
- The robot must be able to keep itself out of harmful situations unless designed otherwise
- More advanced levels of autonomy would feature the robot foraging for its own food (energy) and spare parts, maintaining its operational integrity
- Above everything else, the robot must keep people safe, and sacrifice itself for their safety

https://doi.org/10.52202/075106-0016

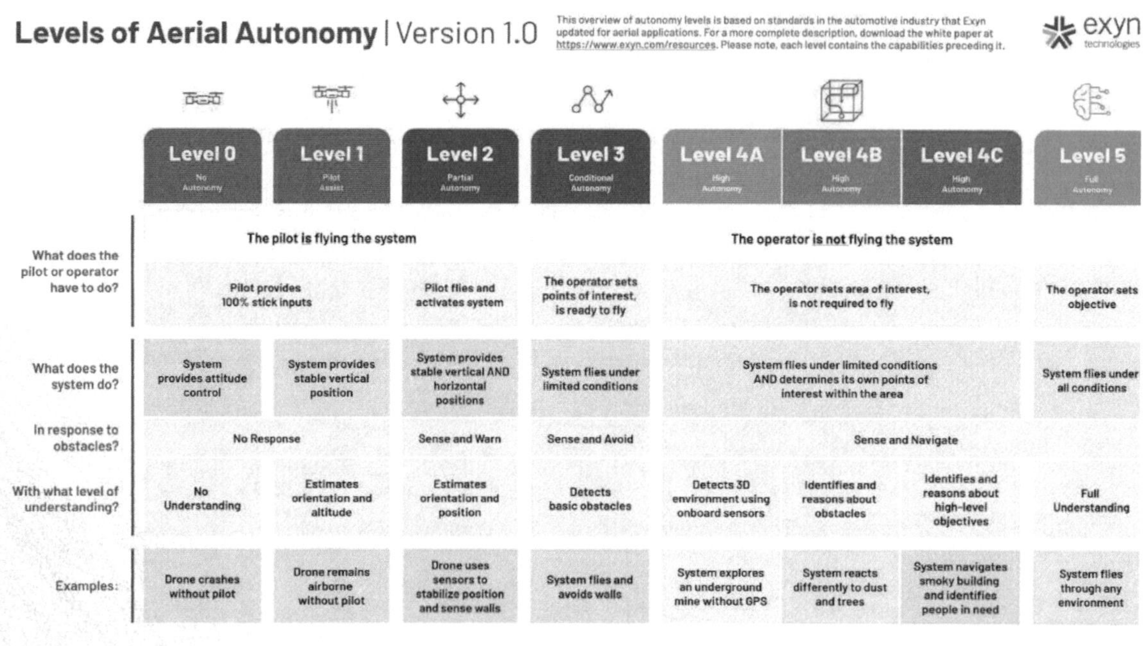

Figure 2. Proposed Levels of Aerial Autonomy

CREATING LEVEL DEFINITIONS AND TAXONOMY FOR AERIAL AUTONOMY

These level definitions, along with additional supporting terms and definitions provided herein, can be used to describe the full range of aerial automation features equipped on airborne vehicles in a functionally consistent and coherent manner.

Table 1. Terms and Definitions

Term	Definition
What is an Operator?	• A general term for referencing the human role in aerial automation
What is an Aerial Vehicle?	• Any vehicle designed to be flown by a pilot or dispatched through an autonomous system • A machine designed to provide conveyance through the air
What is an Aerial Autonomy System (AAS)?	• The hardware and software that are collectively capable of performing part or all of the Dynamic Flight Task (DFT) on a sustained basis; this term is used generically to describe any system capable of level 1-5 aerial autonomy.

What is DFT (Dynamic Flight Task)?	• All of the real-time operational and tactical functions required to operate an aerial vehicle, excluding the strategic functions such as trip scheduling and selection of destination and waypoints, and including without limitation: o Autoflight & hover motion control via rotors (operational); o Yaw, pitch and roll motion control via IMU & rotors (operational); o Monitoring the environment via object and event detection, recognition, classification, and response preparation (operational and tactical); o Object and event response execution (operational and tactical); o Maneuver planning (operational)
What is ODD (Operational Design Domain)? Autonomous Understanding & Reasoning (AUR A, B, C)	• Operational conditions under which a given aerial autonomy system or feature is specifically designed to function, including, but not limited to, environmental, geographical, and time-of-day restrictions, and/or the requisite presence or absence of certain landscape characteristics o Examples ▪ A UAV designed to inspect electrical infrastructure will be bounded to only fly along the electrical grid ▪ A UAV in the mining industry will only map tunnels and shafts when not in use, recharging during operating hours • Autonomy Levels 4 & 5 will require further definitions of the complexities of their environment and at what level the autonomous system can sense and navigate accordingly. Rather than ranking the environmental factors a UAV will need to overcome (light, dust, wind, etc.), complexity will be ranked by how the autonomous system interprets and overcomes obstacles these various environments present. o Examples ▪ A UAV can sense and avoid dynamic obstacles (AUR A) ▪ A UAV can determine an obstacle is dust and fly through it (AUR B) ▪ A UAV can interpret obstacles and use that information to plan aspects of its mission (AUR C)
What is DFT fallback?	• A response by the operator to either perform the DFT or achieve a minimal risk condition after occurrence of a DFT performance-related system failure(s) or upon ODD exit, or the response by an AAS to achieve minimal risk condition, given the same circumstances. o Even at high levels of autonomy (level 3), while the AAS is capable of performing all DFTs within its ODD, it may be unable to perform the DFT fallback in all situations.

https://doi.org/10.52202/075106-0016

What is a minimal risk condition?	• A condition to which an operator or an AAS may bring a vehicle after performing the DFT fallback in order to reduce the risk of a crash when a given trip cannot or should not be completed. o EXAMPLE: When the AAS detects a potentially dangerous wind gust or orientation shift, the AAS or operator (if necessary) would steer the robot to a safe auto hover location to complete the mission (if safe) or return to a safe landing zone.
What is a request to intervene?	• Notification by an AAS to a fallback-ready user indicating that they should promptly perform the DFT fallback, which may entail resuming manual operation of the aerial vehicle, or achieving a minimal risk condition if the vehicle is not operational.
What is a mission?	• A mission is a set of directives that are sent to the robot to complete. This could be as defined as a series of waypoints for the robot to maneuver to, or less specific such as "explore this building" or "find the fire extinguishers in this mine."

IMPACTS OF ENVIRONMENTAL COMPLEXITY ON EVALUATING AUTONOMOUS REASONING

While the steps between Level 0 through Level 3 aerial autonomy are fairly linear, differentiating between High and Complete autonomy (Levels 4 & 5) is remarkably more complex. Robots can achieve large leaps in autonomous intelligence while still remaining inside the Level 4 classification. In order to make these classifications less opaque we must incorporate a subsection of levels outlining the environmental complexities an aerial robot must navigate safely.

Table 2. Autonomous Understanding & Reasoning

	Autonomous Understanding & Reasoning	Autonomy Level
AUR-A	The UAV is capable of sensing & navigating around obstacles in its environment with onboard sensors	4 A
AUR-B	The UAV is capable of sensing & navigating around obstacles in its environment, but also can make determinations about perceived obstacles and how to approach them • Examples o The UAV can delineate between "phantom" obstacles induced by dust while still avoiding actual obstacles.	4 B

AUR-C	The UAV is capable of sensing & navigating around obstacles in its environment, and uses that information to make determinations about how to adjust its mission objectives
	• Example
	  ◦ In a search and rescue operation a robot may have a primary objective of exploring a man-made or natural disaster site for any people or survivors. When the UAV identifies a survivor it can determine whether the person's physical condition warrants immediate attention. If so, the robot can determine whether to pause exploration and instead bring medical support.
	  ◦ The UAV can identify people, doorways, windows through dust & smoke and use that information to execute its mission.
		4 C

LEVELS OF AERIAL AUTONOMY

Operator Performs Part or All of the DFT

Level 0 - No Aerial Autonomy

- Operator (at all times):
  - Performs the entire DFT
- Aerial Autonomy System (if any):
  - Does not perform any part of the DFT on a sustained basis (although other vehicle systems may provide warnings or support, such as visible or auditory cues to the pilot, e.g. low-battery indicator)
  - Flight controller can help support the operator's angular rates
    - EXAMPLE: a FPV racing drone

Level 1 - Pilot Assistance

- Operator (at all times):
  - Performs the remainder of the DFT not performed by the aerial autonomy system
  - Supervises the aerial autonomy system and intervenes as necessary to maintain safe operation of the vehicle
  - Determines whether/when engagement or disengagement of the aerial autonomy system is appropriate
  - Immediately performs the entire DFT whenever required or desired
- Aerial Autonomy System (while engaged):
  - Automatically maintains the specified yaw/pitch/roll rates from the operator
  - Disengages immediately upon operator request
    - EXAMPLE: Commercial drones for photography or recreation

Level 2 - Partial Aerial Autonomy

- Operator (at all times):
  - Performs the remainder of the DFT not performed by the aerial autonomy system

https://doi.org/10.52202/075106-0016

- Supervises the aerial autonomy system and intervenes as necessary to maintain safe operation of the robot
- Determines whether/when engagement or disengagement of the aerial autonomy system is appropriate
- Immediately performs the entire DFT whenever required or desired

- Aerial Autonomy System (while engaged):
  - Automatically maintains the specified yaw/pitch/roll rates from the operator
  - GPS-enabled allowing the robot to return to it's take-off location (in event of a low battery, for example)
  - Disengages immediately upon operator request
    - EXAMPLE: Upper-tier commercial drones for videography

AAS Performs the Entire DFT While Engaged

Level 3 - Conditional Aerial Autonomy

- Operator (while the AAS is not engaged):
  - Verifies operational readiness of the AAS-equipped vehicle
  - Plans waypoint-based mission for the AAS to execute
  - Determines when engagement of AAS is appropriate
  - Becomes the DFT fallback-ready user when the AAS is engaged
- DFT fallback-ready user (while the AAS is engaged):
  - Is receptive to request to intervene and responds by performing DFT fallback in a timely manner
  - Is receptive to DFT performance-relevant system failures in vehicle systems and, upon occurrence, performs DFT fallback in a timely manner
  - Determines whether and how to achieve a minimal risk condition
  - Becomes the operator upon requesting disengagement of the AAS
- AAS (while not engaged):
  - Permits engagement only within its ODD
- AAS (while engaged):
  - Performs the entire DFT
  - Object avoidance enabled
  - Determines whether ODD limits are about to be exceeded and, if so, issues a timely request to intervene to the DFT fallback-ready user
  - Determines whether there is a DFT performance-relevant system failure of the AAS and, of so, uses a timely request to intervene to the DFT fallback-ready user
  - Disengages an appropriate time after issuing a request to intervene
  - Disengages immediately upon operator request
    - EXAMPLE: A UAV equipped with waypoint based autonomy executing a mission with onboard intelligence

Expanding the ODD to Account for Increased Environmental Complexity & Understanding and Reasoning

Level 4 A - High Aerial Autonomy With Moderate Environmental Complexity

- Operator/Dispatcher (while the AAS is not engaged):
  - Verifies operational readiness of the AAS-equipped vehicle
  - Designates area of interest for AAS to explore
  - Determines whether to engage the AAS

- Becomes a safety pilot when the AAS is engaged only if physically present by the vehicle
- Operator/Dispatcher (while the AAS is engaged):
  - Need not perform the DFT or DFT fallback
  - Need not determine whether and how to achieve a minimal risk condition
  - May perform the DFT fallback following a request to intervene
  - May request that the AAS disengage and may achieve a minimal risk condition after it is disengaged
  - May become the operator after a requested disengagement
- AAS (while not engaged):
  - Permits engagement only within its ODD + AUR A
- AAS (while engaged):
  - Performs the entire DFT
  - May issue a timely request to intervene
  - Performs DFT fallback and transitions automatically to a minimal risk condition when:
    - A DFT performance-relevant system failure occurs or
    - An operator does not respond to a request to intervene or
    - An operator requests that it achieve a minimal risk condition
  - Disengages, if appropriate, only after:
    - It achieves a minimal risk condition or
    - An operator is performing the DFT
  - May delay operator-requested disengagement

Level 4 B - High Aerial Autonomy With Increased Environmental Complexity

- Operator/Dispatcher (while the AAS is not engaged):
  - Same as Level 4 A
  - Becomes a passive safety pilot when the AAS is engaged
- Operator/Dispatcher (while the AAS is engaged):
  - Need not perform the DFT or DFT fallback
  - Need not determine whether and how to achieve a minimal risk condition
  - May perform the DFT fallback following a request to intervene
  - May request that the AAS disengage and may achieve a minimal risk condition after it is disengaged
  - May become the operator after a requested disengagement
- AAS (while not engaged):
  - Permits engagement only within ODD + AUR B
- AAS (while engaged):
  - Same as Level 4 A

Level 4 C - High Aerial Autonomy With Severe Environmental Complexity

- Operator/Dispatcher (while the AAS is not engaged):
  - Same as Level 4 A
  - Less severe need for a safety pilot fall-back when AAS is engaged
- Operator/Dispatcher (while the AAS is engaged):
  - Need not perform the DFT or DFT fallback
  - Need not determine whether and how to achieve a minimal risk condition
  - May perform the DFT fallback after disengaging AAS

https://doi.org/10.52202/075106-0016

    - May request that the AAS disengage and may achieve a minimal risk condition after it is disengaged
    - May become the operator after a requested disengagement
- AAS (while not engaged):
    - Permits engagement only within ODD + AUR C
- AAS (while engaged):
    - Same as Level 4 A

AAS Performs the Entire DFT Under Any ODD

Level 5 - Full Aerial Autonomy

- Operator/Dispatcher (while the AAS is not engaged):
    - Verifies operational readiness of the AAS-equipped vehicle
    - Sets a high-level mission objective with the AAS
    - Determines whether to engage the AAS
    - Becomes a non-actor when the AAS is engaged
- Operator/Dispatcher (while the AAS is engaged):
    - Need not perform the DFT or DFT fallback
    - Need not determine whether and how to achieve a minimal risk condition
    - May perform the DFT fallback following a request to intervene
    - May request that the AAS disengage and may achieve a minimal risk condition after it is disengaged
    - May become the operator after a requested disengagement
- AAS (while not engaged):
    - Permits engagement of the AAS under all ODD conditions
    - Complete autonomous understanding and reasoning
- AAS (while engaged):
    - Performs the entire DFT
    - Performs DFT fallback and transitions automatically to a minimal risk condition when:
        - A DFT performance-relevant system failure occurs or
        - An operator does not respond to a request to intervene or
        - An operator requests that it achieve a minimal risk condition
    - Disengages, if appropriate, only after:
        - It achieves a minimal risk condition or
        - An operator is performing the DFT
    - May delay operator-requested disengagement

DIFFERENCES BETWEEN LEVEL 4 AND LEVEL 5 AUTONOMY

In the early levels of aerial autonomy, the main differentiators between them are the capabilities of the system to take over control of the robot. You can see this in how the human operator is receding farther away from overall control, and they're only needed in an emergency situation. But Levels 4 and 5 are remarkably similar. The main differentiator is in the operating environment the robots are capable of flying in.

To achieve Level 5 Autonomy, a UAV must repeatedly demonstrate successful flight in any environment without requiring interaction or assistance from a pilot or operator. Achieving this

https://doi.org/10.52202/075106-0016

Level of autonomy will likely require advances in nearly all elements of a Level 4 stack from motion planning to advanced perception. The goal is to enable a system to exhibit safe behavior through degenerate, degraded visual environments and under high levels of environmental and system uncertainty. It is a formidable challenge.

WHAT COMES NEXT?

While our team put extensive thought and care into the creation of this paper, we acknowledge that they are still incomplete without input from the larger UAV community. If you would like to collaborate with us to improve these levels for the entire autonomous industry, please email us, hello@exyntechnologies.com.

ACKNOWLEDGMENTS

Contributors and editors of this paper include Vijay Kumar, Ph.D, University of Pennsylvania, GRASP Labs, Jason Derenick, Ph.D., Exyn Technologies, Denise Wong, Ph.D., Exyn Technologies, Justin Thomas, Ph.D., Exyn Technologies, Nick Lynch, Exyn Technologies, Alex Burka, Ph.D., Exyn Technologies, Justin Lehmann, Exyn Technologies, and Rachel Appel, Exyn Technologies.

REFERNECES

[1] SAE International. (2018, 12 11). *SAE International Releases Updated Visual Chart for Its "Levels of Driving Automation" Standard for Self-Driving Vehicles*. Retrieved from SAE International: https://www.sae.org/news/press-room/2018/12/sae-international-releases-updated-visual-chart-for-its-%E2%80%9Clevels-of-driving-automation%E2%80%9D-standard-for-self-driving-vehicles

[2] NIST. *Autonomy Levels for Unmanned Systems (ALFUS) Framework.* Retrieved from: https://www.nist.gov/system/files/documents/el/isd/ks/NISTSP_1011_ver_1-1.pdf

[3] *DRONEII: Tech Talk – Unraveling 5 Levels of Drone Autonomy:* Retrieved from: https://dronelife.com/2019/03/11/droneii-tech-talk-unraveling-5-levels-of-drone-autonomy/

[4] Clough, B.T. (2002). Metrics, Schmetrics! How The Heck Do You Determine A UAV's Autonomy Anyway.

[5] Valavanis, K.P., Vachtsevanos, G.J. (2015). UAV Autonomy: Introduction. In: Valavanis, K., Vachtsevanos, G. (eds) Handbook of Unmanned Aerial Vehicles. Springer, Dordrecht. https://doi.org/10.1007/978-90-481-9707-1_144

https://doi.org/10.52202/075106-0016

GPS DENIED - THE NEXT LEVEL IN AUTONOMOUS NAVIGATION TOWARDS FLYING WITHOUT GPS

Georg Puhlfürst[1], Jürgen Hess[1], Stefan Milz[1]

This paper gives an overview of current challenges in GPS-denied drone navigation, highlighting a crucial advancement towards achieving full autonomy in flights devoid of GPS. It emphasizes the essentiality of GPS-denied solutions for ensuring uninterrupted and safe flight operations in environments where GPS signals are unreliable or absent. The exploration is divided into two principal methodologies: managing sudden GPS loss during flight, and initiating flight without GPS reliance. For scenarios involving GPS disruption mid-flight, the study proposes the employment of visual odometry, utilizing both traditional and deep-learning approaches, to sustain navigation accuracy. Alternatively, for flights planned without GPS from the beginning, we rely on a method of aligning live camera images with satellite imagery. This approach not only enhances navigation reliability in GPS-compromised situations but also sets a new standard for autonomous flight technology.

INTRODUCTION

Traditionally, UAVs and drones rely on GPS guidance for navigating their flight paths. This system is popular and serves as the basis of modern drone navigation as it is highly precise and reliable. However, there exist scenarios where utilizing GPS for navigation proves to be challenging if not impossible. This includes situations such as dense urban environments where high buildings create "urban canyons" that disrupt GPS signals, or scenarios where GPS signals are intentionally jammed or spoofed. Under such conditions, GPS can be lost for varied durations, lost permanently, or altogether absent. Without reliable alternatives, drones risk becoming disoriented or ultimately lost, compromising mission success and operational safety. Given these challenges, our objective is to develop robust GPS-denied navigation solutions, ensuring drones can maintain efficacy across a diverse array of operational hurdles.

In this paper, we address the two pivotal use cases of GPS-Denied navigation: experiencing GPS loss mid-flight and operating entirely void of GPS. For the first use case we employ visual odometry techniques, both classical and deep-learning-based. These methods enable drones to estimate their movement continuously, from one time step to the next by analyzing subsequent camera images. For the second use case, navigating entirely without GPS, we combine this visual odometry with an image based global localization. This allows drones to estimate their position within a larger area by matching onboard camera images with pre-existing maps or satellite imagery. Together, these strategies can significantly increase drone autonomy and effectiveness across diverse operational landscapes. Figure 3a shows an example flight.

---

[1] Spleenlab GmbH

https://doi.org/10.52202/075106-0017

APPROACHES FOR GPS-DENIED NAVIGATION

In this section, we detail our solution to the two GPS-Denied use cases, a sudden loss of GPS during flight and navigating entirely without GPS

Use Case 1: Losing GPS During Flight

For drones that lose the GPS signal mid-flight, we advocate the implementation of visual odometry techniques. These methods, which can be either classical or based on deep learning, allow drones to estimate their movement over time by analyzing the visual differences between sequential images captured by onboard cameras.

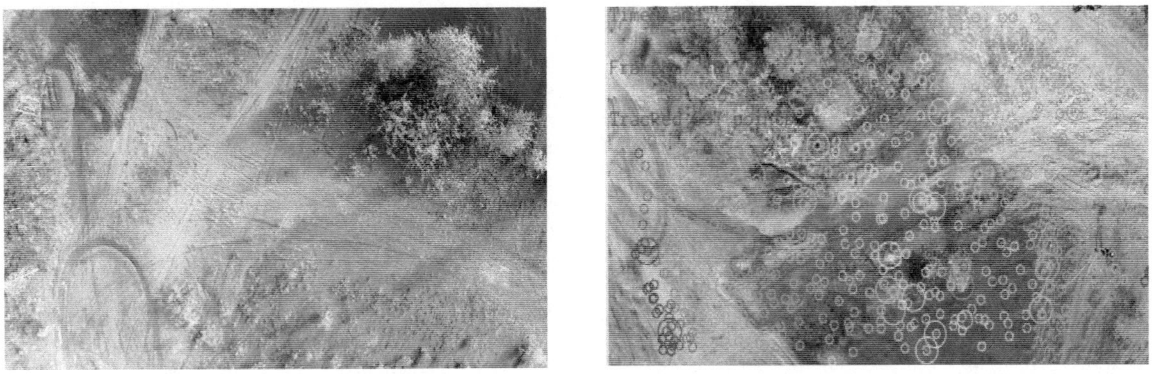

Figure 1: An example image recorded during the flight tracked in Figure 3 (left). An example of image features tracked during flight (right). The color symbolizes the distance to the drone. Dark blue represents features furthest away from the drone camera while dark red mark features closest to the camera.

Our proposed solution employs a monocular camera system for UAV navigation. This approach, however, introduces the challenge of accurately estimating the scale of movement, a task that is inherently complex due to the reliance on a single viewpoint for depth perception. While stereo camera systems might seem like an alternative, offering direct depth information through the use of two cameras, they fall short in aerial applications where the UAV's flight altitude significantly exceeds the baseline, i.e. the distance between the two cameras. In such cases, the stereo system's effectiveness in depth estimation diminishes, rendering it inadequate for large-scale depth perception.

To calibrate the scale for the monocular camera, we utilize GPS data when it is available, e.g., at the start of the flight. While the GPS data is available, we already run and analyze the current scaling of the visual odometry output. This is done by matching the poses that are computed by the visual odometry system to the drone positions as received from GPS. As a result, once the GPS is failing, we can leverage the previously computed scale information together with the continuous visual odometry for calculating an estimated GPS location. An overview of how our approach integrates into the entire navigation system is shown in Figure 2.

To implement monocular visual odometry, we leverage two approaches depending on the use environment the drone is operating in and the computational resources available.

The first system utilizes classical features to compute optical flow between images, such as ORB (Oriented FAST and Rotated BRIEF)[1], which are renowned for their efficiency and low computational cost. ORB excels in identifying and matching distinctive points between

https://doi.org/10.52202/075106-0017

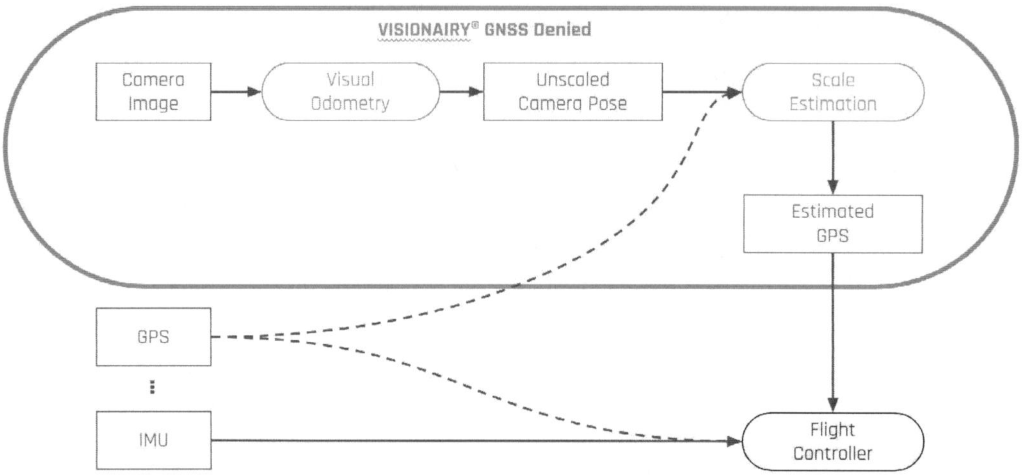

Figure 2: System overview for the use case of losing GPS during the flight. At times where GPS is available, this information can be used to scale the monocular visual odometry. Once the GPS fails, an estimated GPS location can be sent to the flight controller that can also integrate additional information such as IMU or wind sensor data.

consecutive images, enabling the drone to estimate its movement by comparing these points over time. We enhance this with additional steps that help refine the movement and camera position estimation. After obtaining feature correspondences between images, we apply a graph-based optimization together with a keyframe selection scheme to estimate the movement and pose of the camera. This method is particularly effective in environments where the visual features remain consistent and well-defined, allowing for precise movement estimation without the need for intensive computational resources.

Figure 3 shows an example result of applying feature-based visual odometry inside the GPS-Denied system. The blue dot marks the start of the flight. In the beginning, GPS information is still available. We simulate a GPS cutoff at the position of the orange dot. The resulting trajectories are shown in green (true location as received from the receiver) and red (estimated GPS location after the simulated cutoff). As is visible from the figure, the algorithm is highly accurate. While experiencing some expected drift, the estimated trajectory stays close to the measured location.

The second system makes use of neural networks and is based on Deep Pose Visual Odometry (DPVO)[2] to infer the drone's pose and movement from raw image data, offering highly accurate results in complex environments. Unlike classical methods, which rely on explicit feature detection and matching, DPVO learns to recognize patterns and correlations in the visual data, enabling it to adapt to a wider range of visual scenarios. This adaptability makes DPVO particularly suitable for environments where traditional feature-based methods struggle, such as areas with limited distinct visual markers or in varied lighting conditions. However, it does require more computational resources, i.e., the usage of an onboard GPU which also leads to a higher power consumption. In addition, DPVO utilizes dense depth data for training which is hard to obtain, especially for UAV applications. In another work, we have shown how to train DPVO with minimal supervision and thus mitigate this challenge.[5]

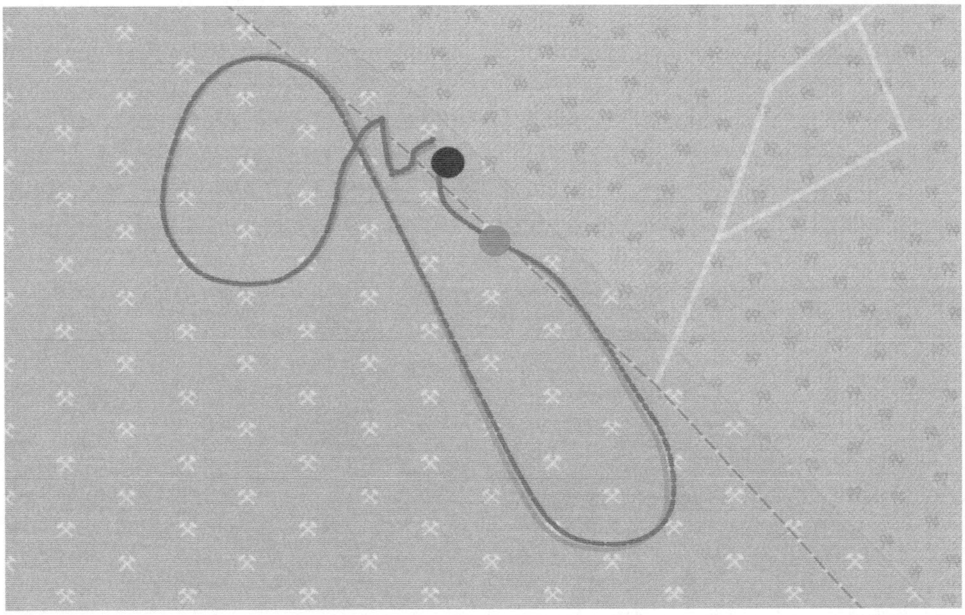

a) Example trajectory flown over a quarry. Plotted are the true GPS trajectory (green) and the estimated trajectory (red). The flight time of the drone was approximately 4.5 minutes. The start is marked in blue, the simulated GPS cutoff is marked in orange. The estimated trajectory stays close to the true trajectory even through the sharp turn at the end.

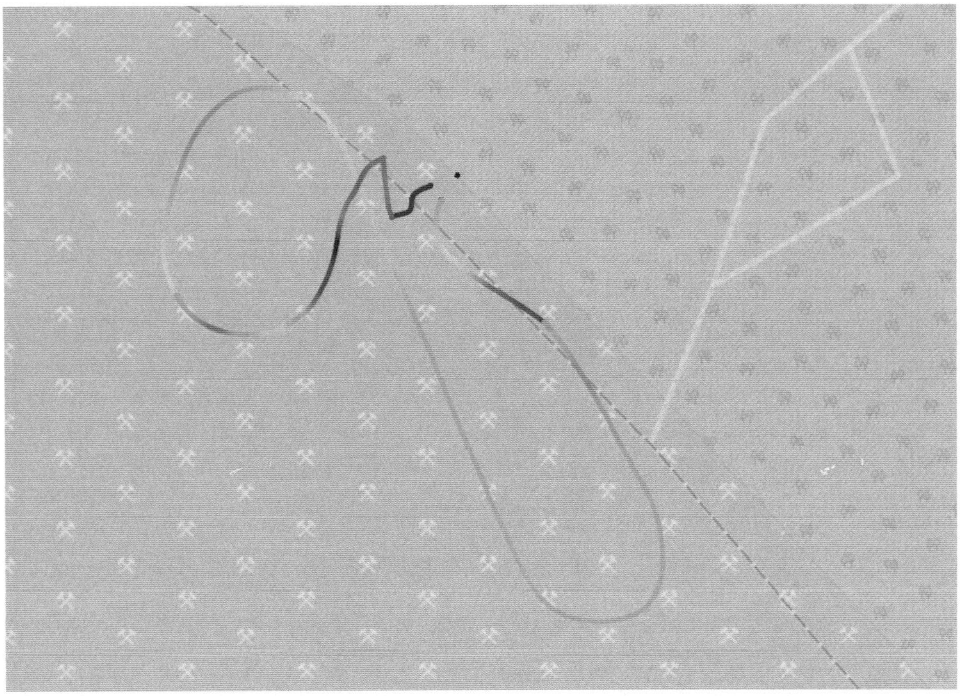

b) Visualization of the absolute trajectory error (ATE) over the entire trajectory. Dark red marks the largest error of about 27m. The mean ATE over the part of the entire flight is ~15m.

Figure 3: Example of a GPS-Denied trajectory flight over a quarry plotted onto an OpenStreetMap.

https://doi.org/10.52202/075106-0017

Use Case 2: Completely Navigating Without GPS

In scenarios where drones are required to operate entirely without GPS, combining visual odometry with global localization techniques offers a comprehensive solution. This dual approach enables drones not only to track their immediate movement through visual odometry but also to locate their position within a larger mapped area. By matching real-time imagery with pre-existing GPS indexed maps or satellite images, drones can achieve a level of precision navigation that would otherwise be unattainable without GPS. Figure 4 shows an example of matching an image obtained from sensory input during the flight to a Google maps aerial image.

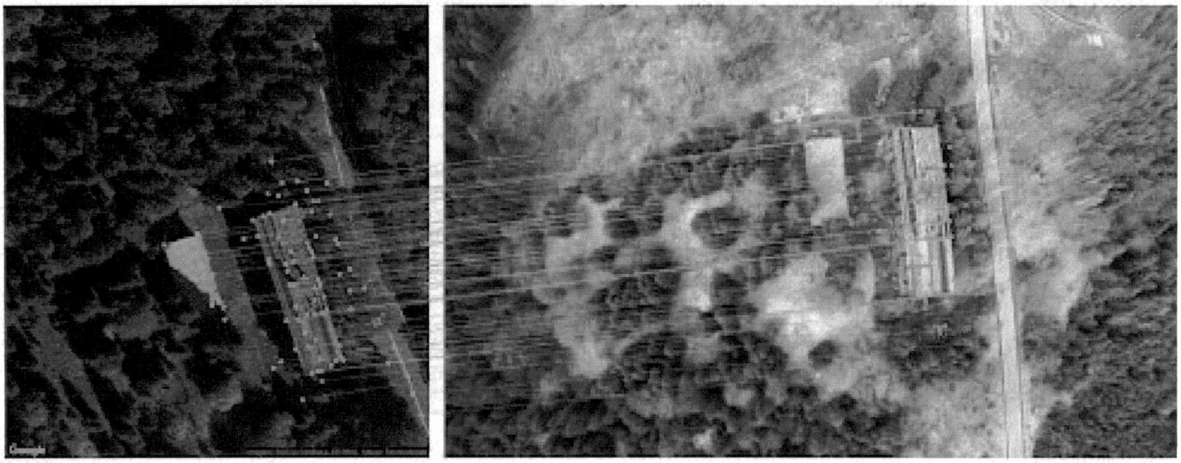

Figure 4: Example of matching visual features of a camera view (right) to Google satellite imaginary (left). The matching proves robust to different viewpoints and even larger vegetation changes.

To implement this matching process, we employ cutting-edge deep-learning based algorithms such as SuperGlue[3] and LightGlue[4], which are exceptional in their ability to identify and match features between the drone's immediate visual input and the corresponding sections of pre-stored maps or satellite images. We have tuned and highly optimized these algorithms for the specific hardware onboard the drone, allowing a processing speed beyond the current state-of-the-art. We then leverage the found feature matches to compute the current GPS position of the drone and thus realize global localization. While the matching algorithm is highly optimized, it yields a lower frame rate than the visual odometry. While visual odometry and global localization can run independently, we propose a hybrid system. This system runs the visual odometry and estimates the relative movement of the drone in addition to the global localization runs to provide a continuous estimate to the flight controller. Figure 5 shows the integration of this approach into the system control view.

This approach, however, does come with its set of challenges. First, its reliance on a GPU for processing means that the energy consumption are factors that must be carefully managed to ensure prolonged operational periods. Furthermore, for the matching process to be effective, the drone needs to maintain a sufficient altitude, ensuring it captures a broad enough view of the surrounding environment to match with the aerial data. This requirement on a high altitude poses limitations on the flight path in densely populated areas, environments with significant vertical obstructions, or other areas where high altitude flight is highly regulated. Additionally, aerial image data, comparable in altitude to that at which the drone is flying, must be available, processed beforehand, and stored onboard. This introduces constraints regarding data storage and

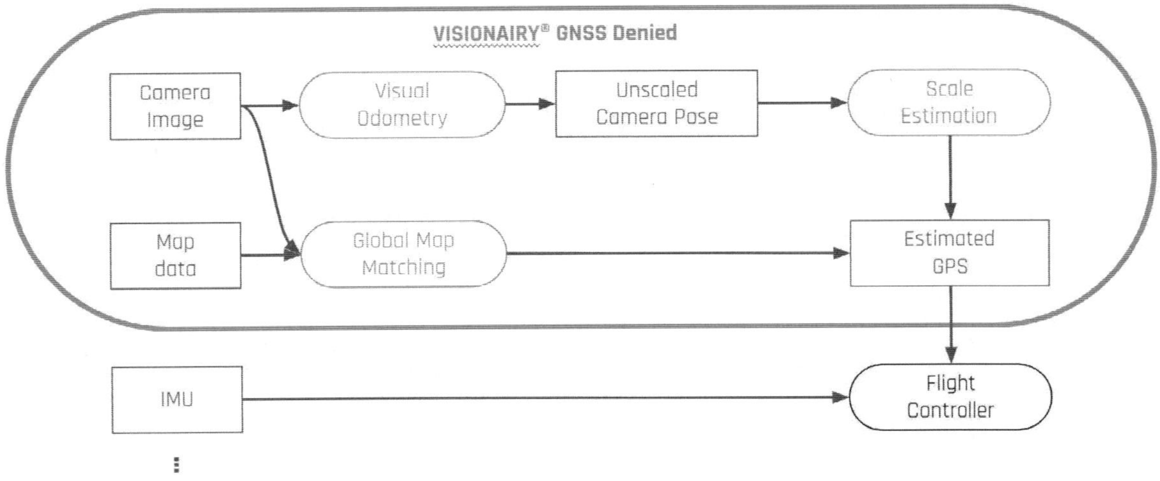

Figure 5: System overview for the use case of navigating completely without GPS. The global map matching typically runs at a lower frequency, providing global GPS locations. The visual odometry estimates the local movement between the global matching runs. The flight controller integrates the estimated GPS locations together with additional sensor information such as IMU data.

management. Despite these challenges, the advantages offered by this method, such as the ability to navigate with high precision in GPS-denied environments, make it a promising avenue for enhancing drone autonomy and operational capabilities.

GPS-DENIED NAVIGATION COMPETITION

To benefit the community with valuable datasets as well as advance our own research, we are launching a competition that challenges participants to navigate drones without GPS by solely relying on visual data from three onboard cameras. Hosted as part of the CVPR OmniCV2024 Workshop, the aim is to develop an algorithm capable of accurately calculating the drone's flight path in GPS-denied scenarios. The competition, named "Omni-BALLOON" in homage to the historic East Germany balloon escape, underscores the transformative potential of camera-assisted navigation. This endeavor not only fosters innovation but also provides a platform for showcasing novel solutions in the realm of autonomous flight. CVPR OmniCV2024 Workshop. We encourage everyone to participate in showcasing and comparing your solution through Kaggle[2].

CONCLUSION

GPS-Denied drone navigation in real-time is extremely challenging. We have presented approaches for two use cases, losing GPS mid-flight and flying the entire mission without GPS. In addition, through our "Omni-BALLOON" competition, part of the CVPR OmniCV2024 Workshop, we offer the community valuable datasets for fostering innovation in autonomous flight technology. There are still a number of challenges ahead, including flying in low-light or night time conditions and flying robustly without GPS for several hours at a time. However, through these efforts, we lay the groundwork for future advancements in drone navigation, moving closer to reliable operations in environments where GPS is compromised or unavailable.

[2] https://www.kaggle.com/competitions/omni-balloon

 https://doi.org/10.52202/075106-0017

REFERENCES

[1] E. Rub2lee, V. Rabaud, K. Konolige and G. Bradski, "ORB: An efficient alternative to SIFT or SURF," *2011 International Conference on Computer Vision*, Barcelona, Spain, 2011, pp. 2564-2571

[2] Teed, Zachary, Lahav Lipson, and Jia Deng. "Deep patch visual odometry." *Advances in Neural Information Processing Systems* 36 (2024).

[3] Sarlin, Paul-Edouard, et al. "Superglue: Learning feature matching with graph neural networks." *Proceedings of the IEEE/CVF conference on computer vision and pattern recognition*. 2020.

[4] Lindenberger, P., Sarlin, P. E., & Pollefeys, M. (2023). Lightglue: Local feature matching at light speed. In *Proceedings of the IEEE/CVF International Conference on Computer Vision* (pp. 17627-17638).

[5] Abouee, Amin, et al. "Weakly Supervised End2End Deep Visual Odometry", *In Proceedings of the 7th International Workshop on Visual Odometry and Computer Vision Applications Based on Location Clues at the CVPR 2024* (accepted, to be published)

A DEEP LEARNING APPROACH TO ESTIMATE GROUND RISK AND PLAN MISSIONS FOR UAS IN URBAN AREAS

Jeffrey Pattison, Ph.D.*

The past decade has seen an explosion in the use of Uncrewed Aerial Systems (UAS). Concerned with safety, regulatory agencies across the globe have imposed regulations on the use of UAS, especially around people in heavily populated areas. As a result of these restrictions, highly populated areas have been slow to adopt UAS. As industry pushes for more autonomous operations and the expansion of UAS applications, additional risk assessment and mitigation techniques are required to ensure the safety of the people and property is maintained as the role of the pilot diminishes. Unlike common risk assessment methods found in literature, this research proposes a novel way to use machine learning to rapidly evaluate the ground risk for a UAS. This rapid risk assessment can be used to determine whether a UAS route exceeds an allowable level of risk, or it can assist in the UAS route planning, guaranteeing the ground risk is minimal. The main contributions are to introduce a new way to evaluate and benchmark UAS ground risk while also introducing an additional risk mitigation technique by being able to rapidly generate low risk UAS routes.

INTRODUCTION

Public interest in Uncrewed Aerial Systems (UAS) has grown in the past few decades as they are being adopted to complete an increasing number of tasks, ranging from surveillance to mapping. However, safety concerns around the use of UAS have slowed the adoption of the UAS into the National Airspace (NAS). As a result, the process of UAS adoption has been incremental, where rules and regulations are slowly relaxed or repealed. Some of the more restrictive regulations forbid operating Beyond Visual Line of Sight (BVLOS) and largely forbid operating over groups of people. While obtaining waivers to operate BVLOS is possible, the ability to operate over people is still heavily regulated due to safety concerns. Hence, these requirements significantly hinder UAS use in urban areas that are dynamic and highly populated.

Urban areas can see large benefits from the capabilities UAS has to offer. This includes decreasing response time from first responders to improving logistics for shipping food and goods, or even delivering medical supplies in record times. These heavily populated areas have made UAS use very difficult with current regulations. Currently, regulatory agencies around the world limit the ability to fly over people depending on the risk levels a UAS presents to those below, so a revision of current regulations would be required to exploit the full benefits of UAS. This change cannot occur until there is sufficient reason to believe the use of UAS in these areas is sufficiently safe. Therefore, realistic and detailed risk analysis for UAS is required to ensure any UAS operation over people does not exceed some acceptable level of risk.

* Founder and President, Pattent, pattisonenterprisellc@gmail.com

https://doi.org/10.52202/075106-0018

With safety being the primary goal of the FAA and similar regulatory agencies, risk evaluation and mitigation has long been required for all air traffic. Due to the long history of crewed aircraft, this risk assessment can be done using historical aviation accident and incident data[2]. Because the introduction of UAS has been more recent compared to crewed aircraft, there is an insufficient collection of data related to operations like flight hours, number of accidents and incidents, and failure rates [2]. Consequently, the use of models and simulations has proven to be the next best option for UAS risk assessment. While the risk assessment for crewed aircraft includes the risk to those onboard, UAS does not pose similar threats, so the primary risk for UAS is to those in other aircraft and to those on the ground. This work will focus on the latter and explore ground risk.

There are several factors that can impact the ground risk levels of UAS, including the surrounding environment and the characteristics of the UAS and its trajectory. Several approaches can be found in literature outlining how to approximate risk for a UAS. From Reference 1, the risk for a UAS is quantified as the expected number of fatalities per flight hour, and this metric can be determined using the kinetic energy of the UAS on impact, the probability of hitting a person, and the availability of shelter that might absorb some of the kinetic energy. The common ground risk assessment using this approach is to take a predefined UAS path and simulate different failure events along the path. For each failure event, a descent trajectory can be approximated using the governing laws of physics to estimate the most probable impact locations. For example, in the event of a power failure, a ballistic trajectory can be approximated to predict where the UAS will land. Finding the probable impact locations is required because there is some uncertainty in the UAS initial position, speed, and aerodynamic characteristics, 4 so several descent trajectories need to be simulated. From the impact location, the kinetic energy can be determined. Combining kinetic energy with the population density, the expected fatality rate can be calculated[1]. As mentioned, this process is repeated along various points of the UAS trajectory. Because the risk assessment is completed for a predefined path, this is of little service if the safety levels for the given path exceed the maximum level of safety. Hence, more information is required to assist in the route planning phase to ensure the path of the UAS does not exceed maximum safety levels.

In Reference 4, the authors build on the ground risk assessment approach previously described but use the approach to compile a risk map. This risk map identifies the high-risk areas and is used to assist in the path planning of a UAS. Using the risk map with a path planning algorithm, the UAS can find an optimal path that minimizes the ground risk. The authors accomplish this by discretizing a flight zone and assessing the ground risk at every discretized location for a given cruising speed and UAS model. Using the high-fidelity probabilistic ground risk approach previously described at every location in the risk map can become quite time consuming and computationally expensive. This is problematic in urban areas where the population density is incredibly dynamic, and the ground risk map needs to be updated frequently. To improve upon the work in Reference 4, this work offers a novel approach for ground risk assessment for uncrewed rotor vehicles using a Machine Learning model in place of the physics-based model to enable rapid ground risk assessment. This risk assessment can then be used by the FAA and others to quantify just how dangerous a UAS operation over people is and enables better decision making when planning a UAS mission. To compare the physics-based approach with the Machine Learning approach, both methods will be used to create a risk map for the campus at the Georgia Institute of Technology to see how closely the Machine Learning method accurately estimates the physics-based model.

With a rapid way of generating a UAS risk map, UAS users can not only determine if a predetermined UAS route exceeds a maximum level of allowable risk, but it can also be used to find a route that minimizes risk. By minimizing the ground risk, an additional layer of safety is added to UAS operations in urban areas, which is required if the role of the pilot is to be reduced. Reducing the need for a pilot benefit all who are looking to use UAS by decreasing the cost of operation.

However, the safety measure alone is not the only benefit since there is a component of public trust as well to using UAS in urban areas. By demonstrating the UAS will be avoiding heavily populated areas, this may build trust in the use of UAS.

PHYSICS-BASED MODEL FOR UAS GROUND RISK ASSESSMENT

As mentioned in the previous section, there is a need for risk evaluation for UAS operations to safely operate over people in urban areas, but with UAS being introduced into the airspace relatively recently, there is insufficient flight data for UAS to make similar risk assessments as for crewed aircraft. For this reason, modeling and simulation has become the best way to estimate UAS risk. The modeling and simulation method used in this work to approximate UAS risk is a physics-based approach based on the state-of-the-art methods outlined in References 1,4,5, and 6. This physics-based approach will then be used to generate training data so a Machine Learning algorithm can approximate the UAS risk in a more time efficient manner.

In Reference 1, the ground risk metric is defined as the expected rate of fatalities, with an acceptable level of risk being 10^{-7} fatalities per flight hour based on equivalent levels of safety seen in crewed aircraft for ground risk. The approach in Reference 4 also uses this metric for ground risk assessment. The equation for the expected rate of fatalities can be seen below for a given location, as shown in Reference 4.

$$f_F = A_{\text{exp}} * D_p * P(fatality|exposure) * f_{gia} \tag{1}$$

In the above equation, f_F is the expected rate of casualties, A_{exp} is the area exposed during the crash, D_p is the population density for the area of the crash, $P(fatality|exposure)$ is the probability of a fatality given the exposure, and f_{gia} is the rate of ground impact accidents.

Area Exposed During a Crash

The term A_{exp} from Equation (1) is the area exposed to a crash for a single person on the ground. From Reference 4, this area can be found with Equation (2) below.

$$A_{exp} = \pi(r_p + r_{uas})^2 sin(\gamma) + 2(r_p + r_{uas})(h_p + r_{uas})cos(\gamma) \tag{2}$$

As seen above, r_p is the radius of the average person, r_{uas} is the radius of the UAS, γ is the glide angle, and h_p is the height of the average person. With this, the area exposed during a crash can be determined.

Population Density Estimation

The population density of an area plays a crucial role in the ground risk for a UAS. Highly populated areas will result in a higher probability of a UAS striking a person in the event of an unplanned and uncontrolled descent. Obtaining accurate estimates on the population density information is critical for enabling UAS ground risk assessment, otherwise the risk assessment is meaningless. Multiple methods for estimating population density have been mentioned in literature.

In Reference 4, city census data is used to estimate the population density. Using census data is adequate for demonstrating proof of concept in an academic setting and is easy to obtain, but this data is static and may not be indicative of how humans move throughout the day. Additionally, census data may be stale and outdated by the time it is available. Another method mentioned in Reference 4 relies on the use of mobile phone data. Mobile phone data may be a good resource for accurate estimates throughout the day, but obtaining this data is difficult. Accessing mobile phone data for a given area may take days to obtain depending on the size of the area of interest. In Reference 14, the LandScan Global Population Database is used from the Oak Ridge National

https://doi.org/10.52202/075106-0018

Laboratory[15]. The LandScan database is open-source and provides high resolution (about 90m x 90m) averages of population density throughout the day and contains population density averages for daytime as well as nighttime. Because the LandScan database offers averages at different times of the day, it was the chosen database for this work to assist in population density estimates. The below images show a heatmap of the population density distribution at the Georgia Institute of Technology for daytime and nighttime created using the 2021 LandScan database. The units for the heatmap are people per square meter with each cell being approximately 10m x 10m.

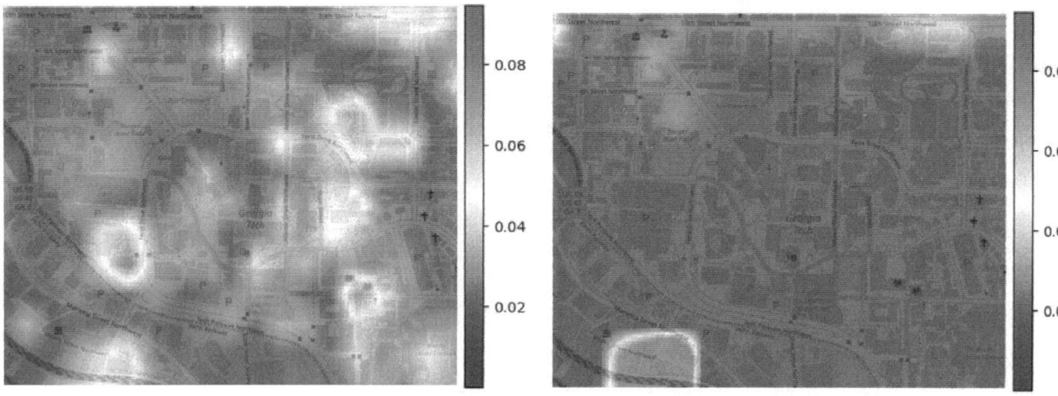

Figure 1: LandScan Population Density at Georgia Tech for Daytime (Left) and Nighttime (Right)

The LandScan database has one of the same disadvantages as the census data in that it is relatively static. Because LandScan is historical data, it fails to account for anomalous events that might result in a change in the normal expected population density. In urban areas, these anomalous events include festivals, concerts, or parades. As a result, any risk assessment solution used in urban areas will need to account for these occurrences that LandScan cannot in order to obtain a more complete picture on the population density as it changes. This work proposes to supplement the use of historical data with a new method for monitoring population density by using social media activity.

The popular social media site Snapchat has accessible information that shows how many posts have been made within a given radius for any given GPS location. Using this method, one can request this information for any location of interest to generate a heatmap of the Snapchat activity. This can become useful to gain insight on events that might draw large numbers of people, which would not be accounted for in the historical data for population densities. To test the validity of this, the social media activity was observed while the Dogwood Festival occurred at Piedmont Park in Atlanta, Georgia from April 15,2023 to April 17, 2023. The two images below show the social media activity at Piedmont Park during different times of the day while the festival took place.

https://doi.org/10.52202/075106-0018

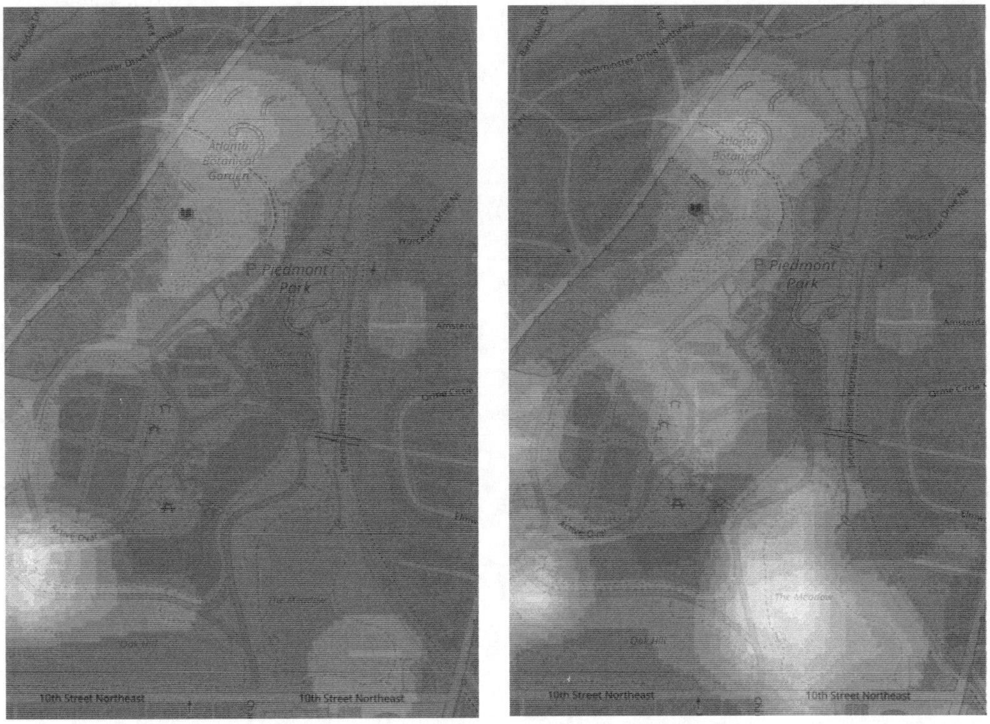

Figure 2: Social Media Activity Generated During the 2023 Atlanta Dogwood Festival at 10:40AM (Right) and 8:30PM (Left)

From the images above, there was more social media activity as the day progressed, suggesting there was a growing population. To show this change in population was abnormal, the social media data was compared with the LandScan data for the same location. This comparison is shown in Figure 3.

https://doi.org/10.52202/075106-0018

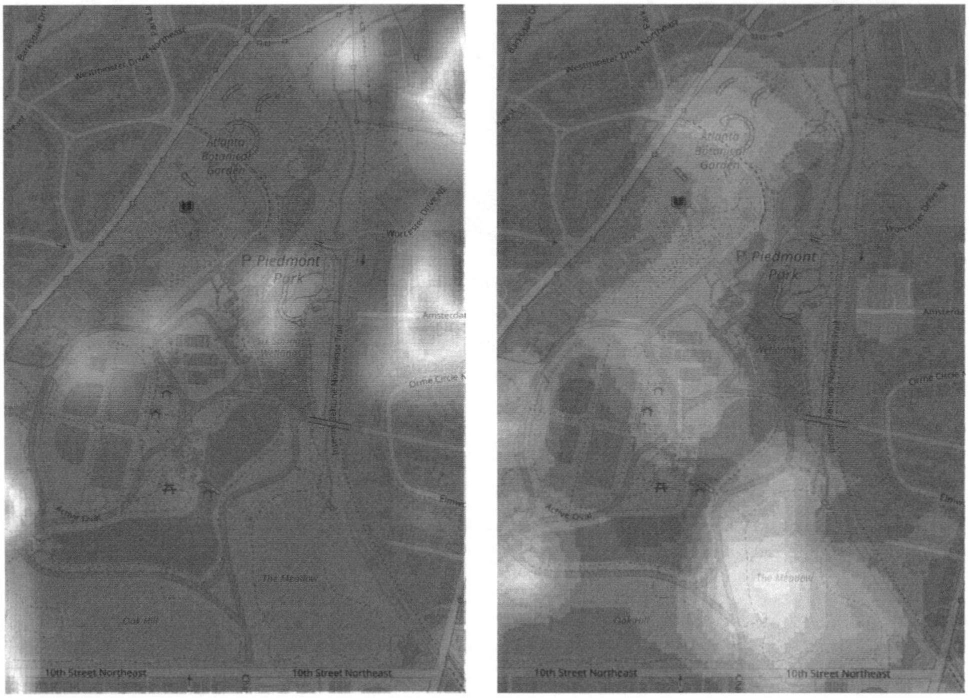

Figure 3: A Comparison Between the LandScan Database (Left) and the Social Media Activity (Right)

The discrepancy between the LandScan population density and the social media data collected confirms the historical data does not properly account for anomalous events like the Dogwood Festival. However, the social media data collected cannot be used by itself because not everyone uses social media, nor is it used everywhere. That is why the two combined can be used to paint a better picture of population densities by taking the maximum value between the two resources. This way the large events that draw crowds and generate social media activity can be taken into consideration while the LandScan database can provide average estimates at every other location that does not generate much social media activity. As better methods for estimating population densities are derived, they can be used in place of the methods proposed in this paper.

Probability of Fatality Given Exposure

The probability of fatality given exposure is the probability of a UAS strike resulting in death if it were to impact a person. One approach outlined in Reference 1 is to map the kinetic energy of the UAS on impact to the probability of resulting in a fatality given the UAS impacts a person. This model takes into account not only kinetic energy, but it also includes a shelter factor that can provide some protection to people on the ground. This shelter factor may be different depending on if the shelter provided is a building, tree, or if there is no shelter at all. The more protection a shelter may provide, the higher this shelter factor, and the less likely a UAS accident is to result in a fatality. No shelter would have a shelter factor of 0 while a building would have a shelter factor of 5, as discussed in Reference 4. The equation for finding probability of fatality given exposure, $P(fatality|exposure)$, can be found below with Equation (3) as seen in Reference 1.

$$P(fatality|exposure) = \frac{1}{1+\sqrt{\frac{a}{b}}\left[\frac{\beta}{E_{imp}}\right]^{\frac{1}{p_s}}} \quad (3)$$

https://doi.org/10.52202/075106-0018

In the above equation, E_{imp} is the kinetic energy at impact and p_s is the sheltering factor to consider surrounding structures that may absorb some of the energy. The α parameter is the impact energy required for a fatality probability of 50% with a sheltering factor of 0.5 while β is the impact energy required for a fatality as the sheltering factor goes to 0. According to Reference 4, acceptable values for α and β are 100 kJ and 34 J, respectively. Common values for the sheltering factor for different types of shelter include 0 for no shelter, 2.5 for sparse trees, and 5 for low buildings[4]. The open-source database OpenStreetMap contains information on the location of buildings that can be used for finding the shelter factor[16]. The image below shows the building coverage layout for the Georgia Institute of Technology, with blue representing the buildings. Every location that did not have any building coverage was assumed to provide no shelter. Each cell in the image below is approximately 10m x 10m.

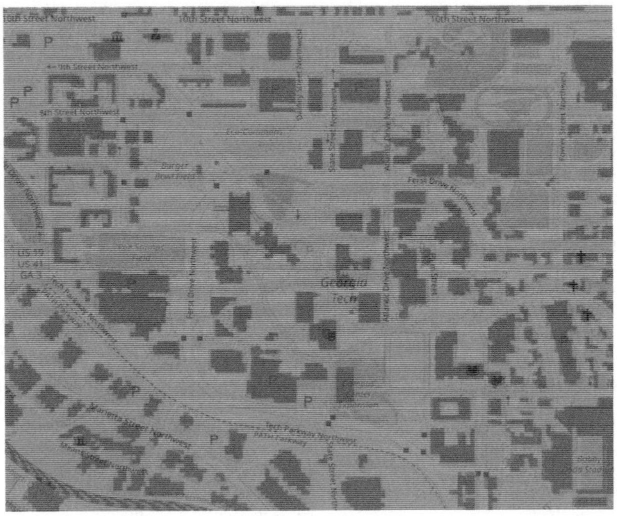

Figure 4: Recorded Building Layouts for Georgia Institute of Technology

The final component of Equation (1) is f_{gia}, which is the rate at which ground impacts occur. This value is measured as number of occurrences per hour, and ideally would be based on flight history data. In Reference 1, the value is estimated to be between 10^{-6} to 10^{-9} accidents per flight hour and is based on the average accident rate involving uncrewed aircraft. However, this failure rate is dependent on each vehicle specifically, and is subject to change as vehicles become safer. For this work, a constant value of 10^{-6} incidents per flight hour is used as the conservative estimate within the range from Reference 1. This value can be updated and changed as better estimates are collected on the true value of the probability of UAS failure.

Determining Probable UAS Impact Locations

The above approach outlined the different components of Equation (1) to quantify the ground risk as expected fatality rates for a UAS in each initial location. With the sheltering factor and population density playing important roles in the ground risk calculation, it is important to know where a UAS is going to land. If a UAS experiences a failure event at a given location, its descent trajectory could land the vehicle at a location far from where the incident occurred depending on the UAS altitude and speed. As a result, estimating the probable impact locations of the UAS is required for proper ground risk assessment to identify the quantities of its impact location, like shelter and population density. There are countless ways that a UAS could fail resulting in an un-planned or uncontrolled descent. Some of these failures include a power outage, a loss of one or more propellers for multi rotor vehicles, or a loss of control from a pilot. Each of these different

https://doi.org/10.52202/075106-0018

failure types result in different descent trajectories. According to Reference 5, the ground risk of a UAS is dominated by a ballistic descent type that might occur if power is lost when compared to other descent trajectories from other failure types, so for this work only a ballistic descent is modeled to find the probable impact locations. The governing laws of physics can be used to find the probable impact locations of a UAS that loses power at some location with some velocity, altitude, and physical characteristics. The governing equations to find the impact locations are shown below.

$$m\ddot{x} = -\frac{1}{2}\rho|\dot{x}|\dot{x}C_d A \tag{4}$$

$$m\ddot{y} = -\frac{1}{2}\rho|\dot{y}|\dot{y}C_d A \tag{5}$$

$$m\ddot{z} = -\frac{1}{2}\rho|\dot{z}|\dot{z}C_d A - mg \tag{6}$$

In the governing equations, m is the UAS mass, $\ddot{x}$, $\ddot{y}$, $\ddot{z}$ are the acceleration of the UAS in the global frame with z being the altitude, ρ is the air density, C_d is the UAS drag coefficient, and A is the frontal area of the UAS. For some given initial location x_0, y_0, z_0 with speeds $\dot{x}_0$, $\dot{y}_0$, $\dot{z}_0$ the governing equations can be solved for when the final altitude z_f is zero to find the values of x_f and y_f , the UAS impact location. However, there is likely going to be some uncertainty in the initial position, velocity, and drag coefficient[7]. To account for the uncertainties of the initial conditions, several descent trajectories are required to find the most likely impact locations. Therefore, for a given descent trajectory i, the initial conditions x_0, y_0, z_0, $\dot{x}_0$, $\dot{y}_0$, $\dot{z}_0$, and C_d are pulled from a normal distribution with a given mean. For example, if the probable impact locations are needed for a UAS flying with a recorded initial speed of 5 m/s at some given location, then the velocities used for simulating the probable impact locations would be taken from a normal distribution centered around 5 m/s. The table below illustrates the normal distributions used for the position, velocity, and drag coefficient[7].

Table 1: UAS Initial Condition Distributions

UAS Parameter	Distribution
x_0	$N(x_0, 0.5)$
y_0	$N(y_0, 0.5)$
z_0	$N(z_0, 0.5)$
$\dot{x}_0$	$N(\dot{x}_0, 2.0)$
$\dot{y}_0$	$N(\dot{y}_0, 2.0)$
$\dot{z}_0$	$N(\dot{z}_0, 2.0)$
C_d	$N(C_d, 0.2)$

To fully account for the uncertainty, 500 descent trajectories were simulated to find the probable impact locations for a UAS traveling with some given initial conditions. Because the objective of the risk assessment is to create a risk map to assist in route planning, no route for the UAS has been determined yet, so the direction the UAS travels has not been specified. To account for this, each of the 500 different descent trajectories is given a different heading, with the heading determined by sampling from a uniform distribution ranging between 0 and 360 degrees. The image below shows the results of 500 simulated descent trajectories to find the most probable impact locations for a UAS with some initial conditions. In the image, it is assumed the UAS initially starts at (0,0) with each blue marker identifying one of the probable impact locations.

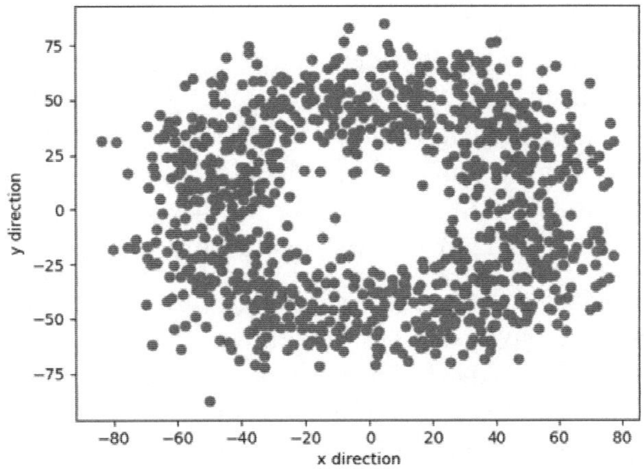

Figure 5: 500 Probable Impact Locations for a UAS

For each of the 500 trajectories simulated at a given location, the kinetic energy of the UAS, population density and sheltering factor were recorded at the location of impact and used to find the expected fatality rate with the equations above. After calculating 500 expected fatality rates for a given UAS starting location, the mean fatality rate of the 500 trajectories was recorded as the fatality rate for that location with the given initial conditions. To create a risk map used to assist in route planning, this process needs to be repeated at various locations. For every location within the map, probable impact locations and expected fatality rates need to be calculated. Because of this, the process to generate a risk map can become quite time consuming as the size of the map increases. The research in Reference 4 and Reference 7 attempt to make simplifying assumptions for determining the probable impact location to reduce the computation time with the trade-off of losing some accuracy. However, adding simplifying assumptions can only go so far before the credibility of the analysis is lost. For this reason, machine learning methods are explored to estimate the risk in a more time efficient manner compared to the high-fidelity physics-based model. As a result, the physics-based model can be made to be as high fidelity as possible with little concern for the computation time because the machine learning model can be used to approximate it within some reasonable degree of accuracy in a fraction of the time required.

MACHINE LEARNING APPROACH FOR UAS GROUND RISK ASSESSMENT

From predictive text and language processing to computer vision and autonomous vehicles, Machine Learning has become quite popular with the rise of Artificial Intelligence (AI). Machine Learning algorithms have shown tremendous capability in learning to recognize patterns in data for problems that are complex for traditional approaches[8]. For this reason, this work explores how well Machine Learning methods would be able to learn and estimate the ground risk of UAS based on the data collected from the high-fidelity physics-based model.

There are many different types of algorithms that fit into the Machine Learning category, with some being better suited for certain applications over others. For the application of estimating UAS

https://doi.org/10.52202/075106-0018

ground risk, the objective is to feed the UAS initial condition parameters and spatial data around a GPS location into the Machine Learning model as inputs, which would then output an expected fatality rate for that given location. This process could be repeated at every location in an area of interest to generate a risk map. One way to do this would be to generate data using the physics-based model to create a database of fatality rates mapped to the input conditions of the UAS. This is a supervised learning approach where the physics model generates the output data, and the Machine Learning model can learn the pattern between the input parameters and the output fatality rate. Because the fatality rates are continuous, the Machine Learning algorithm will need to be used for regression rather than classification. Therefore, a supervised learning algorithm for regression narrows down the list of suitable Machine Learning algorithms to use.

The type of input data will also affect which type of Machine Learning algorithm is suitable. Based on the description of the physics-based model, the type of input data required is mixed between numeric and spatial data. Numeric data would include the mass, speed, frontal area, and altitude of the UAS. All these components affect the kinetic energy upon impact, and therefore affect the expected fatality rate. Beyond the parameters just mentioned, the sheltering factor from the surrounding coverage and the population density also play a role in the fatality rate as well. These values are scalar but based on the probable impact locations seen in Figure 5, it is difficult to pinpoint a single value to use since the population density and building coverage will change based on where the UAS lands. For this reason, the entire area encapsulating the probable impact locations is required as input to accurately estimate the expected fatality rates. Therefore, the Machine Learning model used needs to be able to account for both scalar values like the UAS characteristics and initial conditions and also the spatial inputs like shelter and population density. While some Machine Learning algorithms can handle either numeric data or spatial data, there is no one Machine Learning algorithm well suited for both. The approach proposed in this paper combines two popular Machine Learning algorithms to account for the different input data types. Based on the work in Reference 9, it can be shown that a Multilayer Perceptron (MLP) and a Convolutional Neural Network (CNN) can be combined to account for numeric as well as spatial data. The MLP accounts for the numeric data while the CNN accounts for the spatial data. The outputs of each of these models can then be combined with additional layers to produce a single output, predicting the expected fatality rate. A simple summary of the model can be seen in the image below.

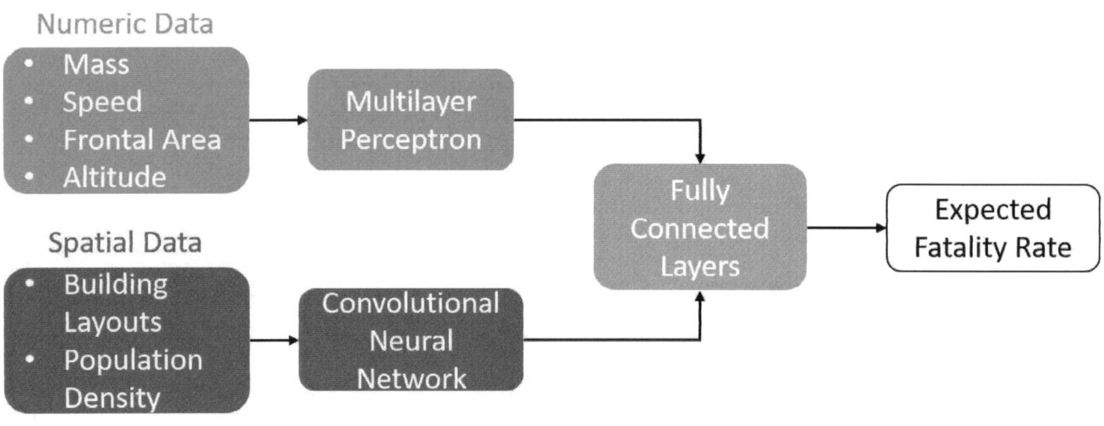

Figure 6: Flowchart for the Machine Learning Model

The Multilayer Perceptron

The Multilayer Perceptron is one of the simplest Artificial Neural Network (ANN) architectures and is comprised of one or more hidden layers between an input layer and an output layer[8]. Each layer consists of several neurons, and each neuron in a layer takes as input all the values of the neurons in the layer before it and then transforms the inputs using an activation function. These values are then passed to all the neurons in the next layer. For traditional regression applications using only a MLP, the final layer is called the output layer and consists of a single neuron, which is the predicted value. The diagram below illustrates a MLP with one input layer and two hidden layers used for this work. The first hidden layer has eight neurons, and the second hidden layer has four neurons, which is the same architecture as found in Reference 9. The input parameters are the UAS characteristics and initial conditions.

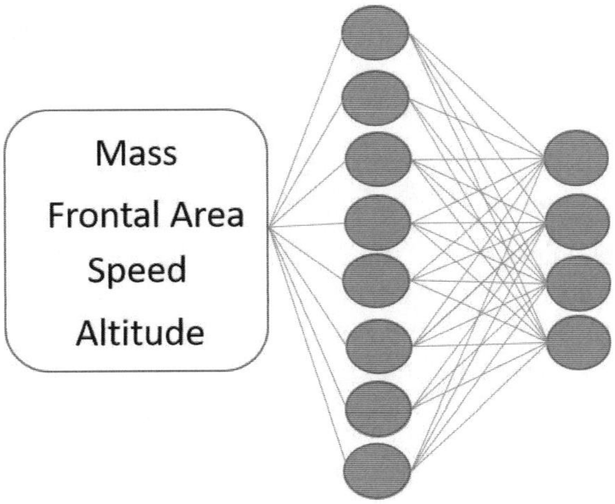

Figure 7: The Multilayer Perceptron of the Machine Learning Model

The Convolutional Neural Network

Convolutional Neural Networks are specialized for processing data with grid-like topology to learn patterns and spatial relationships, which makes them popular when images are inputs since images are grid-like arrays of pixels[10,11]. The main components of the CNN are the convolutional layers and the pooling layers[12]. Most CNN architectures are a combination of convolutional and pooling layers with the final layer being a fully connected layer comprised of all the nodes of an input array unraveled into a single layer.

For this work, the grid-like topology input for the CNN is the building coverage and population density for a given location of interest. This input data takes the form of two arrays, population density and building layouts, of the surrounding area of the location of the UAS. These arrays are intended to only encompass the area of land where the UAS is likely to land. The cells in the array for the building coverage take the value of either 0 and 1 with 0 representing shelter and 1 representing no shelter. The values in the population density array are the number of people per square meter. Based on the CNN architecture in Reference 9, a summary of the CNN can be seen in the image below. The CNN has an initial convolutional layer with 16 filters, then a max pooling layer, then another convolutional layer with 32 filters and max pooling layer, then the CNN is flattened and connected to a hidden layer with 16 neurons, and then finally one last layer with four neurons.

https://doi.org/10.52202/075106-0018

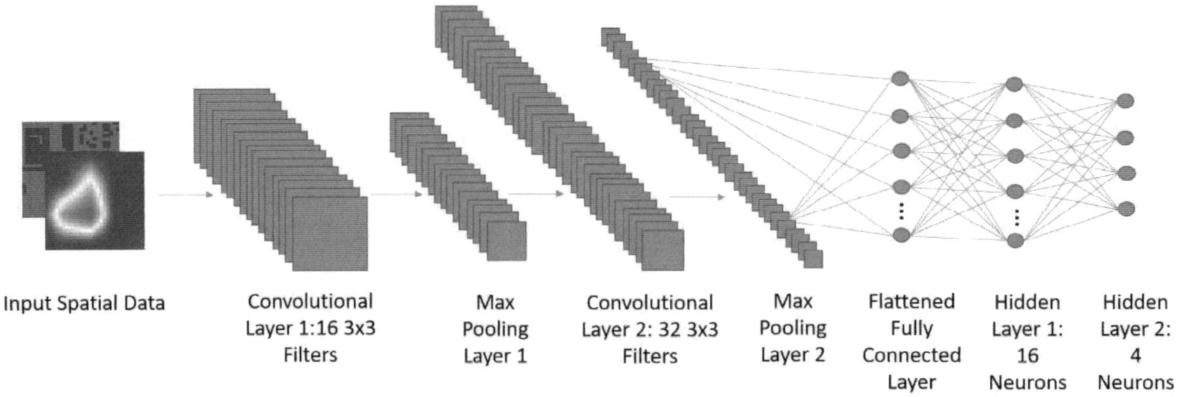

Figure 8: Illustration of the CNN Used

To estimate risk using the MLP and the CNN, the two output layers of each model were then concatenated together. Three additional layers were then added to the combined output. Two hidden layers, one with 10 neurons and one with five neurons, and finally an output layer with one neuron, were added to the model. The final output layer with one neuron is the predicted fatality rate. The image below shows the entire model.

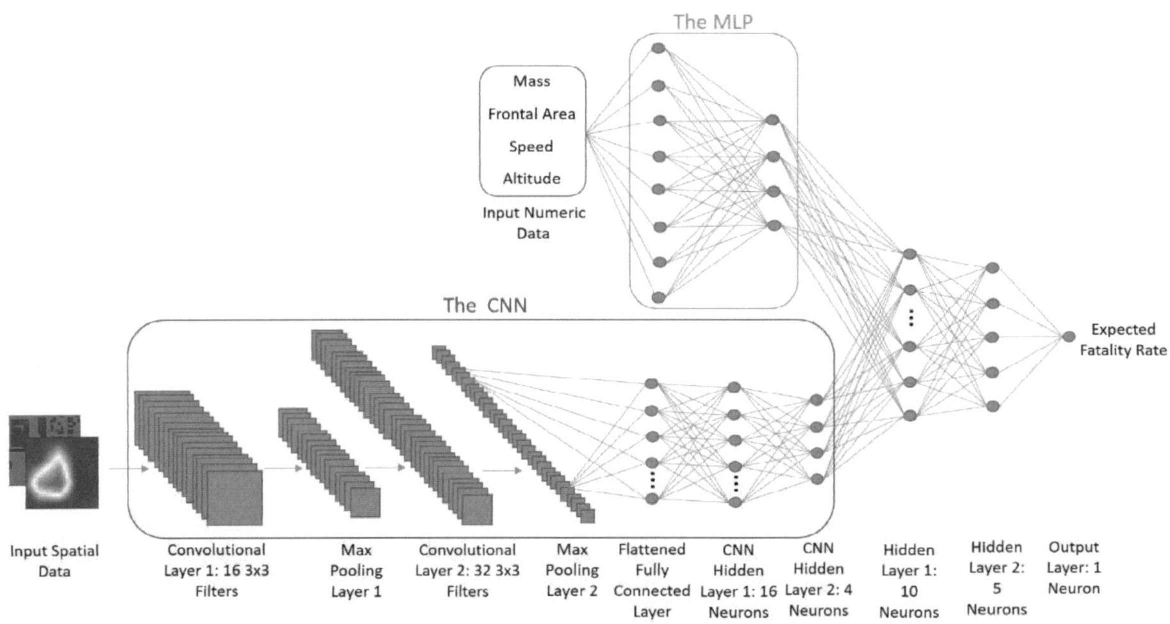

Figure 9: Combination of the MLP and the CNN to Predict Fatality Rates

Incorporating a ground risk assessment is vital to UAS operations to protect people on the ground as UAS becomes more ubiquitous in the rapidly evolving national airspace, especially in highly populated areas. The benefit of using a Machine Learning model like the one proposed over the physics-based models currently used is the reduction in computation time required. However, this benefit comes with the trade-off that there will be some error between the results of the Machine Learning model and the physics-based model. The main challenge with Machine Learning applications is how to reduce the error. Often this error reduction can be accomplished by increasing the data used for training. By leveraging the physics-based model, a significant supply of data can be generated to train the Machine Learning model.

Generating Training Data for the Machine Learning Approach

To train the Machine Learning model, data is required mapping the input conditions to the output fatality rate. Fortunately, the physics-based model can be used to generate this data by using the initial conditions of the UAS and the spatial data at the UAS location to calculate an expected fatality rate. An ideal Machine Learning model used for UAS risk assessment would be able to handle any combination of UAS speed, altitude, mass, frontal area, and velocity along with any distribution of shelter and population density, so the training data would also ideally reflect several of these combinations. However, this is not practical in the real world, so the objective is to include many these combinations by observing the common operating conditions of UAS today. A summary of the ranges used for the UAS characteristics can be seen below and was created using off-the-shelf UAS characteristics.

Table 2: UAS Parameter Ranges

UAS Parameter	Minimum	Maximum
Mass (kg)	2.0	9.0
Frontal Area (m^2)	0.347	0.81
Speed (m/s)	5	35
Altitude (m)	15	140

The minimum limit for the mass was based on the findings from Reference 17 that show a significant difference in the chance of causing a neck injury between the 1.2 kg DJI Phantom 3 and the 3.1 kg DJI Inspire 1. The DJI Phantom 3 had little chance of causing a neck injury while the DJI Inspire 1 had a much higher chance, so a mass of 2 kg between the DJI Phantom and DJI Inspire was used as the minimum value for this work.

With the ranges for the UAS characteristics determined, finding a way to sample many combinations of parameters is still required to train a Machine Learning model to be as robust as possible. For this, a Design of Experiments was used to efficiently explore the design space. A Latin Hypercube Design of Experiments was chosen because it is a space-filling design that creates design points evenly spread throughout the design space[18]. To ensure the design space was being thoroughly sampled, 248,000 data points of different combinations of UAS parameters were generated. However, the design points of the UAS parameters are only a portion of the data required, and spatial data for each of the design points was still needed.

To generate the input training spatial data, the building coverage and population density for the Georgia Institute of Technology were used. In Figure 1 and Figure 4, the population density and building coverage for the Georgia Institute of Technology are shown as 2D arrays, with each 10 m x 10 m cell being a unique location. The total size of the array was 113 rows and 135 columns, resulting in 15,504 unique locations. For each unique location of interest, a 29 cells x 29 cells window of the building coverage and population density was extracted and used as the spatial input data for that location of interest. The figure below demonstrates this concept. In the figure, the expected fatality rate is estimated for the location identified by the marker.

https://doi.org/10.52202/075106-0018

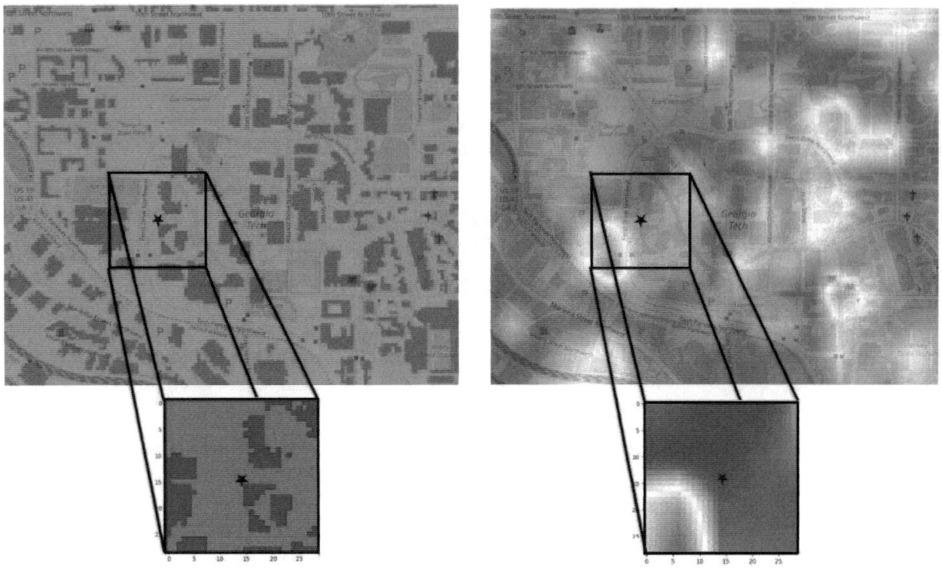

Figure 10: Spatial Data Extracted for a Single Location of Interest with Building Layouts (Left) and Population Density (Right)

To capture the effect of population density and shelter coverage on the fatality rate with the effect of the UAS conditions, 16 different UAS conditions were used at each location of interest to utilize the full data set generated using the Latin Hypercube Design of Experiments. With the combination of the input spatial data and the input numeric data, the physics-based model described in the paper was then used to estimate the expected fatality rate for each of the 248,000 cases.

Training the Machine Learning Model

A key aspect to any machine learning model training is the hyperparameters. Like how the model architecture can alter the performance, so can the hyperparameters set for training. These hyperparameters include the type of loss function used, the learning rate, batch size, and even the number of epochs used for training. The loss function used for this work was the Mean Absolute Percent Error 16 (MAPE). The batch size and number of epochs were 128 and 200, respectively. These values were the same as used in Reference 9. However, an early stop condition was implemented to prevent the model from overfitting the training data if the testing validation accuracy stopped improving.

Aside from the hyperparameters mentioned above, other hyperparameters that affect the performance of the model are those associated with the architecture of the model, like the number of layers or the number of neurons in each layer of the MLP. Changing the architecture can alter the performance, so several different architectures should be explored to find the best one. However, with an unlimited number of possibilities for architectures, it is infeasible to try them all. For this reason, it is recommended to find a suitable model that has decent performance as the baseline and fine tune the baseline model from there. The model architecture described in Reference 9 was used as the baseline architecture of the Machine Learning model. Parameters in the baseline model architecture subject to change include the number of neurons in each of the MLP layers, the number of filters in the CNN convolutional layers, the number of neurons in the CNN fully connected layer, and the number of neurons in the final two fully connected layers after the outputs of the CNN and MLP have been connected. A range of each of these parameters is summarized in the table below.

Table 3: Hyperparameter Ranges for Machine Learning Model Training

Model Parameter	Minimum	Maximum
MLP Hidden Layer 1 Neurons	4	100
MLP Hidden Layer 2 Neurons	4	100
Number of Filters in Convolutional Layer 1	4	32
Number of Filters in Convolutional Layer 2	4	128
CNN Hidden Layer 1 Neurons	4	32
CNN Hidden Layer 2 Neurons	4	32
Hidden Layer 1 Neurons	4	32
Hidden Layer 2 Neurons	4	32

The Machine Learning model training was done in Python using TensorFlow. Compatible with TensorFlow is another Python package, Keras. One of the functionalities of Keras is to automate the random search to find the optimal model configuration, so 200 random model combinations were generated from the ranges above and tested. Because there is some randomness in the actual training process, each model combination was trained three different times. The model with the best performance out of the 200 random combinations was saved.

RESULTS AND DISCUSSION

Evaluating the Machine Learning Model Performance

The baseline Machine Learning model and the Machine Learning model optimized with the hyperparameter random search both showed promising results when observing their MAPE on the training and validation dataset. A summary of their two architectures and their performance can be seen in the table below.

Table 4: Summary of the Machine Learning Model Architectures

Model Parameter	Baseline Model	Optimized Model
MLP Hidden Layer 1 Neurons	8	54
MLP Hidden Layer 2 Neurons	4	54
Number of Filters in Convolutional Layer 1	16	24
Number of Filters in Convolutional Layer 2	32	4
CNN Hidden Layer 1 Neurons	16	32
CNN Hidden Layer 2 Neurons	4	32
Hidden Layer 1 Neurons	10	54
Hidden Layer 2 Neurons	5	12
Training MAPE (%)	16	15
Validation MAPE (%)	22	17

https://doi.org/10.52202/075106-0018

Based on the table above, it is shown that both the baseline model and the optimized model performed similarly with the training data, but the optimized model performed a little better on the validation data. However, the two models still need to be compared to the physics-based model.

To compare the physics-based model with the Machine Learning models, a risk map was created for the campus of the Georgia Institute of Technology using the daytime population density information from LandScan combined with the social media activity collected. A UAS was assumed to have a mass of 6 kg, a frontal area of 0.6 m^2, a flight altitude of 35 m and a flight speed of 25 m/s operating over the campus. With this UAS configuration, a risk map was finally created using the physics-based model after hours required for completion while the Machine Learning models were able to complete the risk map in a matter of seconds with the same UAS conditions. Although the optimized model seemed to do better during training, the baseline model performed better when compared to the risk map created by the physics-based model. When compared to the physics-based risk map, the MAPE for the baseline was 19.3% while it was 22.3% for the optimized model, so the baseline model was selected as the superior model. A comparison between the risk map created using the baseline Machine Learning model and the physics-based risk map can be seen in the image below. As mentioned, it took several hours to create the physics-based risk map and a few seconds to create the Machine Learning risk map. The heat map in each figure represents the expected fatality rate, measured as fatalities per flight hour.

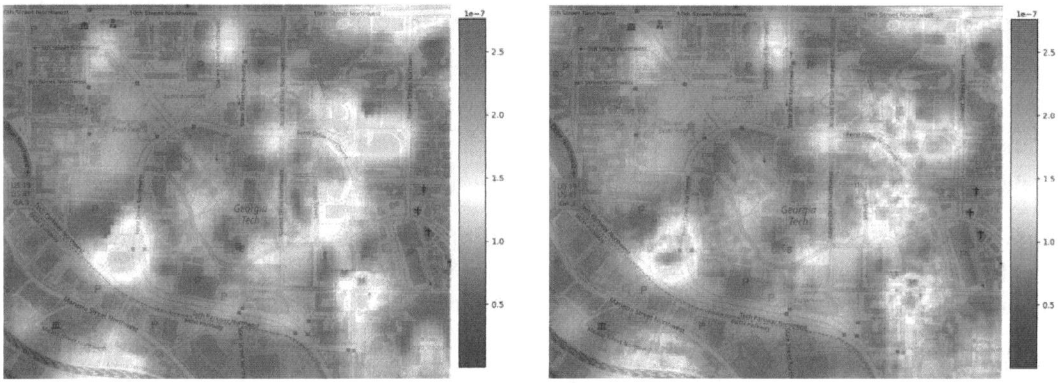

Figure 11: Comparison Between the Machine Learning Model Risk Map (Left) and the Physics-Based Model Risk Map (Right)

The comparison above shows the Machine Learning can identify the same high-risk areas as the physics-based model, although the predicted risk from the Machine Learning model in these high-risk areas appears to be less than that from the physics-based model. However, the MAPE for the Machine Learning model creating this risk map was still only about 19%. This can be interpreted as if the physics-based model determines the risk for a single location is 1×10^{-6} fatalities per flight hour, then the Machine Learning model might predict the risk value to be 0.81×10^{-6} fatalities per flight hour. That is one fatality every 114 years compared to a predicted 0.81 fatalities every 114 years. Because of how low this frequency is, the 19% MAPE was deemed reasonable, although future work will include trying to minimize the MAPE further.

One benefit of using the Machine Learning model is the rapid generation of risk maps for UAS users to identify how the UAS flight conditions might affect risk. The images below show two different risk maps created for two different UAS configurations.

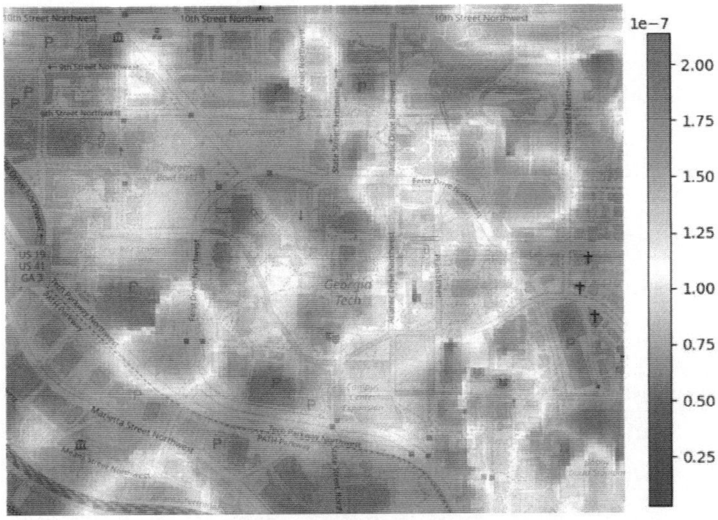

Figure 12: Risk Map Created Using the Machine Learning Model when Mass=8 kg, Frontal Area=0.75 m^2, Speed=30 m/s and Altitude =35 m

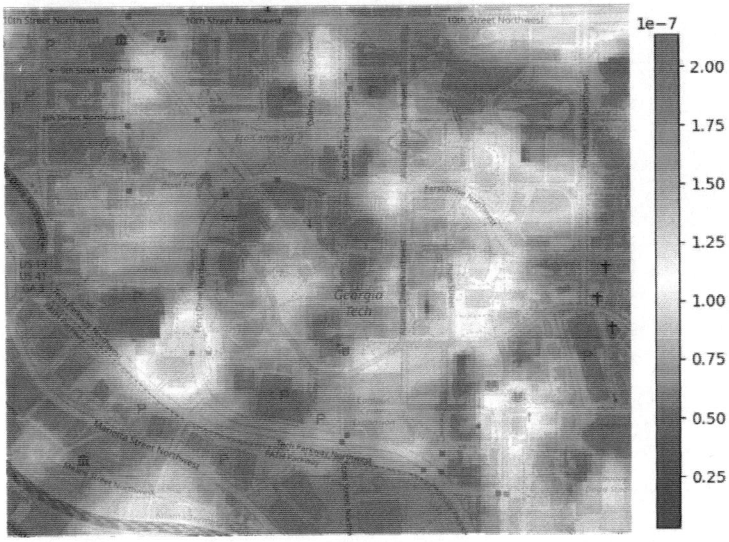

Figure 13: Risk Map Created Using the Machine Learning Model when Mass=2 kg, Frontal Area=0.4 m^2, Speed=25 m/s and Altitude =35 m

By altering the UAS flight conditions and UAS parameters, UAS pilots can observe how the risk map changes. As expected, a larger and faster UAS increases the risk compared to a smaller and slower UAS. This ability to rapidly create risk maps with changing UAS flight parameters is meant to assist UAS pilots in the route planning to ensure the maximum level of allowable safety is not exceeded.

Risk-Informed Route Planning

Using the risk maps, risk-informed routes can be generated using a variety of different route planning methods. One such approach uses a modified version of the common A* algorithm that modifies the heuristic function to become a bi-objective equation comprised of total distance traveled and cumulative ground risk[19]. As a result, the route planning algorithm can be used to find a combination of the safest route or the fastest route as desired. For this work, the Machine Learning model can be used to create a risk map to use with the route planning algorithm. This allows for the algorithm to find the safest route based on minimizing the risk of the path using the risk map.

https://doi.org/10.52202/075106-0018

This route planning algorithm was used to find the safest route between two points at the Georgia Institute of Technology using both risk maps created by the physics-based model and the Machine Learning model to compare the resulting paths. The comparison of the routes created can be seen below where the red route is the route created using the physics-based risk map and the black route is the route created using the Machine Learning model. The heat map in the figure is the risk map associated with the Machine Learning model.

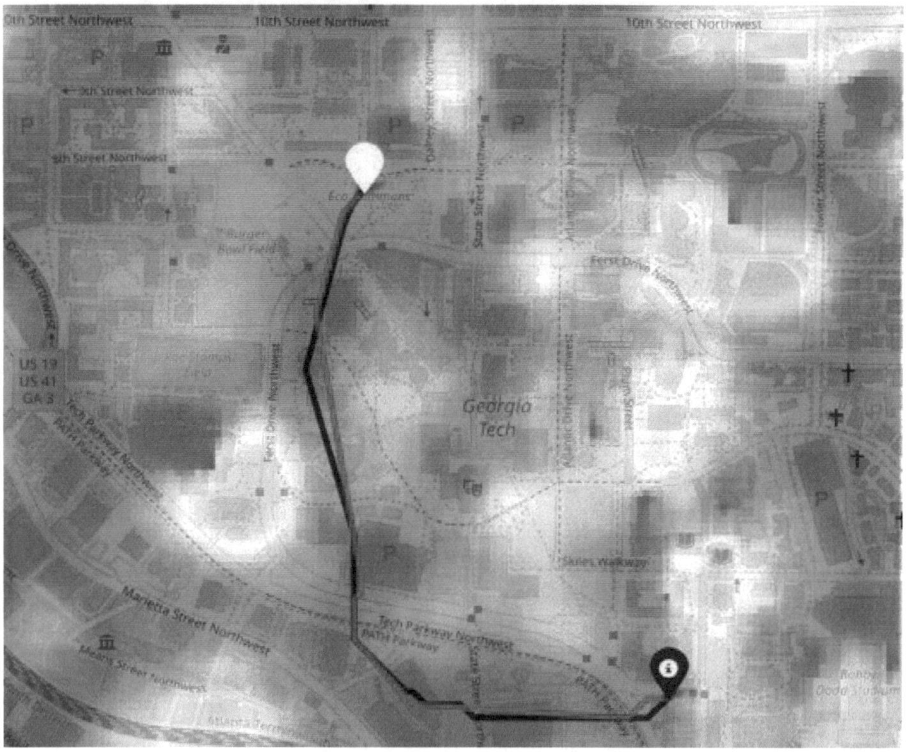

Figure 14: Comparison of Routes Created Using the Machine Learning Risk Map (Black) and Physics-Based Risk Map (Red)

The two routes created using each of the risk maps are strikingly similar, suggesting the Machine Learning model is adequate at replacing the physics-based model for assisting in route planning. Additional confirmation of this is found when comparing the predicted risk and actual risk of the route created using the Machine Learning risk map. The table below summarizes the risk for this route. The predicted risk values and actual risk values are the risk values obtained using the Machine Learning risk map and physics-based risk map, respectively.

Table 5: Summary of Risk Using the Machine Learning Risk Map for Route Planning

Risk	Predicted (fatalities / hour)	Actual (fatalities / hour)	Absolute Percent Error (%)
Maximum Risk	$5.81 * 10^{-8}$	$6.46 * 10^{-8}$	10
Average Risk	$1.60 * 10^{-8}$	$1.64 * 10^{-8}$	2.4

From the table above, the error between the predicted maximum risk level and the actual risk level is greater than the error of the predicted average risk and the actual average risk. As stated previously, the Machine Learning model has a higher error with the higher risk areas, but with the model still able to identify the high-risk regions, it correctly avoids these areas when assisting in route planning. The error between the predicted average risk and the actual average risk is very low, suggesting that if the route planner attempts to avoid the high-risk areas, the predicted risk values will be very close to the actual risk values. It should also be noted that both routes have a maximum risk value less than the acceptable risk level of 10^{-7} fatalities/hour mentioned previously.

Using the Machine Learning model to quickly generate ground risk maps with the route planning algorithm to find acceptably safe routes below a target level of safety, further analysis can be done on how different UAS operating conditions and different target levels of safety impact the route a UAS can take. The below images in Figure 15 and Figure 16 show two different scenarios with the same start and end point, with the only difference being in the UAS operating conditions. In both cases, a route was found below a maximum target level of safety of 10^{-7} fatalities per flight hour. This type of analysis can provide more informed decision making on flight parameters and vehicle parameters and their expected impact on both expected safety levels and available routes to take.

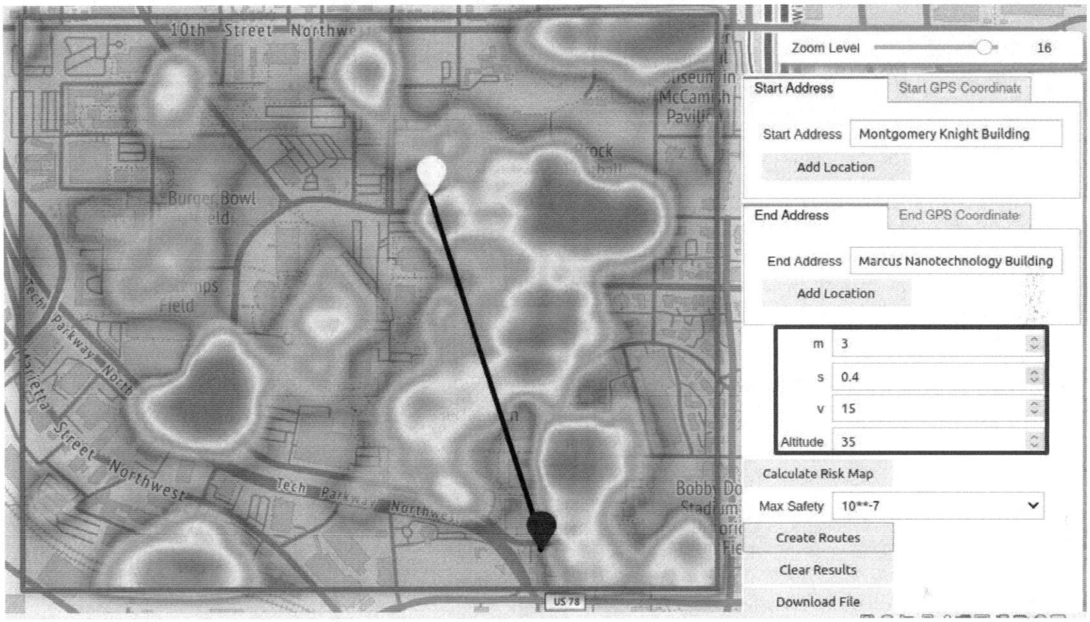

Figure 15: Route Planned for a Low-Risk UAS

https://doi.org/10.52202/075106-0018

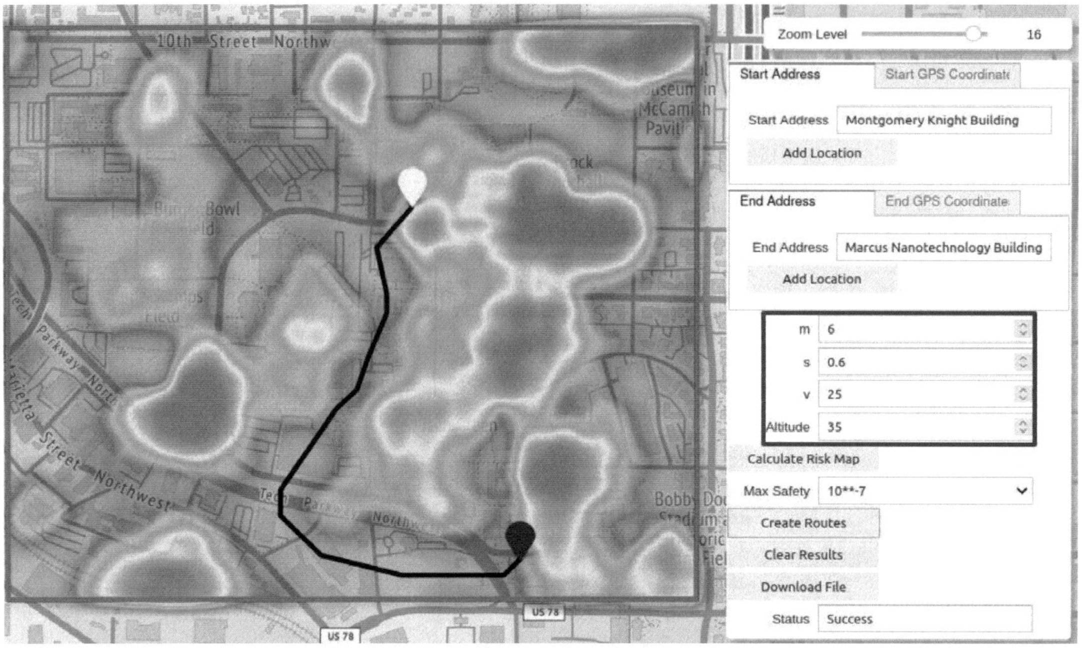

Figure 16: Route Planned for a High-Risk UAS

Alternatively, further analysis can be done to observe the impact of changing the target level of safety requirement. A more relaxed target level of safety may be justified for more urgent use cases, like law enforcement and medical device deliveries. The images below show the resulting routes created for the same UAS configuration with different levels of safety required. As expected, the route that allows for higher rates of fatality takes the more direct path over the high-risk areas.

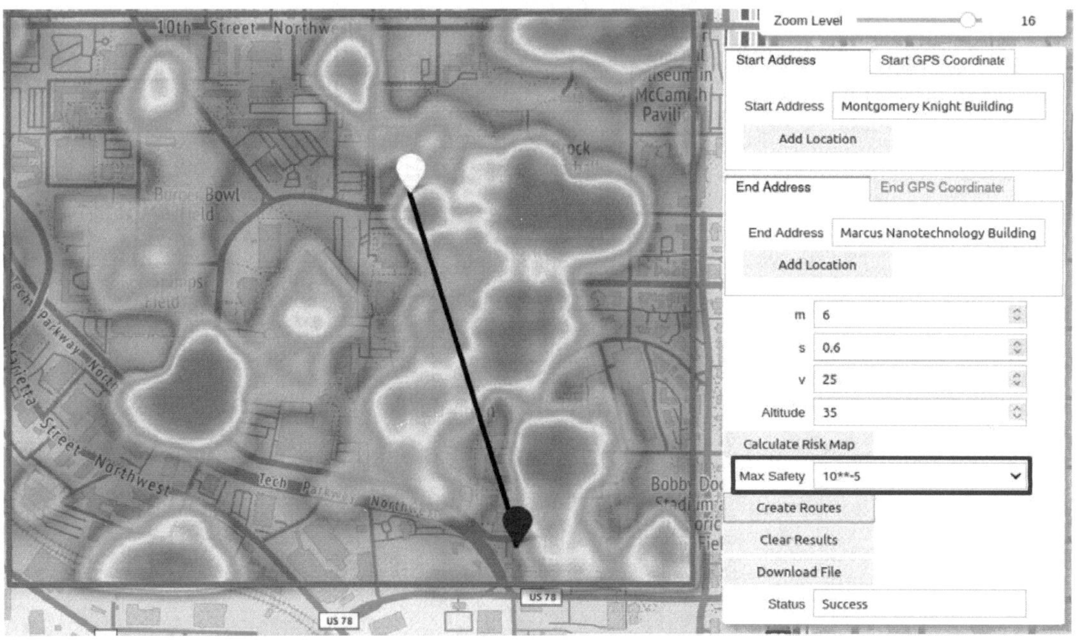

Figure 17: Route Planned with Maximum Target Level of Safety of 10^{-5} Fatalities Per Flight Hour

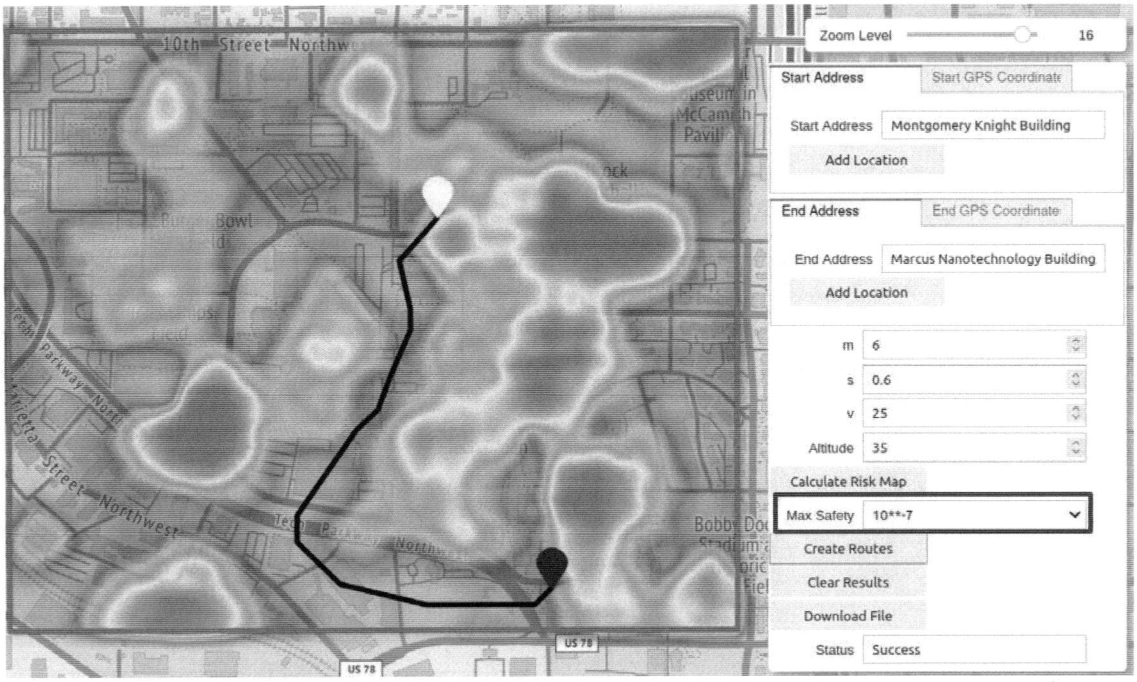

Figure 18: Route Planned with Maximum Target Level of Safety of 10^{-7} Fatalities Per Flight Hour

The objective of combining the risk map with the route planning algorithm was to provide a way to assess UAS ground risk and ensure ground risk values do not exceed some acceptable target level of safety. Whether it is law enforcement or delivery companies, all UAS users can utilize this technology by using it to create safe routes. The key enabler for this is the Machine Learning model that allows for quick risk assessment to ensure any route planned does not exceed a maximum level of allowable safety without taking significant time to determine expected risk values.

CONCLUSION

Like crewed aircraft, UAS require risk assessment to ensure any UAS is operating with some acceptable level of risk. Because the UAS has been introduced relatively recently compared to crewed aircraft, the UAS do not have the comparable historical flight data that would be required for such risk assessment. As a result, modeling and simulation methods have become the next best option to approximate UAS risk levels. The state-of-the-art methods for estimating UAS ground risk rely on physics-based models to determine the descent trajectory for a UAS given a failure and then determine the likelihood of causing a fatality if the UAS were to strike a person. This process can become computationally expensive and is not suitable for dynamic environments like cities. This work shows that Machine Learning methods can be used to replace the slow physics-based methods. Using physics-based methods to generate training data, a Machine Learning model was trained with a 16% error on the training data and a 22% error on the validation data. The Machine Learning model allows for UAS users to rapidly generate a ground risk map based on their desired UAS flight conditions with high-risk areas easily identifiable. With this risk map, it was shown that a route planner can be used to find a path that does not exceed some allowable risk value set by the user. This solution is presented as a flexible web-based application that can be used by any UAS pilot operating in highly populated areas. Future work would include increasing the fidelity of the physics-based model by incorporating additional descent trajectories and the effect of wind. Additionally, more diverse population densities and building layouts should be added to the training to increase robustness.

https://doi.org/10.52202/075106-0018

REFERENCES

[1] K. Dalamagkidis, "On Integrating Unmanned Aircraft Systems into the National Airspace System". SpringerLink, 2009, ISBN: 978-90-481-2096-3

[2] T. Perez, R. A. Clothier, and B. Williams, "Risk-management of UAS robust autonomy for integration into civil aviation safety frameworks."

[3] J. Fortes, R. Fraga, K. Martin, "Safety Analysis for UAS Operation Using Stochastic Fast-Time Simulation."

[4] S. Primatesta, "An Innovative Algorithm to Estimate Risk Optimum Path for Unmanned Aerial Vehicles in Urban Environments" International Conference on Air Transport. 2018.

[5] S. Primatesta, "Ground Risk Map for Unmanned Aircraft in Urban Environments" Journal of Intelligent and Robotic Systems. 2019.

[6] A. Cour-Harbo, "Quantifying Risk of Ground Impact Fatalities for Small Unmanned Aircraft" Journal of Intelligent and Robotic Systems. 2018.

[7] A. Cour-Harbo "Ground Impact Probability Distribution for Small Unmanned Aircraft in Ballistic Descent" 2020 International Conference on Unmanned Aircraft Systems. 2020.

[8] A. Geron, "Hands-On Machine Learning with Scikit-Learn and TensorFlow." Published by O'Reilly Media, Inc. 2017.

[9] A. Rosenbrock, "Keras: Multiple Inputs and Mixed Data" pyimagesearch. 2019. https://pyimagesearch.com/2019/02/04/keras-multiple-inputs-and-mixed-data/

[10] I. Goodfellow, "Deep Learning" MIT Press. 2016.

[11] K. Sangvhi, "Image Classification Techniques" Analytics Vidhya. 2020.

[12] "Convolutional Neural Networks" IBM. https://www.ibm.com/topics/convolutional-neuralnetworks

[13] M. Mishra, "Convolutional Neural Networks, Explained" Towards Data Science. 2020.

[14] J. Breunig et. al. "Modeling Risk-Based Approach for Small Unmanned Aircraft Systems" The MITRE Corporation. 2018.

[15] Oak Ridge National Laboratory LandScan Database URL:https://web.ornl.gov/sci/landscan

[16] OpenStreetMap contributors, "Planet dump retrieved from https://planet.osm.org" 2017. URL: https://www.open-streetmap.org

[17] E. Campolettano et. al. "Ranges of Injury Risk Associated with Impact from Unmanned Aircraft Systems" Annals of Biomedical Engineering. 2017.

[18] R. Myers. "Response Surface Methodology Third Edition" John Wiley and Sons, Inc. 2009.

[19] J. Pattison. "A Deep Learning Approach Using Social Media Data to Estimate Ground Risk of UAS in Urban Areas." 2023 FAA Data Challenge.

THE SOPHISTICATED TECHNOLOGIES POWERING TODAY'S DRONE AND THE CHALLENGES MANUFACTURERS FACE TO PROVIDE THOSE TECHNOLOGIES

Christopher T. Raabe*

This paper explores the significant challenges faced by drone manufacturers in integrating advanced technologies into their products, particularly in a market dominated by a single major player. As the Chief Technology Officer at a leading drone manufacturer, I provide a firsthand perspective on the difficulties of adopting sophisticated technologies such as drone control laws and capabilities, camera payloads, video transmission, and drone batteries. These challenges are primarily due to the high costs of development and the scale required for economical production, which are attainable only by the market leader. However, recent shifts in legislation have begun to disrupt this status quo, providing an opportunity for smaller manufacturers. I propose several strategic solutions, including collaborative research, collective bargaining, development of proprietary technologies, and industry standardization. These strategies are aimed at overcoming the barriers to entry and fostering a more competitive market landscape. Through innovative approaches and strategic partnerships, there is potential for smaller manufacturers to not only compete but thrive in the evolving drone technology sector. This paper underscores the need for industry-wide cooperation and innovation to challenge the dominance of the market leader and to advance the global drone industry.

INTRODUCTION

As the Chief Technology Officer at a prominent drone manufacturer for over seven years, I have observed the dynamic shifts within the drone technology sector, noting the significant hurdles that many manufacturers encounter as they strive to embed sophisticated technologies into their products. The drone market is intensely competitive and is predominantly monopolized by one manufacturer, which holds approximately three-quarters of the total market share. This overwhelming dominance not only stifles competition but also significantly hampers the ability of smaller manufacturers to adopt and implement cutting-edge technologies in their drones. In this editorial, I aim to dissect the integration challenges related to essential drone technologies such as control laws and capabilities, camera payloads, video transmission, and drone batteries. I will also suggest viable strategies to surmount these challenges, helping to level the playing field for all manufacturers

DRONE CONTROL LAWS AND CAPABILITIES

One of the foundational technologies in drone functionality is the control system, which often starts with a basic control law such as the Proportional-Integral-Derivative (PID) controller. While

* Global CTO, ACSL Inc., 710 Lakeway Dr, Ste 200, Sunnyvale, CA 94085.

 https://doi.org/10.52202/075106-0019

the PID controller is widely utilized due to its simplicity, it serves as a prime example of the inherent challenges in developing sophisticated drone autopilots. This linear controller is typically optimized for specific flight conditions like hovering and is not well-suited to the nonlinear dynamics of multicopter drones. Its weaknesses become particularly evident at high speeds or in strong winds, as well as when the drone is modified with various payloads, which are increasingly common as customers demand fast, flexible operations under a variety of environmental conditions.

Building upon the basic control, modern drones require a complex overlay of functionalities to achieve high levels of autonomy. These include smooth autonomous flight, autonomous fail-safe behaviors, obstacle detection and avoidance, smooth landings, and autonomous maneuvers such as survey scans. Each capability introduces additional complexity and demands sophisticated algorithms and significant computational power to effectively manage the drone's interaction with its environment.

Strategic Solution:

To overcome these challenges, drone manufacturers could explore collaborative research and development, possibly forming partnerships with academic institutions that specialize in advanced robotics and artificial intelligence. Such collaborations could foster the development of innovative control systems that are adaptive and robust, capable of performing optimally across various flight conditions and operational demands. Additionally, focusing on specific market segments where advanced control capabilities provide significant value could allow manufacturers to differentiate their products and meet the increasing customer demands for versatile and reliable drone operations.

CAMERA PAYLOADS

The quality of a drone's camera system significantly enhances its market value, especially in sectors that require high-resolution imaging such as environmental monitoring, security, and filmmaking. However, the integration of top-tier camera systems into drones presents substantial challenges for smaller manufacturers. The market leader has consolidated its position further by acquiring a top-tier camera maker, which complicates the availability of advanced camera technologies for other players in the industry. This acquisition not only limits access to high-end camera technology but also places smaller manufacturers at a significant competitive disadvantage.

Smaller drone manufacturers are keen to collaborate with top-tier camera makers to enhance their product offerings. However, these camera manufacturers typically demand production volumes that smaller players cannot guarantee, making such partnerships financially unfeasible. This volume demand stems from the high upfront costs involved in developing and manufacturing advanced camera systems, which camera makers need to recoup through large-scale production agreements.

Strategic Solution:

To address this challenge, smaller manufacturers might consider forming a consortium that can collectively negotiate with camera producers, presenting a united front that guarantees larger volumes and makes collaboration more appealing. Such a consortium could potentially level the playing field, allowing access to advanced camera technologies that are currently monopolized by the market leader. Additionally, investing in modular camera designs that can accommodate various types of camera technologies could also be a strategic move. This approach would not only diversify the product range offered by smaller manufacturers but also attract a broader customer base, thereby increasing the potential sales volume and making partnerships with top-tier camera makers more viable.

VIDEO TRANSMISSION

A critical aspect of drone technology is the ability to transmit high-quality video in real time, which is indispensable for applications such as live surveillance, inspections, and videography. However, the transmission of video data presents significant technical challenges, primarily due to the relatively low data bandwidth of the wireless connection between the drone and its controller. This bandwidth limitation necessitates the efficient compression and decompression of video to ensure rapid and clear transmission without latency or loss of quality.

The need for specialized hardware to handle these compression and decompression tasks poses a further challenge. Much of this technology requires significant investment in research and development, often involving advanced chipsets designed specifically for this purpose. Unfortunately, many of these technologies are produced overseas, which can introduce supply chain vulnerabilities and limit access to preferred sourcing options due to geopolitical and economic factors.

Strategic Solution:

To circumvent these challenges, drone manufacturers could benefit from developing proprietary video compression technologies. This approach would not only reduce dependency on international suppliers but also provide a unique selling point in the drone market. Collaborating with telecommunications and technology companies could also distribute the costs associated with R&D, while accelerating the development and integration of these technologies. Furthermore, exploring advanced software solutions to optimize existing compression techniques could improve performance and reliability of video transmission, enhancing the overall functionality and appeal of the drones.

DRONE BATTERIES

The efficiency and operational capacity of drones are heavily reliant on the performance of their batteries, specifically the gravimetric energy density, which determines how much energy a battery can store per unit of weight. Flight time is directly proportional to this energy density—the higher the gravimetric energy density, the longer a drone can remain airborne without needing a recharge. This is particularly crucial for applications that require extended flight durations such as long-range surveillance, aerial photography, and continuous data collection over large areas.

However, achieving high gravimetric energy density in drone batteries poses significant challenges. The development and production of such high-performance batteries involve substantial costs and are economically viable primarily for manufacturers who can produce at large scales. Smaller drone manufacturers often struggle with this, as they cannot promise the sales volumes necessary to justify or cover the high initial investment required for advanced battery technology development.

Strategic Solution:

Collaborations with innovative energy companies and startups focused on next-generation battery technologies could reduce the financial burden of such ventures. Moreover, advocating for and participating in industry-wide initiatives aimed at standardizing battery technologies could help decrease costs and increase the accessibility of high-performance batteries.

https://doi.org/10.52202/075106-0019

REMOTE ID

Remote Identification (Remote ID) technology serves as an electronic license plate for drones, providing crucial information about the drone's identity and location to other aircraft, regulatory authorities, and potentially the general public. Despite its importance for safety and compliance in increasingly crowded skies, the standardization of Remote ID technology has not been achieved globally. This lack of uniformity forces drone manufacturers to develop their own integrated solutions, leading to significant duplicated efforts across the industry.

Each drone's Remote ID must align with the specific regulatory requirements of the markets in which it operates. This fragmentation means that a drone designed for one market might need significant modifications to its Remote ID system before it can be introduced into a different market, complicating the process for manufacturers who seek global distribution. The necessity to customize these systems for various regulatory environments not only increases the cost and complexity of manufacturing but also delays the time to market for new and innovative drone products.

Strategic Solution:

A collaborative approach among drone manufacturers and regulatory bodies could be instrumental in addressing these challenges. By advocating for and participating in the development of global standards for Remote ID, the industry could reduce redundancy and streamline the integration of this technology across different markets. Establishing a unified standard would not only decrease the regulatory burden on manufacturers but also enhance the interoperability and safety of drone operations worldwide. Additionally, forming partnerships with technology companies specializing in communication and identification systems could provide manufacturers with advanced, adaptable solutions that can be easily modified to meet various international standards, further reducing development time and costs.

CONCLUSION

As the drone industry continues to evolve, the integration of sophisticated technologies presents significant challenges, especially for manufacturers that do not command a large share of the market. This paper has explored some of the complexities associated with integrating advanced drone technologies such as control systems, camera payloads, video transmission, drone batteries, and Remote ID systems. Each of these components is critical to the functionality and marketability of drones but requires substantial investment in research and development, often making them inaccessible for smaller manufacturers.

The strategies proposed throughout this discussion—ranging from collaborative research and collective bargaining to the development of proprietary technologies and advocacy for global standardization—offer practical solutions to these challenges. By forming strategic partnerships and focusing on innovation, smaller manufacturers can overcome the barriers imposed by high development costs and regulatory fragmentation. Furthermore, the push for standardization, particularly in technologies like Remote ID, could significantly reduce the redundancy and complexity currently faced by manufacturers aiming to enter global markets.

Embracing these strategies will not only enable manufacturers to compete more effectively but also contribute to the growth and safety of the drone industry as a whole. As we move forward, it is crucial for industry stakeholders to work together towards a more integrated and standardized approach that will facilitate innovation, ensure compliance, and enhance the operational capabilities of drones worldwide. This collaborative effort is essential for the industry to meet the increasing demands of customers and to navigate the challenges of a technology-driven future.

AUTOMATION ENABLING MULTIPLE UAS AT THE SAME TIME (SAFELY AND LEGALLY)

Wendy Ljungren*, Brent McLaughlin†

The ability to operate multiple Uncrewed Aircraft (UA) simultaneously offers significant commercial potential. However, manual control of such a fleet quickly becomes overwhelming and susceptible to human error. This paper provides an overview of the safety process used to identify and detail the automation needed for m:N Beyond-Visual-Line-of-Sight (BVLOS) operations, where "m" represents the number of human flight crew and "N" signifies the number of deployed UAs. The resulting framework enables safe and efficient control of multiple UAs by a limited crew in BVLOS operations.

INTRODUCTION

As part of a 2023 FAA UAS BAA project ***Standards for Piloting Multiple, Simultaneous UAS BVLOS***‡, the team led by Anzen Unmanned (Au) developed the minimum criteria and open-source flight control software for a Remote Operator (RO) to operate multiple small uncrewed aircraft systems (UAS) simultaneously (also known as m:N, where "m" is the number of flight crew and "N" is the number of UAs in operation) operating Beyond Visual Line of Sight (BVLOS). The accomplishments in this project included defining, developing, and testing the:

- Minimum UA performance and behaviors for normal and off-nominal conditions, including the flight control modes and level of automation necessary to support Remote Operator (RO) responses
    - Included submitting the resulting ArduPilot and MAVLink software updates to the open-source repositories
- RO interfaces (e.g. display, alerts, controls) needed to maintain m:N situational awareness and enable timely, correct RO responses
- Minimum sUAS equipage needed to support the m:N safety case
- Operational pre-flight and flight procedures
- Location checklist that can be scaled nationwide
- Minimum RO qualifications and training
- Organizational practices, including safety management, quality management, configuration management, and training programs
- Safety case to obtain FAA waiver§ under § 107.200 for §107.31 (BVLOS), 107.33b/c2 (VO), and §107.35 (m:N)

---

* Chief Operating Officer, Anzen Unmanned, https://www.anzenunmanned.com/
† Chief Operating Officer, American Robotics, https://asylonrobotics.com/solutions/security-drones/
‡ Au-REP-0048 Anzen Unmanned Final Report for Piloting Multiple, Simultaneous UAS BVLOS
§ FAA 107W-2023-01262 Asylon BVLOS m:N waiver

 https://doi.org/10.52202/075106-0020

Operating multiple UAS BVLOS simultaneously requires a level of automation that aligns with the scope, scale, and complexity of an operation. As the level of automation increases, the RO role shifts from an active hands-on pilot to a hands-off safety monitor.

Defining the level of automation is foundational to understanding the role of the human as a causal factor contributing to performance degradations and safety incidences. Based on the SAE International J3016 standard's Levels of Driving Automation*, UA automation levels were identified as part of this study to ensure human factors were captured within a defined automation context (Table 1).

Table 1 – Automation Levels

Level	Automation Name	Description
0	No automation	Human pilots do all the flight operations.
1	Crew Assistance	UAS is controlled by the crew, but some flight assist features may be included that can assist the RO with telemetry, speed, and altitude.
2	Crew Partial Automation	UAS has combined automated functions, but the crew must remain engaged with the flight tasks and always monitor the environment.
3	Conditional Flight Automation	An automated flight system on the UAS can perform all aspects of the flight tasks under some circumstances. Crew is still required to monitor the status of the UAS in operation. The RO is expected to be takeover-ready to always take control of the UAS with notice.
4	High Flight Automation	The UAS can perform all flight functions under certain conditions. The crew may have the option to control the vehicle.
5	Full Automation	The UAS can perform all flight functions under all conditions. The crew, as a safety monitor, never needs to be actively involved in flight tasks.

During off-nominal conditions, control of multiple UA at once by a single RO is extremely challenging without automation. A key part of the project was defining the minimum automation and technical requirements needed for scalable BVLOS m:N operation.

APPROACH

Most commercially available UAs are already produced with an autopilot for flight stability, control, and waypoint navigation. This allows the UA to follow pre-programmed paths without constant pilot input. Essentially, automation handles the routine aspects of UAS flight, leaving the pilot free to focus on strategic decisions and monitoring overall operations, while being able to immediately intervene in case of unexpected situations.

This project was designed to be at automation level 3 throughout human-in-the-loop operations, as the technology to enable this is readily available. Level 3 still requires that the crew monitor the status of the UAS in operation, and always be ready to take control of the UAS with notice.

* SAE J3016 Taxonomy and Definitions for Terms Related to Driving Automation Systems for On-Road Motor Vehicles

A detailed safety analysis* was performed to identify the tasks requiring RO attention and/or intervention. These tasks were then evaluated to determine whether automation or other means could be used to minimize the RO's workload.

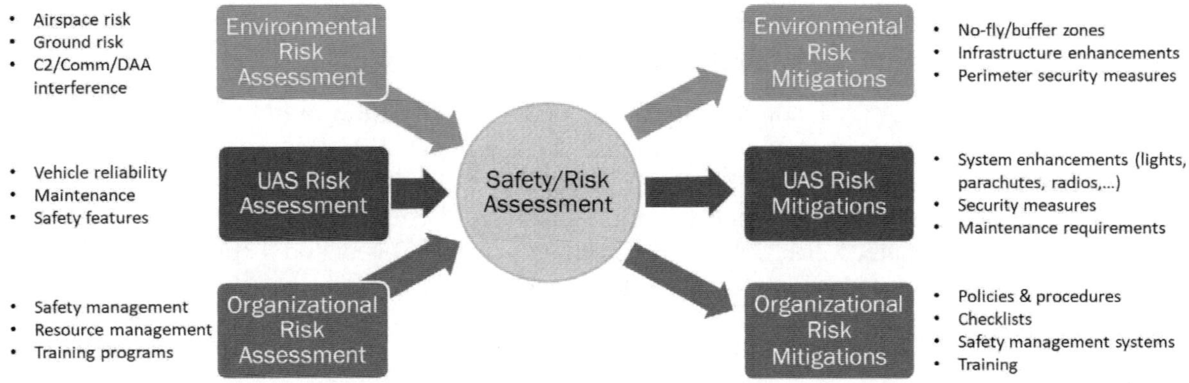

Figure 1 - Safety Risk Mitigations

Given the level of automation and assuming the maximum proportion of time where the RO must focus their attention and be ready to intervene, a task analysis was performed to determine the theoretical maximum number of partially automated UAs that each RO could safely operate BVLOS. The resulting number was verified and validated in both simulations and flight tests with multiple ROs of varying abilities.

MINIMUM BVLOS M:N AUTOMATION

During normal and contingency operations, the RO workload was designed to be low since the UA had the following features:

- Preplanned automated flight routes including take-off and landing (normal and emergency)
- Reliable, secure, real-time transmission of situational and safety data
- Lateral and vertical geofence at operating area perimeter
- Lateral exclusion (no-fly) zones around obstacles and other areas of concern in the operating area
- Deterministic activation of automated contingency actions, including:
    - On-UA lateral and vertical geofence preventing UA excursion from the defined operating area
    - On-UA exclusion zones preventing incursion into high-risk areas (e.g., unavailable Command and Control link (C2), assemblies of people, etc.)
    - On-UA exclusion zones around obstacles above the safe altitude floor (or lower near landing site) and critical infrastructure not involved in the operation
    - C2 anomalies trigger pre-defined Return to Safe Landing Zone (aka Return to Launch, RTL)
    - On-UA ADS-B detection of other aircraft and predictable autonomous avoidance maneuver

---

* SAE 4761A Guidelines for Conducting the Safety Assessment Process on Civil Aircraft, Systems, and Equipment

 https://doi.org/10.52202/075106-0020

- GNSS/GPS anomalies result in landing using secondary/barometric altitude
- Low battery caution automatically triggers pre-defined Return to Safe Landing Zone
- Critical battery warning automatically triggers controlled landing

Automated Controlled Descent and Landings

Since abnormal/emergency landings may be needed away from the initial take-off point, it is important to limit the kinetic energy at low altitudes to less than the Association for the Advancement of Automotive Medicine Abbreviated Injury Scale* level 3 (AIS3) injury criteria. It is also important to ensure the UA can descend fast enough to avoid collisions with Intruding Aircraft. Figure 2 shows the automated vertical speed requirements during Detect and Avoid (DAA) descent and emergency landing that were implemented for this project.

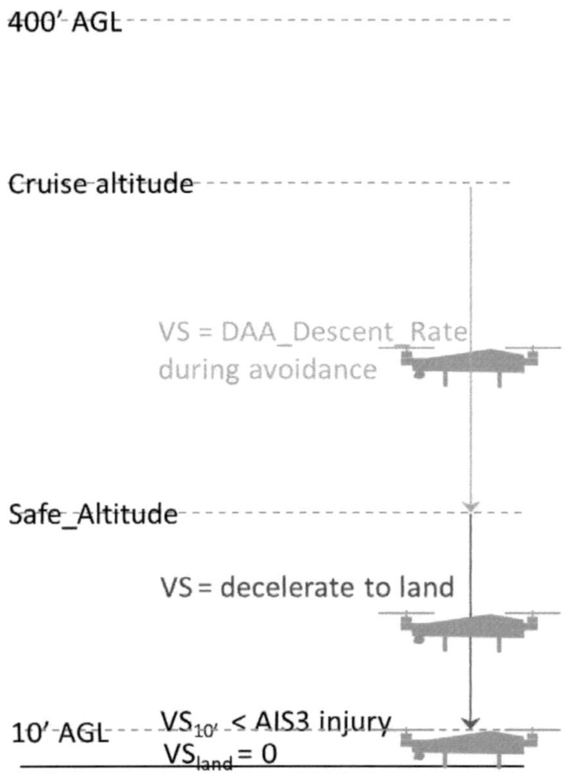

Figure 2 – Controlled Descent and Landings

The project's safe altitude for the operating area was chosen to be both below the typical flight altitude of manned/Intruding Aircraft (~50 AGL) and above the location's obstacles by ~30'.

Vertical descent speeds were automatically set as the situation demanded:

- If an intruding aircraft is detected while the UA is above the safe altitude, the UA performs a controlled descent to the safe altitude as a typical avoidance maneuver for small multicopters. The vertical speed is determined as a function of the maximum stable descent rate, height above safe altitude, and performance characteristics of the DAA

---

* Association for the Advancement of Automotive Medicine Abbreviated Injury Scale

https://doi.org/10.52202/075106-0020

system to ensure the UA remains well clear of the intruder (reference the timing analysis in ASTM F3442* appendix X2).

- In the case of an abnormal/emergency landing, the UA decelerates to minimize the severity of injuries in case there are people underneath. The rate of deceleration must be enough that the UA's kinetic energy is below the AIS3 injury criteria at no less than 10' AGL.

On-site Ground Control Station Automation

The base station provides the on-site GCS infrastructure needed for remote operations. Key safety features include:

- Means for automated safe and accurate takeoff and landings
- Sensors for automated built-in tests and preflight checks (e.g. weather, cameras)
- Secure, reliable communications infrastructure with minimal data latencies to the Remote Operations Center and UA
- DAA Surveillance Sensor, where needed/approved for BVLOS operations

Human/Machine Interface

Since Part 107 UASs typically lack the integrity and availability assurance needed for more complex operations, the RO must be ready to provide situational awareness and some mitigation measures. A detailed safety analysis was performed of the m:N BVLOS operations. Most of the hazards and corresponding safety measures identified are in accordance with BVLOS operations. The primary concern introduced by m:N operation was reducing the risk of hazards resulting from error or inaction by an RO whose attention is divided between multiple UAs. Based on the safety analysis, automated visual and aural alerts were implemented for actions requiring the RO's attention.

Remote Operations Center (ROC)

The ROC houses the equipment needed for the RO and support personnel (e.g., dispatch, supervisor) to monitor and control the UAs. The safety analysis requirements for the ROC include:

- Secure, reliable infrastructure to minimize common failure modes for multiple flight operations (e.g., internet, servers, computers, displays, speakers, controls, failover mechanisms, etc.)
- Automated Task Management software that can manage pre-programmed tasks for each UAS to minimize the pilot's workload.
- Automated, real-time monitoring of the UAS and operational area with caution alerts (visual and aural) for safety issues needing pilot attention/intervention.
- Display of safety related parameters to ensure situational awareness for all UAS
- ROC Supervisor monitoring one or more RO's performance and providing back-up for events (e.g. RO health issue, extended ATC coordination).
- Supervisor control station to monitor multiple ROs' operations and assume control
- Ability to easily coordinate transfer of UA control to/from the supervisor for personal reasons and handling of events.

* Standard Specification for Detect and Avoid System Performance Requirements

https://doi.org/10.52202/075106-0020

ADDITIONAL AUTOMATION TO INCREASE MAXIMUM UA

Level 3 automation was chosen for this project as it is within the scope of many commercially available UAS. To achieve a higher number of simultaneous UAs, higher-level automation should be implemented including:

- Traffic Management system to strategically and tactically deconflict at least UAS and potentially crewed aircraft in the operating area
- Advanced preflight checks
- On-UA DAA system for non-cooperative (e.g. non-ADS-B) detection of other aircraft and birds and then automatic avoidance maneuvers
- On-UA identification and avoidance of people/wildlife on ground
- On-UA obstacle identification and avoidance for areas where current obstacles may be unknown or not precisely mapped
- Higher systems/software/hardware maturity levels for safety and security mitigations such as geofences, exclusion zones, parachute deployment, C2 monitoring
- Redundancy and partitioning of safety mitigations to minimize common mode events

With increased automation, it is expected that the task analysis, simulation, and flight test will confirm that a single RO can safely and effectively operate even more UAs.

CONCLUSION/SUMMARY

Multiple UAs flying simultaneously with a single operator is safe and practical with today's technology. The key is to perform a detailed safety analysis to identify the tasks requiring RO involvement, and then a task analysis to determine the safe number of UAs for a given level of automation.

ACKNOWLEDGEMENTS

This project would not have been successful without the support of strong relationships between Au and its partners:

- Asylon Incorporated for developing the reliable BVLOS m:N DroneCore system, building an impressive ROC simulator, providing most of the ROs, and willingly sharing their lessons learned with the industry.
- New York UAS Test Site operator NUAIR for sharing their expertise, providing ROs with varying levels of experience, giving honest RO feedback during the simulator testing, and hosting the flight test program.
- Purdue University for providing academic rigor during the simulator and flight testing, and much of the analysis for the Human Factors report.
- Anzen Unmanned team who continues to assist the industry with system and safety engineering solutions resulting in precedent setting regulatory approvals for advanced operations.

Please select category below:
Normal Paper ☐
Student Paper ☐
Young Engineer Paper ☒

UAS Aviation Safety Rigour - A Comparative Analysis of Triton Vs Shadow-200

Teddy T. Zvidza[1], Keirin J. Joyce [1], Reuben Strydom [1] and Joshua W. Vicino [1]

[1] *Future Intelligence Surveillance and Reconnaissance System Program Office (FISRSPO), Aerospace Systems Division, Capability Acquisition and Sustainment Group, Defence, PO Box 7922, Canberra BC, ACT*

Abstract

Uncrewed Aircraft Systems (UAS) such as the MQ-4C Triton and the RQ-7B Shadow-200 Tactical Uncrewed Aircraft System (TUAS) each utilise unique approaches to aviation safety. This presents new challenges to both regulators and operators. Whilst each UAS is unique and presents different safety challenges, the fundamentals of UAS safety do not vary simply as a function of the complexity or size of the UAS. This paper compares and considers the drivers for aviation safety effort and rigour between a large UAS like Triton versus smaller UAS such as Shadow-200. To that end, the safety case of both Triton and Shadow-200 will be examined. In doing so, it will be shown that the UAS platforms can have similar levels of aviation safety rigour applied to them. As it stands, Triton and Shadow-200 make use of different risk-based approaches. These approaches are explained and analysed with a view to highlight key differences and argue that despite the scale of each system, a common approach is not only possible, but necessary. This paper will leverage UNCLASSIFIED aviation safety material from the Triton and Shadow-200 programs as prime source material and provide broader insight into the activities that support UAS operations in the ADF. The paper will also consider the differences in the configuration, role and environment (CRE) for the platforms and will highlight the main differences and similarities that support the aviation safety activities.

Keywords: UAS, Safety, Triton, Shadow-200, Uncrewed, Aviation, Risk

Disclaimer

The views, opinions and conclusions drawn from this paper are those of the Authors and do not represent the views of the Australian Defence Force or the Defence Aviation Safety Authority. This paper aims to provide a discussion on the way uncrewed aviation safety is conducted.

Introduction

The operation of Uncrewed Aircraft Systems (UAS) has become vital to modern military operations and other civilian sectors, such as mining, farming, media to name a few. These platforms may be seen as human replacement machines; however, they are introducing new and enhanced operational capabilities than previously realised. Such capabilities include; food and package delivery, intelligence gathering and monitoring, maintenance checks on buildings and on aircraft structures and even providing reconnaissance support to emergency services [1]. The recent war in Ukraine has shown the world the importance of using UAS technology in the theatre of war. Both Russian and Ukrainian forces are using this technology to gather intelligence on troops, deliver small and large munitions, guide troops to where to attack the

https://doi.org/10.52202/075106-0021

opposition [2]. UAS can also reduce the casualties and threats to pilots and combat crews who often would have been used to perform intelligence-gathering activities and would have also been sent out to locations were the enemy may possess more territory. In short, they have become the human on the ground and in the air [1-3].

The reliability and accuracy of UAS technology has greatly increased since their initial use in World War I [3]. There has been an increase in the amount of research and development of these technologies over the past decades. The technological inclusions in these UAS are exceptional. Organisations are developing systems that are fully autonomous, systems that are operated by AI and systems that can be operated on a one-to-many basis [4-5]. Though most UAS can be operated autonomously it is currently considered that there is still need for a human in the loop in their operations to provide a high level of safety assurance.

Designers and operators of these systems need to assure regulators, users, the general population and other airspace users of safety when operating these systems. This is why it is important for the various regulators to set up regulations that are consistent and with set criteria in which these UAS should be operated safely. However, it is to be understood that not all UAS are designed to a specific standard and that most of them have niche characteristics that support their use, which causes challenges for regulators when operators want to bring such systems into service. Most regulatory authorities have adopted risk-based approaches to assist operators in bringing their systems into service. These risk-based approaches ensure safety is at the centre of all operations and that risks to users, general public and other airspace users is reduced So Far As Reasonably Practicable (SFARP).

This paper compares the fundamental level of aviation safety that has been applied to the MQ-4C Triton, coming into Australian service in 2024, and the RQ-7B Shadow-200, a mature system in Australian service for over a decade [5]. These two platforms are military operated UAS, and their mission operations are unique in comparison to other uncrewed systems. Though they have unique operational capability the aviation safety rigour that goes into bringing these UAS into service should not vary in magnitude solely as a result of system complexity. The work presented here will look at the UAS regulations, discuss the configuration, role and environment of the two platforms, and discuss the aviation safety approaches for UAS and any risk methods used to assess them.

UAS Regulations

The two aviation regulators in Australia are the Civil Aviation Safety Authority (CASA) and the Defence Aviation Safety Authority (DASA) [6-7]. They have both developed UAS regulations that guide operators as to how they can operate their RPAS safely without imposing harm to the other airspace and ground users. These regulations at the time of writing this paper are still maturing in the way they regulate and integrate the use of UAS platforms especially those still in development like Triton, MQ-28A Ghost Bat and future platforms that are yet to be developed.

Triton and Shadow-200 operate under the defence aviation safety regulations (DASRs), which has the three categories of Certified, Specific and Open category UAS [6]. Each of these categories aligns with CASA and relevant international and military standards on UAS operations such as STANAG-4671. As they will be military owned and operated UAS, the two platforms are categorised under the Specific category of the DASRs and not under CASA who regulate civilian aircraft. Thus, Triton and Shadow-200 will operate under a DASA

https://doi.org/10.52202/075106-0021

issued UAS Operating Permit (UASOP). The Specific category applies to platforms that have airworthiness standards that applied from initial, continuing to operational airworthiness [6].

Under the Specific category, the UAS have specific restrictions applied to them in order to minimize the risk SFARP on operations. There is specific compliance criteria set out by DASA that should be implemented by UAS operators who wish to operate under a UASOP [6]. Operations under a UASOP have limitations to the classes of airspace in which users may operate UAS. UASOP users fly predominantly in restricted airspace around military locations and any flights in other classes of and over civil airspace require approval from the airspace regulator or other approving authorities. The UASOP application process is outlined by DASA, and it is a risk-based approach that aligns with the 7-Step Safety Risk Management Process [6].

Configuration, Role and Environment Comparison and UAS Function

Shadow-200 by Textron Systems is a tactical uncrewed aircraft system (TUAS) with the ability to provide intelligence, surveillance, target acquisition and reconnaissance (ISTAR) capabilities [11]. The TUAS is also used as a training platform for RPAS operators, mission crew and combat soldiers who need it in the battlefield [11]. It has the capability to operate up to 125km from the UAS operator. This TUAS is designed to provide users with accurate information on activities occurring on the ground in near real-time [9]. Shadow-200 has high-resolution cameras that operate both day and night and can distinguish between a range of tactical vehicles operating on the ground from an altitude of 8000ft and at 3.5km slant range [12]. For an aircraft of its size, Shadow-200 has an endurance of over 8 hours depending on the flight profile and mission objective [12]. Australian operations of Shadow-200 are mainly on military installations, training areas and other restricted airspaces; though, there are instances where it might be flown over sparsely populated areas. One such area was in Afghanistan where Shadow-200 was deployed to replace the missions that the Scan Eagle conducted on behalf the International Security Assistance Force (ISAF) [13]. For such operations, where Shadow-200 is operated over a civilian population, there is usually an assessment of the risk involved from the surrounding infrastructure and airspace and approval is sought from the relevant approving authority to allow flight or missions over high-risk areas. Shadow-200 has a launch system that utilises a pneumatic catapult and recovery system as part of the ground capability [12]. The launcher allows Shadow-200 to be a mobile system that can be deployed to any environment within its operational limits. This provides the users with enhanced capability on the battlefield and in training environments.

The MQ-4C Triton UAS by Northrop Grumman Corporation is a high-altitude long endurance (HALE) aircraft that will be operated by the RAAF for maritime patrol and surveillance. This UAS is equipped with modern real-time ISR capabilities that will support long endurance missions that are over 24 hours operated by the RAAF [13]. The long endurance capability will provide the ADF with excellent Maritime surveillance coverage and the ability to carry out simultaneous and continuous missions over a period. The main areas of operations will be in the North, Southwest Pacific and down to Antarctica [14]. The aircraft has an operational ceiling of up to 50,000ft and has a sensor suite that operates over 2000 Nautical Miles, which then produces a 360-degree view of the surroundings [15]. In Australia, the Triton RPAS will be flown out of RAAF Base Tindal and operated remotely from RAAF Base Edinburgh in Adelaide [15]. The flights out of Tindal reduce the risk SFARP of the UAS as it will fly over less populous areas of Australia. The mission risk for every operation is calculated based on the risk methods that each UAS platform adopts and the operators informed knowledge of the mission or flight profile.

https://doi.org/10.52202/075106-0021

Though both these platforms have varied capabilities and complexities they both have the same basic function of being uncrewed, they are operated by remote pilots located at a specific ground station and they both have the ability to conduct ISR missions.

Risk and Safety Methods

Modern aviation works under a set of assumptions, regulations and legislation to assure safety of crewed and uncrewed aircraft operations [6-7]. One of the fundamental requirements in relation to safety for non-type certified Defence UAS is that risks are eliminated or reduced SFARP [6]. As non-type certified UAS (e.g. Triton and Shadow-200) may not undergo the same level of technical design rigour as certified conventionally piloted aircraft, it is critical to utilise robust risk and safety methods to assess uncrewed aircraft operations. These UAS system safety assessments focus on ground, air and critical infrastructure risk. The risk and safety methods can broadly be categorised into two methodologies: quantitative and qualitative. These system safety methods are used to assess the safety risk of UAS operations based on the CRE, which can be communicated to the relevant decision makers.

Quantitative methods will characterise the risk of UAS operations to people (ground and air) and critical infrastructure [18]. To characterise the ground and air risk effectively, the first step is to gain an understanding of the reliability of the platform to inform human and autonomous decision-making [19-20]. The reliability information then feeds into the risk calculations – including (i) the reachability of the air vehicle in a degraded or failed states [21]; (ii) essential personnel, general public and critical infrastructure distributions [22-23]; (iii) and risk quantification [24-25]. A similar approach can be used for air risk by characterising the airspace density [26] and quantifying the risk of mid-air collisions [27-28].

The limitations of the above quantitative methods are the requirement for significant data or utilising complex uncertainty quantification methods when there is low confidence in the data. In situations where the data is not available, qualitative methods can support the safety case for UAS operations [29-30]. In general, a combination of the quantitative and qualitative methods are used to develop a robust safety case for complex UAS operations. Indeed, both Shadow-200 and Triton are complex UAS, which utilise both qualitative and quantitative risk assessment methods to support a robust safety case.

Discussion of UAS Safety Rigour and Approach

When it comes to uncrewed aircraft safety, every national regulator applies different emphasis on the level of safety that is required. However, for commercial aircraft operations there is a need to comply with ICAO's international policies on safety and to ensure that safe aircraft operations are being carried out worldwide and that they meet a certain standard [32-33]. In the uncrewed environment, safety is still of importance; however, the various regulatory bodies have different approaches to assessing the level of safety and rigour that should be applied to UAS [33-34]. For uncrewed systems very little data has been collected on incident or accident reports, fleet management and other data measures that would be gathered for crewed aircraft to generate mathematical models that can be applied to all uncrewed systems and establish quantitative data for UAS safety [33]. However, it is still worthwhile to note that models have been suggested in literature and some have been used for other platforms. To date, there has not been a consistent quantitative method that meets all the UAS platform requirements. Some of these include tools like the Quantitative Approach and Departure Risk

https://doi.org/10.52202/075106-0021

Assessment (QUADRA) adopted by USN; and the JARUS SORA which has been adopted by CASA when applying for specific operational use that poses different risks [7, 30].

Due to this varied nature of safety analysis, it forces safety authorities in acquisition and design to ask the question "how far should they go when conducting safety analysis for their systems?" Triton and Shadow-200 are unique in the way of their operation, and they provide unique UAS safety perspectives and challenges. However, for Triton's unique design, and operational procedure it provides authorities with a level of assurance on how safe it will be to operate this UAS once it is in service [35-36]. There have only been a few Triton incidents that have been reported, with the main one being caused by a landing gear failure and an inflight mechanical issue that caused it to crash on a Naval base in California [35]. However, Shadow-200 has experienced more crashes during its use in Afghanistan and when being operated by other nations [36-37]. The mature use of Shadow-200 over the years has enabled a greater body of data to be collected on crash incidents compared to Triton, a UAS which is yet to be introduced into Australian Service.

In Australia Triton and Shadow-200 operate in restricted and controlled airspace away from the general population as per their CRE and UASOP limitation. This alters the amount of aviation safety rigour that must be applied to both platforms and it reduces the risks associated with plat platforms crashing over a higher population density. Since both platforms followed the UASOP approval process, the operators and safety authorities would have identified risks and hazards associated with the platforms for various flight configurations and potential mission profiles. These identified risks and hazards advise and inform the risk management systems that are set up for operators when operating the UAS on various missions. This will ensure any known operational and aircraft related risks are reduced before operation.

According to the defence aviation safety regulations (DASR) both Triton and Shadow-200 do not meet the requirements for military type certification [6]. As such, they both need to apply for a Specific Type A category UAS Operating Permit UASOP to fly and carry out mission essential requirements on behalf of the ADF [6]. The UASOP application process and assessments that accompany them will vary in effort as a result of complexity of the system, but they should follow the same phases of the application set out by DASA. However, it should be noted that they both have unique operational requirements and that the levels of risk that they assume are not similar. This is when the variation in aviation safety rigour is seen to differ. Both platforms adopted different safety strategies to satisfy the UASOP application requirements and these strategies are detailed below.

The Triton safety strategy and the level of rigour applied are based on the risk posed by Triton as a large, long range UAS, its cost and complexity, and the approach taken by the USN in approving its operation. As Triton will operate under a UASOP, the safety strategy is based around risk assessment rather than certification against a defined set of requirements. However, in order to leverage the USN airworthiness program and provide defensible argument for the comprehensiveness of the risk assessment, an Airworthiness Criteria Set (ACS) has been developed. The ACS is analogous to a Type Certification Basis in defining criteria that represent best practice in uncrewed system design, but these are used as references rather than strict requirements. Shortfalls against these requirements result in potential hazards which can then be assessed and mitigated to ensure that all risks have been thoroughly identified and reduced SFARP. The outputs of the USN's comprehensive safety program are considered in conjunction with the ACS assessment to develop a comprehensive risk picture for Triton operations. This risk picture is communicated using the bow tie structure described

 https://doi.org/10.52202/075106-0021

in DASA AC 001/2018, providing a consistent framework for assessment and communication of residual risk [38].

For Shadow-200, the safety strategy utilised various mathematical models to calculate the levels of safety risk imposed by the design of the UAS and the exposed population. When Shadow-200 was introduced to service, Defence was still operating under the Technical Airworthiness regulations (TAREGS) that imposed a Safety Target approach to the technical safety evaluation process for UAS [25]. This allowed Shadow-200 to be evaluated under an aviation safety rigour that had an emphasis on an acceptable safety target that they could work to satisfy the probable risks that would be involved. This target would also encompass the various design and operational consideration needed to carry out a mission from a specific area [25]. This method of having safety targets enables UAS operators to assume a certain level of risk when conducting missions. However, this level of risk needs to be defined to ensure the UAS designs can meet the risk standards in operation.

Conclusion

It is evident that for UAS certification and aviation safety procedures there is still a need for standardisation. The regulators, operators and designers need to continually develop these regulations and UAS standards to ensure there is consistent aviation safety efforts being carried out across the uncrewed industry. In a military context it is understood that most platforms have distinct operational needs and having regulations and standards that are consistent allows the military to secure more platforms that satisfy the needs of many users.

The level of complexity of a UAS affects the risks the systems pose to operators, and this is true for both Triton and Shadow-200. The regulations they operate under also influence the rigour of aviation safety in line with the associated risk posed by the UAS. The two UAS presented here, both follow the same fundamental approach to aviation safety as presented by DASA when they apply for the UASOP. However, the risk methods used to establish UAS safety baselines for platforms will differ in effort, but they ought to provide similar level of rigour in application to assure risks are SFARP.

References

1. K. P. Valavanis, Advances in Unmanned Aerial Vehhicles, vol. 33, K. P. Valavanis, Ed., Dordrecht: Springer, 2007.
2. BBC, "Ukraine conflict: How are drones being used?," BBC , 22 August 2022. [Online]. Available: https://www.bbc.com/news/world-62225830. [Accessed 01 October 2022].
3. K. P. Valavanis, Advances in Unmanned Aerial Vehhicles, vol. 33, K. P. Valavanis, Ed., Dordrecht: Springer, 2007.
4. Insider Intelligence, "Drone technology uses and applications for commercial, industrial and military drones in 2021 and the future," Business Insider, 13 January 2021. [Online]. Available: https://www.businessinsider.com/drone-technology-uses-applications.

https://doi.org/10.52202/075106-0021

5. G. Waldron, "Northrop rolls out first Australian MQ-4C Triton," Flight Global, 15 September 2022. [Online]. Available: https://www.flightglobal.com/military-uavs/northrop-rolls-out-first-australian-mq-4c-triton/150201.article#:~:text=Work%20on%20the%20RAAF%27s%20first,to%20Australia%20set%20for%202024.. [Accessed 04 October 2022].
6. Defence Aviation Safety Authority, "Uncrewed Aircraft Systems," Defence , 05 October 2022. [Online]. Available: https://defence.gov.au/DASP/UAS/.
7. Civil Aviation Safety Authority, "Drone Rules," CASA, 05 October 2022. [Online]. Available: https://www.casa.gov.au/drones/drone-rules.
8. Insider Intelligence, "Drone technology uses and applications for commercial, industrial and military drones in 2021 and the future," Business Insider , 13 January 2021. [Online]. Available: https://www.businessinsider.com/drone-technology-uses-applications.
9. Dronenerds, "The Evolution of Drones, Then and Now," [Online]. Available: https://enterprise.dronenerds.com/the-evolution-of-drones/.
10. Akihabara News, "One Pilot Flying Multiple Drones," 29 August 2022. [Online]. Available: https://akihabaranews.com/one-pilot-flying-multiple-drones/.
11. Army , "Unmanned Aerial Vehicles (UAVs) provide Intelligence, Surveillance, Target Acquisition and Reconnaissance (ISTAR) support for land force operations. The Australian Army currently operates the RQ-7B Shadow-200 in this role.," Defence, 2022. [Online]. Available: https://www.army.gov.au/our-work/equipment-uniforms/equipment/surveillance/unmanned-aerial-vehicles.
12. Military.com, "RQ-7B Shadow," 2022. [Online]. Available: https://www.military.com/equipment/rq-7b-shadow.
13. AustralianAviation.com.au, "Shadow-200 Operational In Afghanistan," 7 May 2012. [Online]. Available: shadow-200 operational in Afghanistan.
14. Army-Technology, "Shadow-200 RQ-7 Tactical Unmanned Aircraft System," 13 March 2020. [Online]. Available: https://www.army-technology.com/projects/shadow-200-uav/.
15. A. Thorn, "First RAAF Triton Drones will arrive one year late," Australian Aviation , 30 April 2021. [Online]. Available: https://australianaviation.com.au/2021/04/first-raaf-triton-drones-will-arrive-one-year-late/.
16. L. Reynolds, "Australia commits to next generation Triton Remotely Piloted Aircraft," 18 June 2020. [Online]. Available: https://www.minister.defence.gov.au/media-releases/2020-06-18/australia-commits-next-generation-triton-remotely-piloted-aircraft.

https://doi.org/10.52202/075106-0021

17. AIRFORCE, "MQ-4C Triton Unmanned Aircraft System," 2022. [Online]. Available: https://www.airforce.gov.au/aircraft/mq-4c-triton.
18. J. Rushby, "New Challenges In Certification For Aircraft Software," in Proceedings of the Ninth ACM International Conference On Embedded Software, Taipei, 2011.
19. R. Strydom and A. Buckley, "A proposed framework and review of probabilistic methodologies to characterise UAS operational risks," in AIAC19, 2021.
20. E. Denney, G. Pai and M. Johnson, "Towards a rigorous basis for specific operations risk assessment of UAS," in IEEE/AIAA 37th Digital Avionics Systems Conference (DASC), 2018.
21. K. S. Krishnakumar, P. H. Kopardekar, C. A. Ippolito, J. Melton, V. Stepanyan, S. Sankararaman and B. Nikaido, "Safe autonomous flight environment (safe50) for the notional last "50 ft" of operation of "55 lb" class of uas.," AIAA Information Systems, p. 0445, 2017.
22. E. Rudnick-Cohen, J. W. Herrmann and S. Azarm, "Modeling Unmanned Aerial System (UAS) Risks via Monte Carlo Simulation," in International Conference on Unmanned Aircraft Systems (ICUAS), Atlanta, 2019.
23. Australian Bureau of Statistics, "Australian Population Grid," 2018.
24. A. Rose, J. McKee, K. Sims, E. Bright, A. Reith and M. Urban, "LandScan Global 2020 [Data set]," Oak Ridge National Laborator, 2021.
25. A. L. Mckinnon, "Practical Applications to Evaluate the Safety of UAS in an Australian Defence Context," in Australian International Aerospace Congress, 2015.
26. W. Williams, I. Vuletich, J. Tisato, M. Brett and S. Wilson "Application Of The Range Safety Template Toolkit To Uas Risk Hazard Analyses Supporting Airworthiness Regulation," in ICAS, 2012.
27. A. McFadyen, "Probabilistic determination of maximum safe altitudes for unmanned traffic management," in IEEE/AIAA 38th Digital Avionics Systems Conference (DASC), 2012.
28. T. L. Molloy, J. J. Ford and L. Mejias, "Detection of aircraft below the horizon for vision-based detect and avoid in unmanned aircraft systems," Journal of Field Robotics, pp. 1378-1391, 2017.
29. A. Weinert, S. Campbell, A. Vela, D. Schuldt and J. Kurucar, "Well-clear recommendation for small unmanned aircraft systems based on unmitigated collision risk," Journal of Air Transportation, pp. 113-122, 2018.
30. JARUS, "JARUS guidelines on Specific Operations Risk Assessment (SORA)," Joint Authorities for Rulemaking of Unmanned Systems (JARUS), 2019.

https://doi.org/10.52202/075106-0021

31. J. R. McCall, A Qualitative Journey to Seeing Safety Differently., Saint Louis University, 2019.
32. ICAO, "Safety," 2022. [Online]. Available: https://www.icao.int/safety/Pages/default.aspx
33. H. G. Wolf, Drones Safety Risk Management for the Next Evolution of Flight, London: Routledge, 2017.
34. B. S. Gobin, "Quantitative Approach and Departure Risk Assessment for Unmanned Aerial Systems," Doctoral Dissertation Virginia Tech, 2020.
35. R. Abbott, "MQ-4C Triton crashes During California Flight Test," Avionics Today, 13 September 2018. [Online]. Available: https://www.aviationtoday.com/2018/09/13/mq-4c-triton-crashes-california-flight-test/.
36. E. Slavin, "UAV Crashes near Casey," 26 August 2007. [Online]. Available: https://www.stripes.com/news/uav-crashes-near-casey-1.68056.
37. Australian Defence Magazine, "Shadow tops UAV crash list," 5 November 2010. [Online]. Available: https://www.australiandefence.com.au/news/shadow-tops-uav-crash-list
38. Defence Aviation Safety Authority, "Advisory Circulars (ACs) and Factsheets" Defence , 10 October 2022. [Online]. Available: https://defence.gov.au/DASP/Docs/Manuals/Policy/index.htm#t=ACs_Factsheets.htm

https://doi.org/10.52202/075106-0021

AUTHOR INDEX

Association for Unmanned Vehicle Systems International
2700 South Quincy Street
Suite 400
Arlington, VA 22206

ISBN 978-1-7138-9933-4

CURRAN ASSOCIATES INC.
proceedings.com